T0185584

The Application of Artificial Intelligence

Zoltán Somogyi

The Application of Artificial Intelligence

Step-by-Step Guide from Beginner to Expert

 Springer

Zoltán Somogyi
Antwerp, Belgium

ISBN 978-3-030-60034-1   ISBN 978-3-030-60032-7   (eBook)
https://doi.org/10.1007/978-3-030-60032-7

This Springer imprint is published by the registered company Springer Nature Switzerland AG.
The registered company address is: Gewerbestrasse 11, 6330 Cham, Switzerland

To Hannah and Matthies.

"Big things in life are made not by luck but by hard work and curiosity!"

Zoltán Somogyi

Preface

There are many books in the field of machine learning and artificial intelligence, but most of them are either too complex with a lot of mathematical theory and equations or full of all kinds of scripting (Python, R, etc.) source code. The aim of this book is to present a unique easy-to-understand view of the subject with many practical examples and without scripting and source code. Whether you are a complete novice in the field or an expert I am convinced that you will find a lot of valuable information in this book which is the result of several years of research and hard work. There are many topics which are not properly explained in easily understood lay terms in most of the books out there about machine learning, and this is one of the reasons why I decided to write this book. You will not find a single page in this book which is simply filling the space; everything is designed to help you understand this subject in a short period of time. The book contains several "Expert sections" with more details about specific subjects, where I felt that this was needed, and many practical real-world machine learning application examples explained step by step.

The book is organized as follows:

- **Part I Introduction**

  - The first chapter is a helicopter view of what machine learning and artificial intelligence are and what the different sub-fields supervised learning, unsupervised learning, and reinforcement learning mean.

- **Part II In Depth Machine Learning**

  - The second chapter is a detailed overview of the most important supervised learning, unsupervised learning, and reinforcement learning algorithms.
  - The third chapter contains an extensive overview of machine learning performance evaluation methods and tools.
  - The fourth chapter contains everything you need to know about machine learning data - how to collect, store, clean, and preprocess the data.

- **Part III Automatic Speech Recognition**

  - The fifth chapter is completely dedicated to the subject of automatic speech recognition (ASR). After reading this chapter you will understand how ASR works and how it can be used.

- **Part IV Biometrics Recognition**

  - The sixth and seventh chapters are dedicated to the subject of face and speaker recognition.

- **Part V Machine Learning By Example**

  - The eighth chapter contains numerous real-world machine learning case studies covering applications such as disease recognition, recommendation systems, root cause analysis, anomaly detection, intrusion detection, business process improvement, image recognition, and predictive maintenance. All case studies can be applied with the accompanying AI-TOOLKIT software (no programming or scripting skills are needed).

- **Part VI The AI-TOOLKIT. Machine Learning Made Simple.**

  - The ninth chapter contains a description of the AI-TOOLKIT and useful information about how to use it.

- **Appendix from Regular Expressions to HMM**

  - This appendix contains expert information about hidden Markov models (HMMs) which are often used in natural language processing (NLP) and automatic speech recognition.

The book is accompanied with software called the AI-TOOLKIT which is free for non-commercial purposes, enabling any reader to try out all of the examples in the book without having to invest in other programs. The AI-TOOLKIT contains several easy-to-use machine learning tools. At the time of writing, the following tools are included (which may be extended later): AI-TOOLKIT Professional (the flagship product), DeepAI Educational (a simple neural network learning visualization tool), VoiceBridge (an open source automatic speech recognition tool), VoiceData (a speech data generator for VoiceBridge), and DocumentSummary (a tool for summarizing documents).

The aim of the AI-TOOLKIT is to make the application of machine learning simple and accessible to everyone. No programming or scripting skills are needed except for the fully open source VoiceBridge tool, which is a ready-to-use C++ source code for high-performance speech recognition on MS Windows. The AI-TOOLKIT is compatible with the 64-bit version of MS Windows, and it supports all three major categories of machine learning: supervised, unsupervised, and reinforcement learning, with several types of algorithms per category, and also several machine learning applications. It includes many user-friendly built-in templates for all types of machine learning models.

I hope that you enjoy reading this book and that you find a lot of useful information in it!

Antwerp, Belgium
July, 2020

Zoltán Somogyi

Support Material and Software

You can find all of the data files referred to in the book along with extra information, errata, and all open source software on GitHub: https://github.com/AI-TOOLKIT

You can download the full version of AI-TOOLKIT Professional (MS Windows 64-bit) free of charge for non-commercial purposes here: https://ai-toolkit.blogspot.com/p/download-and-release-notes.html

The direct link to the open source (C++) VoiceBridge is as follows: https://github.com/AI-TOOLKIT/VoiceBridge

Acknowledgments

I would like to thank everyone who has, directly or indirectly, contributed to this book and to the accompanying AI-TOOLKIT software. I have done my best to highlight important contributions with links to the relevant references. It is not possible to mention each individual here, but the references and recommended reading sections in this book and the attribution sections in the software (e.g., on GitHub, and in the "help" and "about" boxes) hopefully provide a complete list. If I have inadvertently overlooked your contribution then please let me know and I will include a reference in future editions or errata of the book!

I would also like to extend my thanks to Springer Nature and, in particular, to Ronan Nugent, executive editor, for his help, support, and hard work and to Amy Hyland for copy editing the book!

Contents

Abbreviations

AI	Artificial intelligence
ALS	Alternating least squares
ARG	Argument
ASR	Automatic speech recognition
AUC	Area under the ROC curve
BPI	Business process improvement
BPR	Bayesian personalized ranking
BSMOTE	Borderline synthetic minority oversampling technique
CB	Content based
CF	Collaborative filtering
CFE	Collaborative filtering with explicit feedback
CFFNN	Convolutional feedforward neural network
CHI	Calinski–Harabasz index
CPU	Central processing unit (processor)
CSV	Comma separated values
CVD	Cardiovascular disease
DFT	Discrete Fourier transform
DOS	Denial-of-service
EGA	Evolutionary genetic algorithm
EXP	Exponential
FFNN	Feedforward neural network
FN	False negative
FNR	False negative rate
FP	False positive
FPR	False positive rate
FSA	Finite state automaton
FST	Finite state transducer
GA	Genetic algorithm
GMM	Gaussian mixture model
GPU	Graphics processing unit

GRU	Gated recurrent unit
HAC	Hierarchical agglomerative clustering
HMM	Hidden Markov model
IDCT	Inverse discrete cosine transform
IDFT	Inverse discrete Fourier transform
IPA	International Phonetic Alphabet
KNN	k-nearest neighbors
LCL	Lower control limit
LDA	Linear discriminant analyses
LM	Language model
LN	Natural logarithm
LSTM	Long short-term memory
MAX	Maximum
MDP	Markov decision process
MFCC	Mel-frequency cepstral coefficient
MIN	Minimum
ML	Machine learning
MLFLOW	Machine learning flow
MLLT	Maximum likelihood linear transform
NLP	Natural language processing
NORMINV	Inverse normal distribution function
OOV	Out of vocabulary word
PCA	Principal component analysis
PMML	Predictive maintenance machine learning
PPV	Positive prediction value
R2L	Unauthorized access from a remote machine
RCA	Root cause analysis
RELU	Rectified linear unit
RGB	Red, green, blue
RL	Reinforcement learning
RLS	Reinforcement learning system
RMSE	Root mean squared error
RNN	Recurrent neural network
ROC	Receiver operating characteristic
RUL	Remaining useful lifetime
SAT	Speaker adaptive training
SER	Sentence error rate
SSC	Sum of squared distances of the centroids
SSE	Sum of squared errors
SVM	Support vector machine
TANH	Tangent hyperbolic
TN	True negative
TNR	True negative rate
TP	True positive

TPR	True positive rate
U2L	Unauthorized access to local super-user
UCL	Upper control limit
UNK	Unknown words
WALS	Weighted alternating least squares
WER	Word error ate
WFST	Weighted finite state transducer
WHO	World Health Organization

List of Figures

List of Tables

Part I
Introduction

Chapter 1
An Introduction to Machine Learning and Artificial Intelligence (AI)

Abstract It is not always clear to people, especially if they are new to the subject, what we mean by machine learning and when and why we need it. A lot of people are aware of artificial intelligence (AI) from science fiction but they may not really understand the reality and the connection to machine learning. This chapter will explain in clear lay terms what machine learning and AI are, and it will also introduce the three major forms of machine learning: supervised, unsupervised and reinforcement learning. The aim is that after reading this chapter you will understand what, exactly, machine learning is and why we need it.

1.1 Introduction

Machine learning is a process in which computers learn and improve in a specific task by using input data and some kind of rules provided to them. Special algorithms, based on mathematical optimization and computational statistics, are combined together in a complex system to make this possible. Artificial intelligence is the combination of several machine learning algorithms which learn and improve in several connected or independent tasks at the same time. At present, we are able to develop parts of a real artificial intelligence but we cannot yet combine these parts to form a general artificial intelligence which could replace humans entirely.

We could also say that learning in this context is the process of converting past experience, represented by the input data, into knowledge.

There are several important questions that arise: To which kind of tasks should we apply machine learning? What is the necessary input data? How can the learning be automated? How can we evaluate the success of the learning? Why don't we just directly program the computer with this knowledge instead of providing the input data?

Let us start with answering the last question first. There are three main reasons why we need machine learning instead of just using computer programming:

© Springer Nature Switzerland AG 2021
Z. Somogyi, *The Application of Artificial Intelligence*,
https://doi.org/10.1007/978-3-030-60032-7_1

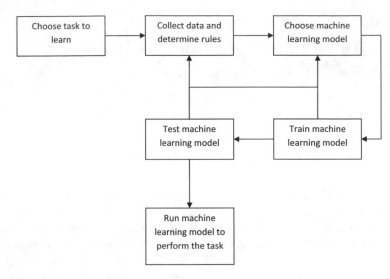

Fig. 1.1 A typical machine learning process

1. After a computer program is made it is difficult to change it every time the task changes. Machine learning adapts automatically to changes in the input data/task. As an example after software has been programmed to filter out spam e-mails, it cannot handle new types of spam without re-programming. A machine learning system will adapt automatically to the new spam e-mails.
2. If the input is too complex, e.g. with unknown patterns and/or too many data points it is not possible to write a computer program to handle the task.
3. Learning without programming may often be very useful.

In order to be able to answer the other questions, let us first look at a typical machine learning process as represented on Fig. 1.1. First we need to decide which task to teach to a machine learning model considering the three reasons mentioned above. Next we need to decide which data and rules we need to feed to our machine learning model. Then we need to choose a machine learning model, train the model (this is when the learning takes place) and test the model to see if the learning is correct. Collecting the data, choosing the model, training and testing are all recursive tasks (note the arrows going back to former steps) because if the model cannot be adequately trained then we often need to change the input data, add more data or choose another machine learning model.

Machine learning tasks can be classified into three main categories:

1. Supervised learning
2. Unsupervised learning
3. Reinforcement learning

In the next sections we will see what machine learning means in more detail, get to understand these three categories and discover what some of the real-world applications are.

1.2 Understanding Machine Learning

The concept of machine learning, as we have discussed previously, is quite abstract and if you are new to the subject then you may wonder how it works and what it really means. In order to answer these questions and make things more tangible let us look at one of the most simple machine learning techniques called linear regression. Linear regression should be familiar to most people since it is typically part of a basic mathematical course. Real-world machine learning algorithms are of course much more complex than linear regression, but if you understand this machine learning adapted explanation of linear regression then you understand how machine learning works!

The well-known mathematical expression of linear regression can be seen in Eq. (1.1).

$$\hat{y} = w \cdot x + b \qquad (1.1)$$

What is the aim of linear regression? There is a set of x and y values as input data. We want to model their relationship in such a way that we can predict future y values for any given x value. There are two parameters in this model, 'w' which we could call weight and 'b' which we could call bias or error. As we know from our basic mathematical studies the so called 'weight' parameter controls the slope of the regression line (see Fig. 1.2) and the 'bias' parameter controls where the regression line will intercept the y axis instead of going through zero. You probably understand already that we have chosen to use the terms *weight* and *bias* deliberately because they are special machine learning terms.

Fig. 1.2 Linear regression

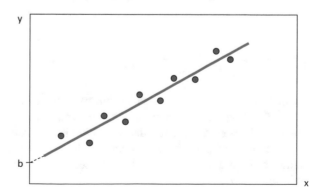

The performance of our simple machine learning model can be measured by calculating the mean squared error of the deviations of the predictions from the original points.

It is important to mention at this point that we make a significant distinction between how well the machine learning model performs (success of learning) on a learning (training) dataset and on a test dataset which is not used during the learning phase (this will be explained in detail in the next section about Accuracy and Generalization)! For this reason as a first step let us divide the input (x, y) points into two sets. One set will be used for learning (training) and one set will be used for testing. We will see in later chapters how to select the training and test datasets, for now let us just assume that from the ten points on Fig. 1.2 we select the first eight points as training data and the last two points as test data.

The next step (which will not be explained here because it is the simple linear regression method) is the estimation of the values of 'w' and 'b' by minimizing the mean squared error on the training set. Then we can calculate the final mean squared error on the training and test sets (after applying the regression line to the test set) with the well-known formulas presented in Eq. (1.2).

$$
\begin{aligned}
MSE_{training} &= \frac{1}{n_{training}} \sum_{i=1}^{n_{training}} \left(y_i - \hat{y}_i \right)^2_{training} \\
MSE_{test} &= \frac{1}{n_{test}} \sum_{i=1}^{n_{test}} \left(y_i - \hat{y}_i \right)^2_{test}
\end{aligned}
\tag{1.2}
$$

These two mean squared error (MSE) parameters provide the performance measures of our simple machine learning model. The MSE on the training dataset and on the test dataset are both important! The MSE on the test dataset is often called the *generalization error* in machine learning. Generalization means that the machine learning model is able to handle data which was not seen during the learning phase. This is often important in real-world applications because we want to train our machine learning model with a dataset collected in the past but we want to use the model with data which will be collected in the future! We will look at accuracy and generalization in more detail in the next section.

1.2.1   Accuracy and Generalization Error

As we have seen in the previous section we make a significant distinction between how well the machine learning model performs (success of learning) on a learning (training) dataset and on a test dataset which is not used during the learning phase!

Depending on the difference between the accuracy (and error) on the training dataset and the accuracy on the test dataset we say that the model is under-fitted, well-fitted or over-fitted.

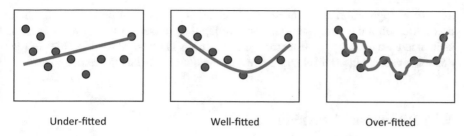

Under-fitted   Well-fitted   Over-fitted

Fig. 1.3 Under-fitted, over-fitted and well-fitted machine learning models

Under-fitted means that the machine learning algorithm failed to learn the relationships (patterns, knowledge) in the training data which resulted in a low accuracy on the training data and will also cause a low accuracy on the test data.

If the accuracy of the machine learning model on the training data is much higher than the accuracy on the test data then we say that the model is over-fitted. In other words the machine learning algorithm is fitted too closely to the training data and it does not generalize well.

We want a good fit and a good accuracy on both datasets (training and test) and we often sacrifice accuracy for a better generalization! A good generalization in the case of a good fit thus means that the machine learning algorithm is good at handling data which it has not seen during the learning phase.

Figure 1.3 shows how these three forms of fitting can be visualized and the importance of model selection because if we modeled this dataset with linear regression (straight line) then we would have the under-fitting problem!

This last thought leads us to the question of how to positively influence the accuracy of our machine learning model on the test dataset? First of all the selection of the training and test sets are of crucial importance! Both datasets must be independent from each other and must be identically distributed! If one of these requirements is not met then we cannot adequately measure the generalization performance. Furthermore, the complexity of the machine learning model is also of crucial importance (as mentioned previously in the discussion about linear regression and Fig. 1.3). If the model is too complex then most probably over-fitting will occur (see Fig. 1.3). If the model is too simple then under-fitting will occur. One way of causing over-fitting in our simple linear regression example is by using a polynomial regression model instead of a linear one. But if the input data is more complex, which is best modeled with polynomial regression, and we use linear regression then under-fitting will occur. Machine learning model selection is often a process of trial and error in which we try several models (or model parameters) and check the training and test (generalization) errors or accuracy.

In the case of over-fitting, increasing the number of input data points may also help!

In the case of a small dataset (when no more data is available), the so-called k-fold cross validation procedure may be used in order to get a statistically better estimate of the errors. Just dividing a small dataset into training and test sets would not leave us with enough information in the data for learning. The k-fold cross validation procedure splits the dataset into k non-overlapping subsets. The test error is then estimated by averaging the test error across k-trials. On trial 'i' the ith subset of the dataset is used as the test set and the rest of the data is used as the training set.

1.3 Supervised Learning

We speak about *supervised learning* when the input to the machine learning model contains extra knowledge (supervision) about the task modeled in the form of a kind of label (identification). For example in the case of an e-mail spam filter the extra knowledge could be labeling whether each e-mail is spam or not. The machine learning algorithm then receives a collection of e-mails labeled spam or not spam and through this we supervise the learning algorithm. Or in the case of a machine learning based speech recognition system the label is a sequence of words (transcribed sentences). Or another example could be the labeling of a collection of images about animals for an animal identification task. With the extra knowledge of which picture contains which animal the learning algorithm is supervised.

It is not always easy to provide this extra knowledge and label the data. For example, if there is too much data or if we just do not know which data belongs to which label. In this case unsupervised learning will help, which will be explained in the next section.

It is interesting to note at this point that core machine learning algorithms work with numbers. All kinds of input data must first be converted to numbers—for example, an image is converted to color codes per pixel—and for the same reason the label is also defined as a number. For example, in the case of the aforementioned spam filter an e-mail which is not spam could be labeled with '0' and spam with '1'. We often call these labels *classes* and the reason for this will be explained in the next section (Fig. 1.4).

There are two forms of supervised learning:

1. **Classification**—when there are a discrete number of labels (classes), e.g. 0,1,2,3...
2. **Regression**—when the labels contain continuous values, e.g. 0.1, 0.23, 0.15...

In both cases the machine learning algorithms must learn which data record belongs to which label by identifying patterns in the data; and in both cases the algorithms are very similar, but the evaluation of the success of the learning is different. As we have seen previously, we first train the machine learning model and then test it. Testing is done by inference on testing data. Inference means that we feed the testing data to the trained machine learning model and ask it to decide which label belongs to each record. We can obviously easily count the number of correct labels in the case of classification, and the so-called error on the estimate is the percentage of wrongly identified labels. In the case of regression where the labels are in a continuous range we must do something else; we consider the mean squared

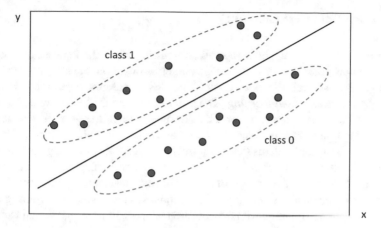

Fig. 1.4 Simple supervised learning

error—or in other words, the average of a set of errors—on the estimate. This is very similar to the performance evaluation of simple regression!

In the case of classification we often use the term accuracy instead of error. Accuracy is the opposite of error—the percentage of well identified labels.

There are several types of supervised learning algorithms and each of them has its advantages and disadvantages. In the next chapter (Chap. 2) we will look at some of these algorithms in more detail.

1.3.1   Supervised Learning Applications

There are already many real-world supervised learning applications and many more will be added in the future. Some of the existing applications are as follows:

- E-mail spam detection based on a collection of messages labeled spam and not-spam.
- Voice recognition based on a collection of labeled voice recordings. The labels identify the person who speaks.
- Speech recognition (part of comprehension) based on a collection of labeled voice recordings where the labels are the transcription of sentences.
- Automatic image classification based on a collection of labeled images.
- Face recognition based on a collection of labeled photos. The labels identify which photo belongs to which person.
- Determining whether a patient has a disease or not based on a collection of personal data (temperature, blood pressure, blood composition, x-ray photo, etc.).
- Predicting whether a machine (auto, airplane, manufacturing, etc.) will break down (and when it will break down—for predictive maintenance) based on a collection of labeled data from past experience.

1.4   Unsupervised Learning

Remember that we speak about *supervised learning* when the input to the machine learning model contains extra knowledge (supervision) about the task modeled in the form of a kind of label. When we do not have this extra knowledge or label then we speak about *unsupervised learning*. The aim of unsupervised learning is the identification of this extra knowledge or label. In other words, the goal of unsupervised learning is to find hidden patterns in the data and classify or label unlabeled data and use this to group similar items (similar properties and/or features) together, and thus put dissimilar items into different groups. Another name for unsupervised learning is clustering (grouping). An example of a two-dimensional (there are only two features or columns in the data) clustering problem can be seen in Fig. 1.5. Clustering can of course be applied to datasets with many more features (dimensions) which cannot be easily visualized.

It is always better to classify (label) your data manually but this is not always possible (e.g., too much data, not easy to identify the classes, etc.) and then unsupervised learning can be very useful.

There are many types of clustering algorithms and each of them has its advantages and disadvantages depending on the input data. In the next chapter (Chap. 2) we will look at some of these algorithms in more detail. Each clustering algorithm uses some kind of similarity criterion and strategy to join items together in one group. Applying several clustering algorithms to the same dataset may yield very different results.

After labeling an unlabeled dataset with unsupervised learning we can of course apply supervised learning!

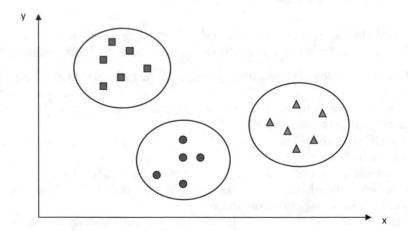

Fig. 1.5 Unsupervised learning—clustering in three groups

1.4.1   Unsupervised Learning Applications

There are already many real-world unsupervised learning applications and many more may be added in the future. Some of the existing applications are as follows:

- Grouping shoppers together based on past purchases and other personal properties; for example, as part of a recommendation system.
- Market segmentation based on chosen properties, e.g., for marketing applications.
- Segmentation of a social network or a group of people, e.g., for connecting people together (as on a dating site).
- Detecting fraud or abuse (by applying unsupervised learning to better understand complex patterns in the data).
- Grouping songs together based on different properties of the music, e.g., on streaming platforms.
- Grouping news articles together depending on the contents or keywords, e.g., as part of a news recommendation application.

1.5   Reinforcement Learning

We could define reinforcement learning as a *general purpose decision making machine learning framework used for learning to control a system.* There are several important keywords in this definition which need some explanation. *General purpose* means that reinforcement learning can be applied to an unlimited number of different fields and problems; from very complex problems such as driving an autonomous vehicle to less complex problems such as business process automation, logistics, etc. *Decision making* means carrying out any kind of decision/action depending on the specific problem, for example, accelerating a car, taking a step forward, initiating an action, buying stocks, etc. *Controlling a system* means taking actions in order to reach a specific goal, where the specific goal depends on the problem (e.g., reaching a destination, having profit, being in balance, etc.).

Reinforcement learning and supervised learning are similar but there are two important differences. Remember that in supervised learning the machine learning model receives labeled data which is used to supervise the learning algorithm. However, in the case of reinforcement learning the model does not receive external data at all but generates the data itself (there are some exceptions to this when the data is generated externally and passed to the reinforcement learning system—for example, when images are used from a video game to learn how to play a game). The second difference is that reinforcement learning uses a reward signal instead of labeled data. It is called a *reward* signal because we tell the machine learning model whether each action taken was successful (positive reward) or not (negative reward or penalty). Giving a reward can also be called positive *reinforcement* and this is where the name reinforcement learning comes from. Both the data and the reward signal are generated by the reinforcement learning system based on predefined rules.

Fig. 1.6 Reinforcement
learning system

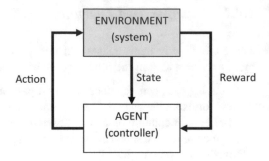

There are several questions arising from this about how to generate the data, how to generate the reward signal and how to design and operate such a system, which we will now consider.

A reinforcement learning system can be symbolized by the interaction between a so-called *Environment* and an *Agent* as you can see on Fig. 1.6. The environment is sometimes also called the *system* and the agent is sometimes also called the *controller*. The environment can mean many different things and can be as detailed as needed. For example, if you want to teach a computer to drive a car then you can place the car into a very simple environment or into a complex real environment with a lot of properties. Many reinforcement learning applications train models in a virtual environment where the model plays a simulation over and over again and observes success and failure while trying different actions (trial and error). This is, for example, how autonomous vehicles are initially trained.

The reinforcement learning system typically starts to operate by initializing the *Environment* to a random state, but it could also start with a specific state. The *state* can mean many different things and depends on the problem. For example, it can be the speed of a car or it can also be several properties at a specific times step such as the speed, the direction, etc. The state is then passed to the *Agent* which calculates an appropriate action (for example, in the case of a car this could be increasing the speed or braking). Each time an action is taken and passed back to the *Environment* a reward is calculated for the last state transition and passed back to the *Agent* (reward signal). This is how the *Agent* knows if the action was good or wrong. This cycle is repeated until an end goal is reached, e.g., the system reaches an expected state such as reaching a destination, winning or losing a game, etc. We call this the end of an *Episode*. After an *Episode* ends the system is reset to a new (random) initial state and a new *Episode* begins. The reinforcement learning system cycles through many *Episodes* during which it learns which actions are more likely to lead to the desired outcome or goal by optimizing the long term reward (a numerical performance measure). We will discuss long term rewards in more detail in Chap. 2 and look at several examples.

We will also see in Chap. 2 how the *Agent*'s actions affect the long term behavior of the environment (when the *Agent* takes an action it does not know immediately whether the action is effective or not on the long run) and how the *Agent* uses the so-called *Exploitation* and *Exploration* strategy to select actions. Exploitation means that the *Agent* leans towards actions which lead to positive results and avoid actions that do not. Exploration means that the *Agent* must find out which actions are beneficial by trialing them despite the risk of getting a negative reward (penalty). Exploitation and exploration must be well balanced in a reinforcement learning system!

1.5.1   Reinforcement Learning Applications

There are currently many real-world reinforcement learning applications and no doubt more will be developed in the future. Some of the existing applications are as follows:

- Self-driving cars. A control system based on reinforcement learning is used to adjust acceleration, braking and steering.
- Automated financial trading. The reward is based on the profit or loss for each trade. The reinforcement learning *Environment* is built using historical stock prices.
- Recommendation systems. The reward is given when, for example, the users click on an item. Real-time learning improves the machine learning model or recommendation systems are trained on historical data.
- Traffic light control.
- Logistics and supply chain optimization.
- Control and industrial applications, e.g., for optimizing energy consumption, efficient equipment tuning, etc.
- Optimizing treatment policies or medication dosage in healthcare.
- Advertising optimization.
- Various types of automation.
- Robotics.
- Automated game play.

Part II
An In-Depth Overview of Machine Learning

Chapter 2
Machine Learning Algorithms

Abstract The first chapter of this book explained what machine learning is and why it is needed. This chapter now gives an in-depth overview of the subject. The most important machine learning algorithms (models) are explained in detail and several important questions are answered: Which algorithm should we select for the task? What are the advantages and disadvantages of the model? This chapter focuses on the practical uses of machine learning; the mathematical background is only explained when it is really necessary—typically in separate 'expert sections' to aid comprehension and to allow interested readers to dive deeper into the subject. Several examples are provided to help explain the different applications of machine learning.

2.1 Introduction

In this chapter we will explore how various machine learning algorithms work and look at several examples. The most frequently used supervised learning, unsupervised learning and reinforcement learning algorithms will be explained in more detail with a focus on practical use. More complex mathematical theory will only be explained when it is really necessary for the understanding of the subject or the practical application. After reading this chapter you will be able to apply each of the machine learning algorithms to real-world problems, e.g., by using the accompanying AI-TOOLKIT software in which all of these algorithms are available!

2.2 Supervised Learning Algorithms

Remember that we speak about *supervised learning* when the input to the machine learning model contains extra knowledge (supervision) about the task modeled in the form of a kind of label (class identification). There are two forms of supervised learning: classification and regression. The machine learning algorithms must learn, in both cases, which data record belongs to which label by identifying patterns in the

© Springer Nature Switzerland AG 2021   17
Z. Somogyi, *The Application of Artificial Intelligence*,
https://doi.org/10.1007/978-3-030-60032-7_2

data. The algorithms are therefore very similar but the evaluation of the success of the learning is different. In the case of classification, the so-called error on the estimate is the percentage of wrongly identified labels. In the case of regression, we consider the mean squared error—or in other words, the average of a set of errors—on the estimate. For classification we often use the term accuracy instead of error; accuracy is the opposite of error—the percentage of correctly identified labels.

2.2.1   Support Vector Machines (SVMs)

A support vector machine (SVM) is a good example of a supervised machine learning algorithm. It is in fact, next to a neural network, one of the most commonly used and useful supervised machine learning algorithms!

SVM is applicable to problems with both linear and non-linear features in the dataset and this makes it very effective. It is also an algorithm with very few parameters and therefore it is easy to optimize for high accuracy and not difficult to use, even for beginners in machine learning. To help our understanding of the algorithm let us start with a simple linear SVM problem before we extend it to a non-linear problem.

Let us assume that we have a dataset with two columns (two feature vectors) which can be easily visualized in a 2D plot. Let us also assume that a third column contains the classification of each data record and that there are only 2 classes (labels) designated with 0 (c0) and 1 (c1). One data record would then e.g. look like "x1, x2, c0". The goal of the SVM algorithm is to find (learn) the best hyperplane which separates the two groups of data points. In the case of a linear problem the hyperplane is a simple line as shown on Fig. 2.1.

There is always a boundary region or margin in which there are only a few data points. We call these points "support vectors" (because they become vectors in higher dimensions). You can see two support vectors (one for each class) on Fig. 2.1 on the boundary hyperplanes (the two dashed lines). The SVM finds the best separating hyperplane by maximizing the *Distance* (see Fig. 2.1) between the two boundary hyperplanes on each side of the separating hyperplane. This is in very simple terms how the SVM algorithm works. One of the advantages of this method is that by maximizing the boundary region (distance) it maximizes the distance of the separating hyperplane (decision boundary) from the data points, which results in a good generalization performance (see generalization in Sect. 1.2.1)!

Most real-world problems are non-linear; therefore, let us extend our linear SVM to more complex non-linear problems. A non-linear SVM works in exactly the same way as a linear one, but it utilizes a pre-processing step that transforms the original data points by projecting them into a higher dimensional space. The reason for this is that the points acquired in this way are often easily separable in the higher dimensional space. This pre-processing step is achieved with a so-called *kernel function*.

Figure 2.2 shows how this pre-processing, or kernel mapping, and then back-mapping to the original space works.

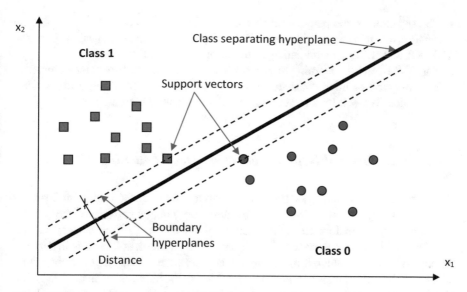

Fig. 2.1 Linear support vector machine (SVM) example

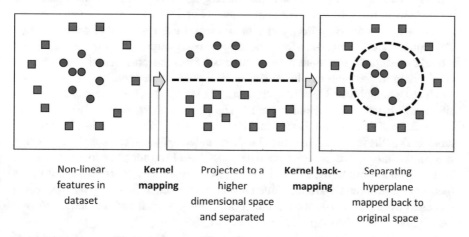

Fig. 2.2 Support vector machines kernel mapping

There are several kernel functions available which can be used with different types of data. Because the choice of the kernel function is important the equation for each function is shown below. Please note that x_i, x_j are the vectors containing the data. It is assumed here that you are somewhat familiar with vector notation (e.g. the transpose of a vector is designated with 'T', etc.).

- Linear kernel (no projection): $K(x_i, x_j) = x_i^T.x_j$
- Polynomial kernel: $K(x_i, x_j) = (\gamma.x_i^T.x_j + coef_0)^{degree}$ (where $\gamma > 0$)
- Radial basis function (rbf): $K(x_i, x_j) = \exp.(-\gamma.|x_i - x_j|^2)$ (where $\gamma > 0$)
- Sigmoid: $K(x_i, x_j) = \tanh(\gamma.x_i^T.x_j + coef_0)$ (where $\gamma > 0$)

The selection of the parameters gamma (γ), degree and coef$_0$ can be done with trial and error or by using past experience. Some software packages, such as the AI-TOOLKIT, offer an automatic parameter optimization module.

The SVM machine learning algorithm has one drawback: it becomes slower to train in the case of huge datasets, in which case neural network algorithms are a better choice. We will look at neural networks in the next section.

2.2.2   Feedforward Neural Networks: Deep Learning

Feedforward neural networks (FFNNs) are one of the most important learning algorithms today next to SVMs. They became very famous because of the success of convolutional feedforward neural networks (CFFNNs) for image classification (thanks to CFFNNs we have self-driving cars). We will look at CFFNNs in detail in the next section. Another form of neural network, the so-called recurrent neural network (where the data is not only flowing through the network, as in an FFNN, but also in a time dependent direction—it feeds its outputs back into its own inputs), has recently had considerable success in natural language processing but it is much more computational resource intensive than an FFNN or SVM.

A neural network contains a series of connected elements, called neurons (often also called nodes), which transform the input into the output and in the process learn the relationships in the input data. Remember from the previous chapter that the relationship in the input can be as simple as a linear regression, but it can also be much more complex and hidden to humans. Figure 2.3 shows a schematic representation of a feedforward neural network. The network starts with a series of input nodes ($X_0 \ldots X_n$). The input is split into its components, this may be just the features (columns) of the input or it may also be an extended feature set filtered by a function or a combination of features (for example $\sin(X_0)$ could be added as an extra feature). Then the data flows to the first hidden layer (1) which also contains several nodes (neurons). There can be several hidden layers. It is called a '*hidden layer*' because it is hidden from the outside world, which only sees the input and the output. Finally the data flows into the output layer ($Y_0 \ldots Y_K$).

Each neuron in the network is connected to all other neurons. The number of input nodes depends on the input data and the number of output nodes depends on the model. For example, in the case of classification the output may be a probability value for each class (the class with the highest probability is the selected class or decision) or just one label (class). Each hidden layer may contain an arbitrary number of neurons (even hundreds of them) depending on the modeled problem.

You may ask yourself the question, why do we need to add these hidden layers? By adding hidden layers in combination with so-called activation functions (see Sect. 2.2.2.1) we can represent a wider range of complex patterns (functions) in the input data! This is the reason why neural networks can represent any kind of complex function! We often call a hidden layer with an activation function an *activation layer*.

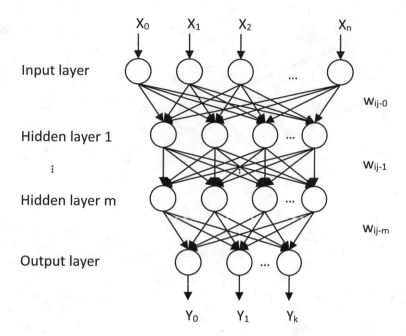

Fig. 2.3 Feedforward neural network

We will discuss the w_{ij-m} (weight) property of each connection between the neurons later (see Fig. 2.3). Let us just note for now that there are weights associated with each connection and these weights are the neural network parameters which are adjusted in the learning process. The neural network learns these weights! Remember the discussion about linear regression and the weight (slope) parameter in Sect. 1.2 of Chap. 1!

Each neuron can be represented by several weighted (w_{ij-m}) input signals (coming from all neurons in the previous layer), a mathematical equation which transforms the input ($x_0 \ldots x_n$) into the output (y), and the calculated output signal (y) going to all neurons (weighted!) in the next layer (see Fig. 2.4). The output is calculated by summing up all of the weighted inputs ($x_i w_i$) optionally extended with a bias (b) and finally filtered by a so-called activation function (F_A).

The optional bias can be thought of as an extra weight connected to a unit (1) input and it is used as a special adjustment for the learning per neuron. Remember how the 'b' term or bias modifies the linear regression model as discussed in Sect. 1.2 of Chap. 1? It shifts the line up or down and determines where the line crosses the vertical axis. The bias in our more complex neural network model has a very similar functionality!

Because the activation function is an important element we will discuss it in more detail in the next section.

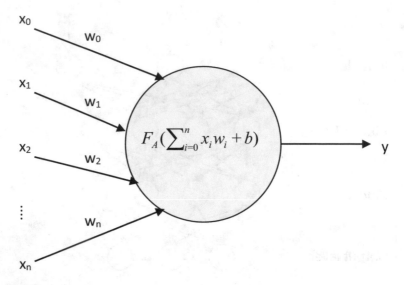

Fig. 2.4 Artificial neuron

2.2.2.1   The Activation Function

The aim of the activation function is to introduce non-linearity in the model by transforming the weighted input (Fig. 2.4). Without this the output would just be a linear function and we would not be able to model non-linear features in the input data. Do you remember how we defined our simple linear regression machine learning model in the first chapter in Eq. (1.1)? It is very similar to the inside of the artificial neuron on Fig. 2.4 except for the activation function! With the help of the activation function the neural network can learn and represent any complex function or relationship in the data instead of just a linear one.

There are many types of activation functions and it is important to know what the advantages and disadvantages of using each of them are in order to be able to make a good decision. Table 2.1 summarizes some of the well-known activation functions and their properties. A variety of helpful and not so helpful activation functions are shown in order to explain the difference! The best choices from the table are the Tangent Hyperbolic (TanH), the ReLU and Leaky ReLU functions! You may, however, experiment with any type of function as long as you take the advantages and disadvantages into account! Other types of activation functions may be developed in the future. You may also be interested to read the expert sections about some of the important properties of activation functions (Expert Sect. 2.1 and Expert Sect. 2.2)!

Table 2.1 Activation functions

Name	Chart	Equation	Range
Identity		$f(x) = x$	$(-\infty, \infty)$
	This does not affect the input.		
Step		$f(x) = 0, x < 0$ $f(x) - 1, x \geq 0$	$(0, 1)$
	– Disadvantage: This tends to switch off (0) specific neurons which then stay 0 (lost). More specifically neurons which are initially not activated (left side) will cause a zero gradient and switch off the signal through those neurons (data loss!). – Disadvantage: Not zero centered.[a]		
Sigmoid		$f(x) = \frac{1}{1+e^{-x}}$	$(0, 1)$
	– Disadvantage: Not zero centered.[a] – Disadvantage: Saturated boundaries.[a] NOTE: this activation function is more common in recurrent neural networks (RNNs) owing to some of the additional requirements of RNNs (see Sect. 2.2.4)!		
TanH		$f(x) = \; tanh\,(x)$	$(-1, 1)$

(continued)

Table 2.1 (continued)

Name	Chart	Equation	Range
	– Advantage: Zero centered.[a] – Disadvantage: Saturated boundaries.[a]		
ReLU		$f(x) = 0, x < 0$ $f(x) = x, x \geq 0$	$[0, \infty)$
	– Disadvantage: This tends to switch off (0) specific neurons which stay 0 (lost). More specifically neurons which are initially not activated (left side) will cause a zero gradient and switch off the signal through those neurons (data loss!). – Advantage: Computationally efficient. – Advantage: Accelerates the convergence of the optimization (gradient descent)!		
Leaky ReLU		$f(x) = c \cdot x, x < 0, c < 1$ $f(x) = x, x \geq 0$	$(-\infty, \infty)$
	– Advantage: This overcomes the switch off (0) problem in ReLU. Usually c is 0.01 (a small number) but it may also be parameterized and learned by the network as an extra hyper parameter ($c < 1$).[a] – Advantage: No saturation problem.[a] – Advantage: Computationally efficient. – Advantage: Accelerates the convergence of the optimization (gradient descent)!		
SoftPlus		$f(x) = \ln (1 + e^x)$	$(0, \infty)$
	– Advantage: This is similar to ReLU but there is no hard cut off and thus also no switch off problem.[a] – Disadvantage: saturated on the left side.[a]		

[a]Advanced information is available in Expert Sect. 2.1 and Expert Sect. 2.2

Fig. 2.5 'Zigzagging'
during gradient update

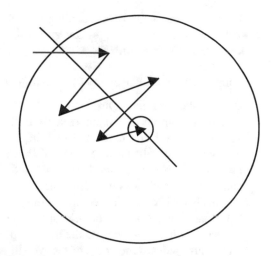

Expert Sect. 2.1 The Importance of Zero-Centered Activation Functions
Neurons without a zero-centered activation function (e.g. sigmoid) transmit
signals which are also not zero-centered to the receiving layers. This has a
negative effect on the solution search (gradient descent) because all gradients
during the backpropagation phase will have the same sign, and this will cause
very inefficient 'zigzagging' in the search for the solution. Zigzagging means
that the gradient updates go too far in different directions which makes
optimization more difficult. Backpropagation and solution search are
discussed in more detail in Sect. 2.2.2.3. Please refer to this section later if
you do not know yet what these terms mean (Fig. 2.5).

**Expert Sect. 2.2 The Importance of Non-saturated Boundaries
in Activation Functions**
If the activation function is getting saturated at the boundaries (e.g. sigmoid)
then the local gradients will be near zero in these regions. As a result during
backpropagation these near-zero gradients will turn off the neurons which
means that no signal (data) will flow through them. This problem is actually
less severe than using a non-zero centered activation function because there
are some other positive effects which partially neutralize the problem during
batch processing.

2.2.2.2 Neural Network Layer and Connection Types

We can categorize the different types of layers and the different types of connections
between the neurons. Each layer contains the same type of neurons and different
layers may contain different types of neurons.

Layers can be categorized according to their activation function (the activation function determines the type of layer); however, there are some additional layer types which also transform the input data but in a different way than activation functions.

Some of the special layer types are as follows:

- **Batch normalization layer.**

    This transforms the input data to give zero mean and unit variance, by subtracting the batch mean and dividing by the batch standard deviation, and then scales and shifts the data. The scale and shift become learnable parameters in the network. Normalization is done for individual training cases, and the mean and standard deviations are computed across the batch.

    This layer is usually added after a fully connected or convolutional layer (see Sect. 2.2.3) but before a non-linearity layer (activation functions). The batch normalization layer speeds up the training process by allowing a higher learning rate. It also reduces the effect of the weight initialization of the network (before training the best weight parameters aren't known) and it has a positive influence on generalization.

    More information about batch normalization can be found in Expert Sect. 2.3!

- **Dropout layer.**

    The aim of the dropout layer is to improve generalization (preventing over-fitting—see Sect. 1.2.1)!

    It has one parameter: the probability, p (default value $= 0.5$). It randomly regularizes the input by setting the connection values with p probability to 0. Setting the connection values to zero means that some neurons will be switched off (no data will flow through them) which also means that data or information will be lost (neurons are either dropped with probability p or kept with probability $1 - p$).

- **Layer normalization layer.**

    This transforms the input data to give zero mean and unit variance and then scales and shifts the data. Normalization is done for individual training cases, and the mean and standard deviations are computed across the layer dimensions. It is very similar to batch normalization, and it should be used instead of batch normalization in recurrent neural networks.

The connections between the layers can be categorized according to how the data is transferred from one layer to another. Remember that each connection contains an input and its weight ($x_i w_i$) and that there can be an extra bias as a kind of weight and a unitary input ($1 b_m$), see Fig. 2.4. According to this we can define the following types of connections (this may not be an exhaustive list):

- **Fully connected (linear) connection.**

    This connection transfers all data from one layer to another and it also adds the bias. Remember that the bias can be thought of as an extra weight connected to a unit (1) input (extra connection) and it is used as a special adjustment for the learning per neuron.

- **Fully connected (linear) connection without bias.**
  This connection transfers all data from one layer to another but the bias is not added.
- **Drop connected.**
  This has the same effect as the dropout layer but instead of zeroing out activations it zeros out weights on the connections.

Expert Sect. 2.3 Batch Normalization

"Training Deep Neural Networks is complicated by the fact that the distribution of each layer's inputs changes during training, as the parameters of the previous layers change. This slows down the training by requiring lower learning rates and careful parameter initialization, and makes it notoriously hard to train models with saturating nonlinearities."[1].

The authors refer to this phenomenon as internal covariate shift, and address the problem by normalizing layer inputs using batch normalization, "by whitening the inputs to each layer, we would take a step towards achieving the fixed distributions of inputs that would remove the ill effects of the internal covariate shift." [1].

How does it work? The batch normalization layer normalizes its input by subtracting the batch mean and dividing by the batch standard deviation. Two trainable parameters are then added to the layer. One of the parameters (gamma) is multiplied with the normalized input (calculated in the first step) and the other parameter (beta) is added to this product. The resultant signal is the new transformed (batch-normalized) output of the batch normalization layer.

Let us assume that we have a mini-batch with $B = \{x_1, x_2 \ldots x_m\}$. The mean of the data in the mini-batch is then.

$$\mu_B = \frac{1}{m} \sum_{i=1}^{m} x_i.$$

The standard deviation can be calculated with

$$\sigma_B = \sqrt{\frac{1}{m} \sum_{i=1}^{m} (x_i - \mu_B)^2}$$

The normalized input then can be calculated with

$$\widehat{x}_i = \frac{x_i - \mu_B}{\sigma_B + \varepsilon}$$

Where ε is a small number to prevent division by zero.

(continued)

Expert Sect. 2.3 (continued)
The final batch-normalized signal can then be written as

$$y_i = \gamma \widehat{x}_i + \beta$$

Gamma (γ) and beta (β) become additional training parameters and the neural network learns their optimal value.
An example can be seen on Fig. 2.6.

2.2.2.3 The Learning Process

Training a supervised feedforward neural network (the learning process) involves a complex combination of several algorithms. The aim of this section is to give a comprehensive overview of the learning steps, focusing on the details which are necessary for the application of neural networks. The complex mathematics behind these steps is only required if you want to implement or further develop/improve these algorithms. We will keep things simple with the minimum number of mathematical formulas. After reading this section you will understand how neural networks learn and you will be able to use this knowledge while applying neural networks in your own machine learning task.

Fig. 2.6 Neural network setup with batch normalization

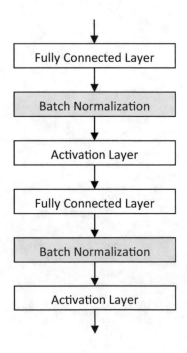

If we have an input vector **x** (a series of numbers or features; one row or data record) and a target label y (one number) we want the neural network to learn the relationship between **x** and y. A well trained neural network model will provide a prediction ŷ that is equal to or close to the target value y for any given **x**. Training the neural network involves searching for a value of w (weight) that produces a function that fits the input data well. In order to be able to measure how well the neural network performs this learning task an error or objective function (sometimes called the cost function) is defined as a function of w. This objective function sums up for each input/target pair {**x**, y} the error measuring how close the prediction ŷ(**x**; w) is to the target y. The neural network training process minimizes the objective function (i.e. minimizes the total error) by adjusting w. The function minimization algorithm also makes use of the gradient of the objective function with respect to the parameter w, which is calculated by the so-called backpropagation algorithm (more about this a little bit later).

The reason we need to make use of the gradient is that the hidden layers complicate the error calculation (objective function). The error at the output layer is easy to calculate by comparing the target y with the prediction ŷ but the error at each hidden layer is unknown! Without knowing the error at each hidden layer we cannot adjust the weights. But by back-propagating the error from the output layer to each hidden layer we can solve this problem. When we know the errors at each hidden layer we can update the weights of each connection during the training process. The idea is that each node is responsible for a fraction of the error in the output nodes it is connected to. The error is distributed according to the strength of the connection between the hidden nodes and the output nodes and is back-propagated to provide the errors for all hidden layers.

The objective function can be defined based on the output of the neural network (after a forward propagation) with general terms:

$$O(w) = \frac{1}{n} \sum_{i=1}^{n} L\left[\hat{y}^{(i)}\left(x^{(i)}; w\right), y^{(i)} \right]$$

In this equation $\hat{y}^{(i)}\left(x^{(i)}; w\right)$ is the prediction result (based on the input $x^{(i)}$ and weight w), and $y^{(i)}$ is the actual value (label), where (i) represents the ith data record and there are n numbers of records in the dataset. The letter L defines a so-called loss (error) function which depends on the type of learning, classification or regression, and on both parameters—the prediction and the actual value. Please note that x in this equation is a matrix and y is a vector because we are representing the whole dataset here instead of just one record! For the same reason $x^{(i)}$ is a vector (one row of the data) and $y^{(i)}$ is a number (label).

Let us simplify for the next equations the term $\hat{y}^{(i)}\left(x^{(i)}; w\right)$ with just $\hat{y}^{(i)}$.

In the case of classification when the model output is a probability between 0 and 1 (the probability for each class) this loss function is called the *cross entropy loss* and the term L (see above) can be replaced in the *objective function* as follows:

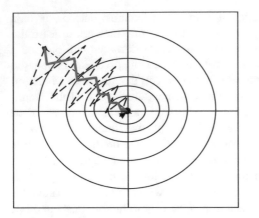

Gradient descent   – – –→

Gradient descent + adaptive
learning rate algorithm (Adam)   ——→

Fig. 2.7 Gradient descent algorithm + adaptive learning rate

$$O(w) = \frac{1}{n} \sum_{i=1}^{n} \left[y^{(i)} \log\left(\widehat{y}^{(i)}\right) + \left(1 - y^{(i)}\right) \log\left(1 - \widehat{y}^{(i)}\right) \right]$$

In the case of regression (outputs are continuous numbers) the so-called *mean squared error* Loss is used:

$$O(w) = \frac{1}{n} \sum_{i=1}^{n} \left[y^{(i)} - \widehat{y}^{(i)} \right]^2$$

This is exactly the same as what we saw in Sect. 1.2 of Chap. 1 while discussing our simple linear regression machine learning model (mean squared error)!

The objective function optimization (minimization) is then performed with the so-called gradient descent algorithm. To put it very simply, the gradient descent algorithm first calculates the gradients and then updates the weights in the network and repeats these two steps until the required convergence is reached (the optimum minimum value of the function is found). The gradients are calculated with backpropagation as explained previously. A graphical representation of how gradient descent works can be seen on Fig. 2.7 (a 2D topology view of a complex function optimization). There is also a so-called stochastic gradient descent algorithm which is very similar but it works with mini batches of the data instead of each data point. Stochastic gradient descent is more efficient (it has better accuracy and is faster to train) and therefore it is used more often.

Using a mathematical formula we can show that the weights are estimated by **minimizing** our objective function:

$$\widehat{w} = \underset{w}{argmin} O(w)$$

The optimization of the objective function is a difficult task because of well-known optimization problems (e.g. false local minima) and also because of the so-called learning rate problem. The learning rate is used in the gradient descent algorithm to adjust the weights (the steps on Fig. 2.7!). If steps are too large or too small then the optimization may fail or may be too slow or may even never converge. The new weights are calculated with the following simplified formula:

$$\widehat{w} = w - \eta \frac{\partial O(w)}{\partial w}$$

Where η is the learning rate and $\frac{\partial O(w)}{\partial w}$ is the partial derivative (gradient) of the objective function in respect of w (weight). It does not matter if you do not understand the mathematics behind the partial derivative, you can think about it as the gradient of the objective function in respect of w (weight)—where gradient means the direction of fastest decrease!

The solution to the learning rate problem is using adaptive learning rates instead of fixed learning rates (see also Fig. 2.7). The learning rate is then calculated based on the value of the gradient, the speed of the learning, the value of the weights, etc. depending on the algorithm. There are many adaptive learning rate algorithms including the following:

- Adam
- SARAHPlus
- Adadelta
- RMSProp
- Adagrad

All of these algorithms work in a similar way but some of them are improved versions. It is interesting to know that RMSProp is the improved version of Adagrad [40]. Adam [38] is the improved version of RMSProp and one of the best and most used algorithms today. They are all of mini-batch based optimizers. Optimizer development is an active research area and thus new optimizers are likely to appear in the future.

It is not the aim of this book to explain in detail how these optimizers work but just to help you understand the main principles and to help you choose the best optimizer. Based on experience the Adam optimizer is recommended but you can also try SARAHPlus [41] and Adadelta [39] which are other good optimizers. The AI-TOOLKIT supports these three adaptive learning rate optimizers with Adam being the default choice!

2.2.2.4   How to Design a Neural Network

There are no exact rules for designing a neural network because the number of
hidden layers and the number of hidden units (neurons or nodes) depend on a
complex combination of many parameters including the following:

- The training data: number of records, the complexity of the patterns (information)
  in the data which must be learned by the neural network, the noise in the data, etc.
- The number of input and output nodes in the neural network which also depend
  on the input data.
- The selection of the activation functions in the network.
- The learning algorithm and its parameters.
- Etc.

If the architecture of the neural network is not complex enough (too few hidden
layers and nodes) then the model will not be able to learn the information in the data
which will result in poor model performance (a high error rate). If the architecture of
the neural network is too complex (too many hidden layers and nodes) the model
may perform well but may also have a high generalization error (a high error on the
unknown test dataset), or the model may not converge to the final solution at all.

The only way to design an effective neural network model is by trial and error—
setting up and training several neural networks and evaluating their performance.
This trial and error process may be cleverly automated but to date the general results
in this field have not been very promising (this is the subject of ongoing research).
There will be more about this later in the next Section!

With regard to the manual trial and error process, there are some basic directions
(rules of thumb) that may help:

- The number of input and output nodes is known and can give an indication of the
  optimal number of hidden nodes. The number of input nodes is equal to the
  number of features (columns) in the input dataset plus one more unit for the bias
  term which usually exists in most neural networks. The number of output nodes
  depends on the neural network model (algorithm) and it is usually equal to 1 in
  the case of a regression model and to the number of unique decision values
  (classes or labels) in the decision variable in the case of a classification model.
  Some classification models may have only one output if they just predict the class
  instead of the probabilities of each class. The AI-TOOLKIT works with probabilities per class label, and therefore the number of output nodes equals the
  number of unique class labels.
- For many problems one hidden layer may be sufficient. Adding more hidden
  layers can help learn more complex functions (information) in the data but this
  will increase the complexity and training time and could also result in over-fitting.
  In most cases two or three hidden layers are sufficient. It is best to start with one
  hidden layer and then add one at a time if this is not sufficient. There is also
  Kolmogorov's theorem which states that a neural network containing three
  hidden layers can learn any function in the data [42, 43]. However, the theorem

does not say how many hidden nodes are necessary or how to determine their number in each layer.

- Concerning the number of hidden nodes (H) a good approach is as follows:

  - Try the mean of the input (I) and output (O) node counts.

$$H = \frac{I + O}{2}$$

  - Try two times the output or input if there is only one output.

$$H = 2 \cdot O$$

  - Try

$$H = \frac{2}{3} \cdot I + O$$

  - If there is more than one hidden layer then try to set the hidden layer just before the output layer to one of the above values and increase the number of nodes in each subsequent layer, e.g. by doubling (or even 3x, 4x...) the number of nodes. Try to end just before the input layer with a number of nodes that easily connects to the input.
  - You can also try to double (or 3x, 4x...) the input and output in the hidden layers just before the output and just after the input and work your way towards the middle to obtain easy connections there (you can see an example of this in Sect. 8.5.7.1).

- Choosing the right activation function and its parameters for each hidden layer is also not trivial. Section 2.2.2.1 explains the advantages and disadvantages of several activation functions. Start with LeakyReLU and TanH activation functions and if these do not work well then try other types, try other combinations of layers, or add special layers such as dropout, batch normalization, etc. Section 2.2.2.2 shows how to build some of the basic architectures.
- There are several other parameters which are also important in a neural network model such as the batch size, step size (learning rate), number of iterations, type of optimizer, etc. All of these parameters are model and algorithm dependent. Concerning the AI-TOOLKIT, refer to the different model types in Sect. 9.2.3 for a detailed explanation of all of these parameters and directions about how to choose their values.

Constructive, Pruning and Evolutionary Genetic Algorithms for Neural Network Design

The architecture of a neural network contains the hidden layers and the hidden nodes (neurons) on these hidden layers, and several other parameters such as the initial weights, the activation functions with their parameters, the learning rate, etc. may also be included. There are two types of methods which may be used to automatically determine the neural network architecture:

- Automated trial and error constructive and pruning methods.
- Evolutionary genetic algorithms.

The constructive trial and error method starts with a very small network (e.g. a single hidden layer and a few nodes) and it adds hidden layers and hidden nodes and reevaluates the model (training and performance evaluation) until a pre-defined performance is reached. The pruning trial and error method is similar to the constructive method but instead of increasing the number of hidden layers and nodes it starts with a larger network and removes (prunes) hidden nodes and layers step by step. Both methods are very time consuming because of the many parameters which must be optimized, and if you have a lot of data and/or you need a large neural network it may be useful to use a supercomputer, grid or cloud computing. It may be much less work to design the network manually with some clever guesses!

Evolutionary genetic algorithms (EGAs) got their name from biological evolution theory, which they try to replicate. The terms EGA and biological evolution might seem complex at first but these algorithms are actually simpler than they sound (as will see later); they are a kind of search optimization method. There are many types of evolutionary algorithms, and the so-called genetic algorithms are used for neural network architecture optimization (design). EGAs have been the subject of much research but there is not yet a standard method for designing neural network architecture efficiently. One of the problems is that EGAs do not know anything about the problem they are optimizing. They are general optimization methods and this, in combination with the many parameters in neural network architectures, makes things very complex and time consuming. Neural networks often do not evolve in the right direction or the generated neural network architecture is not trainable using standard training methods due to invalid connections. Using grid or cloud computing is a good idea while working with EGAs.

An evolutionary genetic algorithm (EGA) for neural network design works as follows:

- **STEP 1**: <u>Random</u> generation of M number of initial neural network architectures which are then encoded into a binary string. There are two types of encoding schemes, *direct encoding* and *indirect encoding*. Direct encoding creates an N × N matrix with all of the nodes in the network, where N is the total number of nodes. The matrix may just contain values of '1' (if there is a connection between two nodes) and '0' (if there is no connection) or it may also contain the weights on the connections. The encoded binary string contains the concatenation of the elements in the N × N matrix (e.g. row wise)—thus it may be a very big string! Indirect encoding just encodes some of the selected (most important) characteristics of the neural network architecture (the number of hidden layers, the number of nodes, the connections, etc.) but it doesn't produce directly trainable architectures.
- **STEP 2**: Training and evaluation of the performance of the chosen neural networks.
- **STEP 3**: Selection of the best performing architectures for reproduction (generating new architectures).
- **STEP 4**: <u>Random</u> generation of new architectures from those selected in Step 3 through "crossover" and "mutation" operations (the terms come from biological evolution theory). We will see a bit later a simple example of what these operations do.
- **STEP 5**: Training and evaluation of the performance of the new architectures.
- **STEP 6**: Replacement of the former architectures with the newly created ones.
- **This procedure is repeated from Step 3 to Step 6 until a pre-defined performance is obtained.** The performance of the neural network model is the objective function which is maximized during the optimization.

As you can appreciate the EGA process is straightforward and simple but it may take a long time to execute because a number of neural network models (architectures) are trained and evaluated. A good representation and encoding scheme is crucial and there have been many attempts to find a good encoding strategy (strings, parse trees, graphs, etc.) but some researchers think that perhaps the best method is not to use encoding at all [44]. This also indicates that EGAs are not yet really ready for automatically designing neural network architectures.

Expert Sect. 2.4 A Simple Genetic Algorithm (GA) Example
Optimizing the architecture of a neural network also means optimizing the performance of the trained model (for example maximizing the accuracy). Let us assume for simplicity that we do not have to take per-class and other performance metrics into account, only the accuracy of the model. Remember that maximizing the accuracy means minimizing the mean squared error (MSE) in the case of a regression problem and maximizing the class prediction accuracy in the case of a classification problem.

(continued)

Expert Sect. 2.4 (continued)

Let us therefore simulate in this very simple example the maximization (optimization) of a simple function $f(x) = 2 \cdot x$ by using a genetic algorithm (GA). Let us assume that the value of x may vary between 1 and 15.

The first step in the GA is to choose how to represent and encode the problem and randomly generate a population. In this very simple example f (x) = 2·x will vary if the value of x changes (between 1 and 15), and therefore we will represent this problem with the variable x. For the encoding we will simply use binary encoding. The numbers between 1 and 15 can be represented with a maximum of four digits; therefore, we will use a four digit sequence for each value of x. For example, 1 can be binary encoded as 0001 and 15 can be binary encoded as 1111. We extend each binary value which is shorter than four digits with zeros in the front. You can calculate the binary value of any integer on a calculator or in MS Excel.

We have decided the representation and the encoding; the next step is to generate a population (some cases or examples). Let us assume that we need five cases.

Table 2.2 is a simple Excel sheet with the input data and calculations for this example. In the upper part of the table random x values are generated, the binary code is determined and the f(x) function is calculated for all five cases. The reproduction probability (p-reproduction) is the probability of using the case in further reproduction, and it is calculated by dividing each f(x) by the sum of all f(x) values (48). Since we are maximizing f(x), higher values will have higher probability and their case is more likely to be reproduced. This is the aim of this probability. The number of times each case is reproduced (count-reproduction) is calculated by dividing each f(x) value by the average of all f(x) values and then rounding the result. Cases which should not be reproduced (shown by a 0 value) are eliminated and not used further.

The crossover operation is shown in the lower part of the table. We randomly pair up two cases and calculate a random crossover-site. Note that case 2 is not included (it has been eliminated)! The crossover-site is the point that separates the digits that will remain unaltered from the digits that will be swapped (using a zero based index). For example, taking the first row (1|001, 0|011, from the second digit (counting starts from zero at the beginning of the binary value and the crossover-site is equal to 1) until the end of the binary value we swap each number in the base item code with the corresponding number in the paired item code (the binary numbers from the paired up cases are shown in the last column of the lower table). We obtain two new binary numbers in this way (1011 and 0001 in the first row).

Next we decode the binary strings and calculate the f(x) values and replace the former population with the new one. The new x values are 11 (1011),

(continued)

Expert Sect. 2.4 (continued)
9 (1001), 5 (0101), 3 (0011) and 1 (0001). The x values are already starting to increase from the first iteration step!

We will not use mutation in this example, but if there was mutation then we would just need to flip each binary digit ($0 \rightarrow 1$, $1 \rightarrow 0$); in some cases this is determined by the mutation probability. Mutation usually has a very low probability. If, for example, the mutation probability is 1% and we generate 5 items which all have 4 binary digits then $0.01 \cdot 5 \cdot 4 = 0.2$ items must be mutated (the digit flipped)—note that because this number is below 1 no items will be mutated.

The whole procedure outlined above is then repeated until we find the maximum value for f(x). Remember that in the case of a neural network f(x) would be the accuracy of the trained model.

This example has been inspired by an example in [45].

2.2.3 Feedforward Convolutional Neural Networks: Deep Learning

Convolutional feedforward neural networks (CFFNNs) are very similar to general feedforward neural networks (FFNNs) but they contain some special layers. They are mostly used in image classification (visual recognition) but may also be used for other purposes. It is thanks to CFFNN that neural networks have become so successful and they have facilitated a breakthrough in machine learning. Some of the very useful applications are, for example, autonomous vehicles, face recognition, automatic image classification, etc.

The genius idea of CFFNN (the name comes from the special *convolution layer* which we will see later) came from studying the animal visual cortex and how animals (including humans) very quickly recognize objects by just perceiving the patterns (edges, curves . . .) in an image. Each object has some unique features and without analyzing the whole object humans can quickly make a distinction. By using past experience we can further improve on this recognition (learning). A CFFNN looks for these edges and curves and then builds up a more complex object through a series of convolutional layers and filters.

Most of the features (structure, activation functions, learning, optimization, etc.) of general FFNNs that we saw in the previous section are also true of CFFNNs. For this reason we will only focus here on the features which are special to CFFNNs, the layer types and their purpose and parameters, and some CFFNN architectures.

The following three layer types are the special layers present in a CFFNN:

- Input layer.
- Convolution layer—unique to CFFNN.
- Pooling layer—unique to CFFNN.

Table 2.2 Genetic algorithm example

min(x)	1
max(x)	15

n	x	code	f(x)=2x	p-reproduction	Count-reproduction	Note
1	5	0101	10	20.8%	1	1x
2	2	0010	4	8.3%	0	Eliminated!
3	9	1001	18	37.5%	2	2x
4	3	0011	6	12.5%	1	1x
5	5	0101	10	20.8%	1	1x
		sum	48	100.0%	5	
		max.	18			
		min.	4			
		avg.	9.6			

base item			paired up item			crossover-site	new item
n	x	code	n	x	code		
3	9	1001	4	3	0011	1	1|001 0|011 1011 0001
1	5	0101	5	5	0101	2	01|01 01|01 0101 0101
3	9	1001	1	5	0101	2	10|01 01|01 1001 0101
4	3	0011	5	5	0101	3	001|1 010|1 0011 0101
5	5	0101	3	9	1001	2	01|01 10|01 0101 1001

Each of these layer types will be explained in detail in the following sections.

There can of course be activation layers, fully connected layers, etc. in a CFFNN, as in a general FFNN, but they function in the same way as we saw earlier in this chapter.

Since convolutional neural networks are used most often in problems involving images or video we will concentrate on images as the input to the network.

2.2.3.1 Input Layer

This holds the raw pixels of the input image. For example, a fully colored image of size 48×48 has $48 \times 48 \times 3 = 6912$ pixel values. The color of a pixel is defined by three color values—red, green and blue—which range between 0 and 255. An image (for example from a video feed) could have a size of 600×600 which in full color would mean 1,080,000 pixel values. This simple calculation exercise shows how much computational resource is needed for CFFNNs! For this reason we often first convert a color image to grayscale so that we only have one color component! Colors (RGB) usually do not hold important information for recognizing forms in an image and grayscale images are sufficient.

2.2.3.2   Convolution Layer

The convolution layer does most of the work in a CFFNN. It has several filters (sometimes called kernels) which all look at different features of the input image. Each filter window slides (convolves) through the input image and multiplies the input image pixel values (in the case of a grayscale image it is a value between 0 and 255) with the values (weights) in the filter (see Fig. 2.8). A very simple mathematical calculation! The resultant matrix of values contains a so-called feature map (or activation map) which is actually an edge or curve in the original input image. Refer to Expert Sect. 2.5 for an example of what a convolution filter does to an image!

Every convolution layer acts as a detection filter for the presence of specific features or patterns present in the input image. The first convolution layers in a CFFNN detect large features that can be recognized and interpreted relatively easy (e.g. large contour lines of objects). Subsequent layers detect increasingly smaller features that are more abstract (and may be present in the larger features detected by earlier layers). The final layer of the CFFNN (a fully connected layer) makes a classification by combining all of the specific features detected by the previous layers in the input data.

Each convolutional filter (remember that there can be several filters in a convolutional layer) has a number of weights (these are the values in the filter)

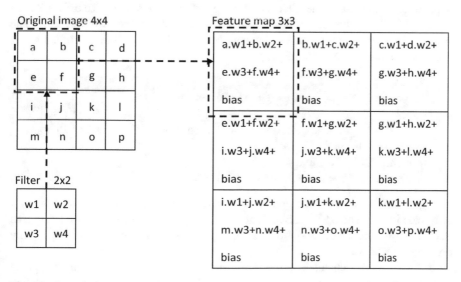

Fig. 2.8   Convolution operation

Table 2.3 Convolution filter
(3 × 3)

−1	−1	−1
−1	8	−1
−1	−1	−1

Fig. 2.9 Example image

depending on the size of the filter. For example if we have a filter size of 2×2 then there are four weights in the filter distributed in a table form (see Fig. 2.8). The weights in each filter are being learned by the CFFNN while minimizing the objective function (see Sect. 2.2.2.3).

Expert Sect. 2.5 Convolution Operation on an Image
If we apply the 3×3 convolution filter in Table 2.3 to Fig. 2.9 the result is Fig. 2.10. As you can see this simple filter, by applying a convolution operation to the original image (as explained in Fig. 2.8), finds the most important edges of the objects in the image.

Figure 2.8 shows how the convolution operation is performed in the case of a grayscale image. The optional bias is also indicated in the resultant feature map (see Sect. 1.2 for more details about the bias). A grayscale image has a depth of 1 (only one color channel).

If there are three channels in the input image (i.e., a color image with red, green and blue components) then we say that the image has a depth of 3. The first

Fig. 2.10 Convolution filtered example image

convolution layer must also have filters with a depth of 3! The convolution operation
is then applied to all channels. The operation still results in one number in the feature
map. For the example in Fig. 2.8, if each pixel has a depth of 3 the top-left value in
the feature map can be calculated as follows:

$$a_r.w1_r + b_r.w2_r + e_r.w3_r + f_r.w4_r+$$

$$a_g.w1_g + b_g.w2_g + e_g.w3_g + f_g.w4_g+$$

$$a_b.w1_b + b_b.w2_b + e_b.w3_b + f_b.w4_b + \text{bias}$$

The subscripts r, g, b indicate the variables relating to the three color components.
As you can see, not only does the number of pixels increase substantially in the case
of a full color image but also the number of weights!

Remember that the weights in each filter are being learned by the CFFNN while
minimizing the objective function (see Sect. 2.2.2.3). The weights in all of the filters
are usually initialized to a random number before starting the learning process.

Each convolutional layer has three important parameters: the size of the filter, the
so-called stride and the padding. We will discuss these in more detail below.

Filter Size (W × H)

Filters are usually square, i.e., both sides are of equal size. Remember that every
convolution layer acts as a detection filter for the presence of specific features or

patterns present in the input image. The initial convolution layers in a CFFNN detect large features. The size of the detectable features depends on the size of the filter! For this reason it is important to choose the right filter size according to the input image size and also according to the size(s) of the object(s) in the image. For example, if we have an input image of 200×200 and we choose a filter size of 10×10, then the largest features the filter in the first convolution layer will be able to find occupy a maximum of 0.25% of the input image area ($10 \times 10/200 \times 200$)! Remember that further convolution layers find smaller and smaller features.

Stride (S)

The stride of a convolution filter is the number of steps, in pixels, the filter jumps while sliding (convolving) over the image. The stride depends on the size of the filter and on the input image size because some combinations are not feasible! For example, if the input image size is 5×5, the filter size is 2×2 and the stride is chosen as 2, then the filter cannot slide over the end of the image because there are not enough pixels (see Fig. 2.11)! We must always make sure that the stride is feasible!

By choosing a higher value for the stride, the layer size can be decreased (decreasing the number of weights) but finer details will be lost!

Padding (P)

Padding means adding a margin of 0 values around the input image to increase the size of the image. There are two reasons for applying padding. The first reason is to prevent the reduction of the image size while applying convolutional layers (see Fig. 2.12) and the second reason is to be able to apply a stride larger than 1 (see the previous section). *The decrease in image size, without using padding, can become a*

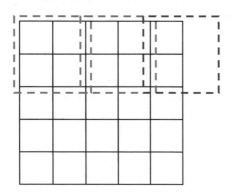

Image size = 5 x 5

Filter size = 2 x 2

Stride = 2

NOT FEASIBLE! Stride must be decreased to 1 or zero padding must be added (see the next section)!

Fig. 2.11 Unfeasible stride

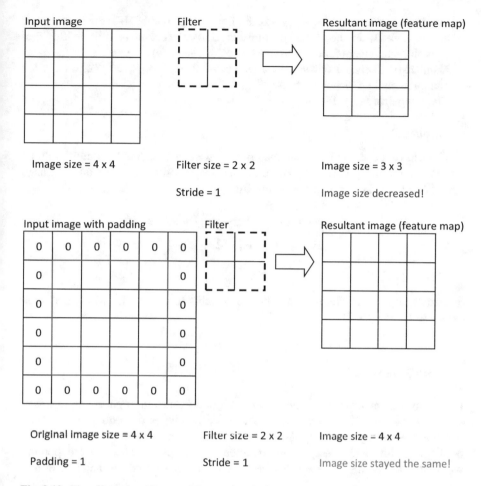

Fig. 2.12 The effect of padding = no image size decrease

huge problem if you apply many convolutional layers, one after the other, because after several convolution operations the image may even disappear (size 0 × 0)!

Calculating the Output Size of a Convolutional layer

In many applications the input size of each layer must be given. When using convolutional layers with possible stride and padding it is not always immediately obvious what the output size from a layer is (the input to the next layer). For this reason let us look at the mathematical equations for calculating the output size from a convolutional layer. Please note that the AI-TOOLKIT has a built-in calculator which helps you to do this very easily.

The input parameters to a convolution layer are as follows:

- Input volume: $W_1 \times H_1 \times D_1$ (W_1 is the input image width, H_1 is the input image height and D_1 is the depth or number of color channels if it is the first convolutional layer (depth = number of color channels)
- Number of filters: N (chosen as a power of 2, e.g., 4, 8, 16, 32, 64 ... for computational efficiency reasons)
- The filter size: $F_W \times F_H$
- Stride: S
- Padding: P

If we have the above input parameters then we can calculate the output volume size $W_2 \times H_2 \times D_2$ (where W_2 is the output image width, H_2 is the output image height and D_2 is the depth of the output image):

$$W_2 = (W_1 - F_w + 2P)/S + 1$$
$$H_2 = (H_1 - F_h + 2P)/S + 1$$
$$D_2 = N$$

Usually $F_w = F_h$! The total number of weights in a convolution layer is then equals to $F_w \times F_h \times D_1 \times N$.

2.2.3.3 Pooling Layer

In this layer a small window (the pooling window) slides through the input image and down samples it (see Fig. 2.13). There are two types of pooling layers: max pooling and average pooling. Max pooling takes the maximum pixel value in the pooling window and average pooling averages the pixel values in the pooling window. Max pooling preserves detected features (edges, curves...) and is therefore

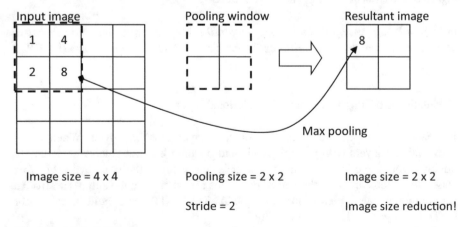

Fig. 2.13 Max pooling

the most commonly used. A pooling layer is always situated between two convolutional layers; it reduces the amount of data in the neural network, and thus the necessary computational resources.

The pooling layer is not very popular anymore because it is not actually very effective and can even be replaced by just using a larger stride.

The input parameters to a pooling layer are as follows:

- Input volume: $W_1 \times H_1 \times D_1$ (W_1 is the input image width, H_1 is the input image height and D_1 is the depth)
- The pooling size: $F_W \times F_H$
- Stride: S

If we have the above input parameters then we can calculate the output volume size $W_2 \times H_2 \times D_2$ (where W_2 is the output image width, H_2 is the output image height and D_2 is the depth of the output image):

$$W_2 = (W_1 - F_w)/S + 1$$
$$H_2 = (H_1 - F_h)/S + 1$$
$$D_2 = D_1$$

Usually $Fw = Fh$! The most commonly used input parameter combinations are $Fw = Fh = 2$ and $S = 2$ (stride) or $Fw = Fh = 3$ and $S = 2$.

2.2.4 Recurrent Neural Networks

Traditional neural networks and machine learning models in general are not able to take sequential information into account. For example, in language processing we often need to know a few of the previous words in a sentence in order to decide what the next word should be, or in the case of event processing we often need to know a few previous events in order to be able to perform a specific task. The only way to handle sequential or time series information with traditional neural networks is to aggregate the data in a time window and use it as input to the neural network (see Sect. 8.7.2.1 of Chap. 8).

Recurrent neural networks (RNNs) solve this problem by making information in a previous time step available in the next time steps by using a feedback loop. It is a feedback loop because the next time step is a back-loop in the network; at each time step the data is pushed through the network again and again. This will be clearer when we draw the RNN structure later.

RNNs were invented for processing sequential (time series) data and they were first introduced in natural language processing (NLP) for learning sentences and language rules. A sentence in a natural language is a sequence of words and each word is a sequence of letters or a sequence of phones (see Chap. 5).

We can represent a recurrent dynamical system with the following equation (modified from [16]):

$$s_t = f(s_{t-1}; W)$$

where s_t is the state of the system at time step t and W is a parameter vector (W refers to weight). We call this system recurrent because it refers back at time step t to the state at time step t-1. It is possible to simplify or unfold this equation so that the final equation does not involve recurrence in the case of a finite number of times steps. This unfolding is the main principle which is applied in the computations of RNNs and it enables us to use the forward and backward propagation algorithms employed in traditional feedforward neural networks. The state equation (s_t) for three times steps can be unfolded and written as follows:

$$s_3 = f(s_2; W) = f(f(s_1; W); W)$$

The state of an RNN is equal to the state of its hidden units (hidden layers and nodes) and therefore the previous state equation can be rewritten as follows:

$$h_t = f(h_{t-1}, x_t; W)$$

where h_t is the state of the hidden units at time step t, f is the activation function (e.g. tangent hyperbolic), x_t is the input vector at time step t, and W is a parameter vector containing the weights. The same activation functions and the same parameters are used at every time step!

A schematic representation of an RNN and of the unfolded RNN can be seen on Fig. 2.14.

The delay shown on Fig. 2.14 needs to be introduced in order to wait until the activations are processed at the next time step. L_t is the loss or objective function for each time step which is optimized.

We often refer to an 'RNN cell' when we talk about an RNN, which can be graphically represented as shown on Fig. 2.15 (all the symbols are the same as on Fig. 2.14; F_A is the activation function).

Depending on the architecture of the RNN different applications are possible. The most important architectures can be seen on Fig. 2.16:

Notice on Fig. 2.16 that there are some RNN outputs (at some time steps) which are ignored. It depends on the required output of the whole network as to which RNN outputs go through at the end of the network.

2.2.4.1 LSTM Cells

Because of the repetitive calculations in RNNs with the same parameters (weight matrix) due to the unfolding of the network, RNNs are sensitive to some problems with the calculated gradients (vanishing gradient and exploding gradient problems)

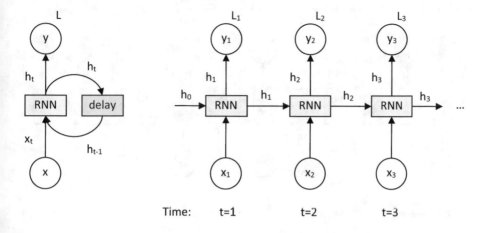

Fig. 2.14 RNN and unfolded RNN

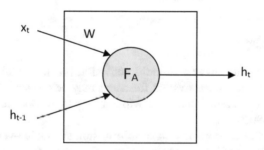

Fig. 2.15 RNN cell

[16]. In order to correct these issues, and in order to build-in long term memory into the RNNs, the so-called *Long Short Term Memory* (LSTM) cell was invented.

Simple RNNs can use information from a few previous time steps. For example, if we try to predict the next word in a simple sentence such as "The leaves fall off the ..." then an RNN would be able to learn that the next word is "tree". But as the gap between the relevant information and the target word grows (e.g., several sentences further) RNNs have more and more problems with learning to connect the information [56]. The solution to this problem is a kind of long term memory which keeps the necessary information in the network for a longer period of time.

The graphical representation of an LSTM cell is shown on Fig. 2.17.

Figure 2.17 with the LSTM cell may look a bit intimidating at first sight, but let us look at the building blocks step by step. An LSTM cell contains an input gate (i_t), an output gate (o_t), a forget gate (f_t), a cell state (C_t), an activation for the input (F_{AI}), an activation for the cell's state (F_{AC}) and three multiplication operators. The input-,

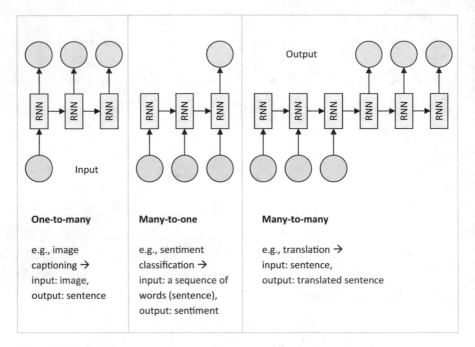

Fig. 2.16 RNN architectures

output- and forget-gates have a sigmoid activation function, which is the σ in the semi circles. The other two activation functions may be chosen by the user, usually the tangent hyperbolic function is used. More information about activation functions can be found in Sect. 2.2.2.1.

An LSTM is able to store and make information available over a long period of time. For example, if the input gate stays closed (i.e., the activation is close to zero as it is a sigmoid function) the value (state) of the cell (C_t) will not be overwritten and may be available in a later time step (further in the sequence) if the output gate is open (i.e., the activation function is close to 1). Once the information has been used and is not needed anymore then it can be erased (set to zero) by using the forget gate. The aim is, of course, not to manually store and erase information in the cell (and network) but for the network to learn when to store and when to erase information automatically! Note also the delayed feedback loop from C_t weighted by the forget gate, this keeps the information (value) available until required (i.e., by not erasing it with $f_t = 0$ until it is no longer needed).

LSTM cells may be connected to each other in an RNN network and they may replace hidden layers in a neural network. The LSTM cell replaces the simple RNN cell shown on Fig. 2.15.

We can write the simplified equations for an LSTM cell (similar to a traditional neural network) as follows (modified from [16, 56]):

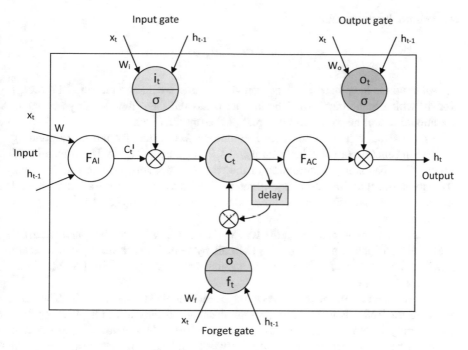

Fig. 2.17 LSTM cell

- The forget gate:

$$f_t = \sigma\left[W_f(h_{t\ 1} + x_t) + b_f\right]$$

- The input gate:

$$i_t = \sigma\left[W_i(h_{t-1} + x_t) + b_i\right]$$

- The cell's state before the input gate:

$$C_t^I = F_{AI}\left[W(h_{t-1} + x_t) + b\right]$$

- The cell's state:

$$C_t = f_t \cdot C_{t-1} + i_t \cdot C_t^I$$

- The output gate:

$$o_t = \sigma\left[W_o(h_{t-1} + x_t) + b_o\right]$$

- And finally the output:

$$h_t = o_t \cdot F_{AC}(C_t)$$

All symbols have been explained previously and are shown on Fig. 2.17 except for 'b' which is the bias term. The bias term has the same functionality here as in traditional neural networks and it is part of the optimization.

Most of the values in the above equations are scaled by the corresponding activation function (a value between 0 and 1 in the case of sigmoid). As mentioned previously, thanks to the unfolding of the recurrent network we can use the same (or very similar) optimization algorithms for RNNs as we use for traditional neural networks. The main difference is how the RNN layer (e.g., the LSTM cell above) is calculated.

There are many variations on the LSTM theme; for example, the gated recurrent unit (GRU) which simplifies the LSTM cell by combining several gates together (such as the input and forget gates into an 'update' gate). RNNs and LSTMs are an active research area.

This section introduced the basics of recurrent neural networks (RNNs). The main use case for RNNs is natural language processing but they have also been applied successfully to other fields such as image captioning. In general we can say that RNNs increase the complexity in neural networks and therefore they also slow down processing (training).

2.2.5   Random Forest (Decision Tree)

Decision tree and random forest (i.e., with several decision trees) algorithms are less widely used than SVMs or neural networks, but for some datasets they may provide good results and because of their simplicity may be preferred. They are also more transparent with regard to the calculation procedure and model structure (except for random forests with a lot of trees), which is less obvious in the case of SVMs and neural networks.

A decision tree can be thought of as a series of appropriately developed follow-up questions and answers. The tree starts with a root node ('question') and continues with several nodes ('answers') after which a new node/question is added (an internal node) at all of the leaf nodes (answers) which become an internal root node and so on. The model building algorithm reads the training data and, step by step, builds the decision tree by carefully deciding on the different nodes with a top-down divide and conquer strategy (i.e., a root node, internal nodes and their hierarchy). A statistical property (the so-called information gain) is used to classify the attributes and to build the decision tree from the data. Every time a node needs to be chosen, the attribute with the highest information gain (the largest split over the input space) is used. This procedure is repeated until a final leaf node is reached which represents the classification target we are looking for.

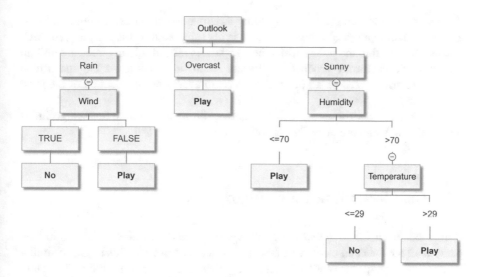

Fig. 2.18 A simple decision tree for deciding whether to play a specific sport

Table 2.4 A simple dataset for deciding whether to play a specific sport

Outlook	Temperature	Humidity	Wind	Decision
Sunny	29	85	FALSE	No
Sunny	27	90	TRUE	No
Overcast	28	78	FALSE	Play
Rain	21	96	FALSE	Play
Rain	20	80	FALSE	Play
Rain	18	70	TRUE	No
Overcast	18	65	TRUE	Play
Sunny	22	95	FALSE	No
Sunny	21	70	FALSE	Play
Rain	24	80	FALSE	Play
Sunny	24	70	TRUE	Play
Overcast	22	90	TRUE	Play
Overcast	27	75	FALSE	Play
Rain	22	80	TRUE	No

The information gain can be calculated in many different ways with the help of the so-called entropy, the Gini index or the gain ratio, etc. The explanation of the detailed calculation procedure and equations is beyond the scope of this book.

A simple example of a decision tree and input data can be seen on Fig. 2.18 and in Table 2.4. Please note that some non-numerical textual data is used for simplicity. In most of the decision tree or random forest algorithms numerical values are used (text labels can be converted to numbers).

A random forest is a set of decision trees built on random samples of the data (random subsampling using a uniform distribution). Instead of using the best policy, the maximum information gain for splitting as explained previously, a random forest algorithm uses random splitting rules (uniformly sampled information gain) for each tree and tries to find the policy which best separates the data. Many trees are built in this way and each may produce a different result or decision. At the end a final decision is made based on all of the results (with averaging or selection of the most voted class).

Decision tree algorithms are sensitive to over-fitting but random forest algorithms reduce this risk and are better at generalization.

2.3 Unsupervised Learning Algorithms

Remember from Chap. 1 that the goal of unsupervised learning is to find hidden patterns in the data and classify or label unlabeled data and use this to group similar items (similar properties and/or features) together, and thus put dissimilar items into different groups. Another name for unsupervised learning is clustering.

In this section a selection of effective and commonly used clustering algorithms will be explained in detail. There are many clustering algorithms available but there is no all-in-one solution which is successful with all types of data. It is difficult to handle all types of data in one algorithm because the data can be low or high dimensional (features or columns), the data may or may not contain noise, the clusters may or may not have varying density, the data may or may not contain hierarchical features, etc. You will often have to use trial and error when you want to apply unsupervised learning to an unknown dataset and try several algorithms. The AI-TOOLKIT includes all of the following algorithms and they can be applied very easily to any dataset.

2.3.1 k-Means Clustering

One of the most commonly used clustering algorithms is k-Means because of its simplicity (it has only one parameter: the number of clusters), and its high performance (accuracy and speed) in the case of several types of datasets. The k-Means algorithm has the advantage that it can handle not only low but also high dimensional (features or columns) data. The accuracy of k-Means increases with the size of the dataset; more data will result in higher accuracy!

The base algorithm, which will be explained hereunder, has been extended and further optimized to provide even better results in more advanced versions of the k-Means algorithm (see the AI-TOOLKIT). One of the most important improvements concerns how the centroids are initialized. The center of a cluster is called a centroid. A refined approach for choosing the initial centroids for k-Means clustering has been introduced by [2]. This approach involves running k-Means several times on random subsets of the data, and then clustering those solutions to select refined initial cluster assignments.

The k-Means algorithm is very simple and can be summarized as follows:

1. It first selects k initial centroids or means (k is the number of desired classes as an input parameter). The basic initial centroid selection is performed by random partitioning. The initial selection method is important, and this is one of the areas in which k-Mean has improved in more advanced algorithms (see above).
2. For each data point it calculates the distance between the data point and each centroid. The data point is then assigned to the nearest centroid.
3. A new set of centroids is computed and all distances are recalculated. This iterative process (points 2 and 3 are repeated) continues until a specific criterion (see Expert Sect. 2.6) is met and the centroids become stable.

Some of the disadvantages of k-Means are as follows:

- The user must provide the number of k classes (clusters) as an input parameter. This may not always be available in the case of an unknown dataset.
- It is sensitive to noise in the data.
- It is sensitive to outliers in the data. It is recommended that the dataset is analyzed and outliers removed before k-Means is applied.

It is not easy to guess the number of classes in an unknown dataset. There are some other algorithms which make an educated guess about the number of classes, e.g., MeanShift clustering (see Sect. 2.3.2) but they require other parameters which are also not easy to estimate. There are also some techniques to estimate the number of classes by other means such as using the so-called elbow method (see Sect. 8.2 for more information about this subject).

In Expert Sect. 2.6 further details regarding the k-Means algorithm will be explained. These details are not necessary for using k-Means, but they may be of general interest.

Expert Sect. 2.6 k-Means Internals
An important part of the k-Means algorithm is the distance calculation between data points and centroids. There are several different distance measures, for example, the so-called Euclidean distance or the Manhattan (or city block) distance. In practice the Euclidean distance is used most often and it can be calculated as follows:

$$d(x_i, c_j) = \sqrt{\sum_{f=1}^{n} \left(x_i^{(f)} - c_j^{(f)} \right)^2}$$

where x_i is the ith data point, c_j is the jth centroid and $d(x_i, c_j)$ is the distance between them; f is the number of features (columns or dimensions) in the dataset.

(continued)

Expert Sect. 2.6 (continued)
The measure of cluster compactness (how near the data points in a specific cluster are to the cluster centroid) is used in the stopping criterion for the k-Means iteration and it is called the sum of squared error (SSE):

$$SSE = \sum_{j=1}^{k} \sum_{i=1}^{m} d(x_i, c_j)^2$$

where k is the number of clusters or classes; m is the number of data points in the jth cluster or group. x_i is the ith data point in the jth cluster; c_j is the centroid of the jth cluster; and $d(x_i, c_j)$ is the Euclidean distance of x_i and c_j (see above).

The stopping criterion of the k-Means iteration is a minimum decrease in SSE. If the SSE does not decrease with a minimum pre-defined ε small value, then the iteration stops (i.e., there is no significant improvement). It is also possible to specify the number of data points that may still change group (i.e., by defining a maximum number—below which the iteration may stop) or the number of centroids that may still change (i.e., by defining a maximum Euclidean distance change which is still acceptable—below which the iteration may stop).

2.3.2 MeanShift Clustering

MeanShift is also a centroid-based algorithm (remember that the center of a cluster is called a centroid) like k-Means but instead of minimizing the distance between the points and the centroids, it slides a window with a given radius (parameter) towards regions of higher density where the centroid is located. The highest density location is simply a sub-area of the sliding window where the most points are located.

The algorithm is very simple:

- It selects a random data point and calculates the so-called mean shift vector in the sliding window with a given radius. The mean shift vector points to the highest density location in the sliding window. How this mean shift vector is calculated is the most important part of the algorithm (see later).
- It moves the center of the sliding window by the mean shift vector.
- It repeats the above steps for each point until convergence.

One of the major advantages of this algorithm is that it does not need the number of classes (clusters) as an input parameter because it discovers it automatically! The algorithm has only one input parameter—the radius of the sliding window. The radius has an important effect on the results (see Expert Sect. 2.7) and deciding its value is not trivial. There are, however, some methods which can be used to estimate the radius (see later).

There is a more advanced version of the MeanShift algorithm (used in the AI-TOOLKIT) in which the convergence of the algorithm is improved by varying the radius of the sliding window per data point. The radius is calculated as follows:

- For each point in the dataset it selects the m nearest points (with the k-nearest neighbor (KNN) method) and it calculates the m distances to the m nearest points—m is chosen as $m = c * n$, where c is a ratio of the n total number of points (e.g., 0.2 (20%)).
- Then for each data point it selects the maximum distance from the m distances. The sliding window radius estimate is then the average of the maximum distances of all n data points.

MeanShift is not only used in clustering; because of its effectiveness, simplicity and speed it also has many applications in machine vision such as color and geometric segmentation, object tracking, etc.

Expert Sect. 2.7 MeanShift Clustering Algorithm
In order to make the explanation of the MeanShift algorithm very simple let us take a one dimensional clustering example. Figure 2.19 shows some data points distributed on a horizontal line. The aim is to group the data points into two clusters. We can draw the distribution of the points denoted with f (x) above the horizontal line (the image is just an illustration and not precise). The MeanShift algorithm will try to find the local maximum (highest density location) of the distribution which is the centroid of a cluster and it will assign all points in the neighborhood to this cluster. This is a so-called gradient ascent method.

To do this we need to estimate the distribution of the points with a continuous function f(x). The MeanShift algorithm does this by defining and applying a so-called *kernel function* to each data point and combining all of the kernel functions in one continuous function. Remember that we are only in 1D at the moment and that our algorithm must scale up to higher dimensions!

Figure 2.20 shows three simple symmetric distribution functions which can be used as kernel functions. In practice the Gaussian kernel is used most often.

Figure 2.21 shows the Gaussian kernel applied to each point (solid lines) and the final continuous distribution (dashed line) which is just the combination (sum) of all of the kernel functions. Think about the kernel function as a statistical trick to be able to define a continuous distribution function.

The width of the symmetrical kernel function (two times the radius r) is an important parameter because if it is too small then too many clusters will be found and if it is too big then too few clusters will be found. Figure 2.22 illustrates this effect—the left side shows a kernel width that is too small and the right side shows a kernel width that is too large.

(continued)

Expert Sect. 2.7 (continued)

The Gaussian kernel function (a normal distribution) can be described with the following mathematical equation:

$$K_N(X) = c.e^{-\frac{\|X\|^2}{2}}$$

The parameter r is the radius of the symmetrical kernel function (half of the width of the kernel function!) which is actually identical to the radius of the aforementioned sliding window used by MeanShift! In the case $r = 1$ then we get the standard normal distribution! The parameter c is a constant which controls the height of the kernel function. The double lines around a variable denote the absolute value of a vector.

The approximate f(X) distribution of our data points in this example (with a kernel K) can be calculated by simply adding the kernel functions at each point together:

$$f(X) = \frac{1}{N} \sum_{i=1}^{N} K(X - X_i)$$

We can replace the general kernel term with, for example, the Gaussian kernel (K_N) function defined above:

$$f(X) = \frac{c}{N} \sum_{i=1}^{N} e^{-\frac{\left\|\frac{X-X_i}{r}\right\|^2}{2}}$$

The gradient (derivative) of this distribution function points in the direction we have to move in order to find the centroid of the cluster (or higher density region; see the beginning of this explanation)! If we denote the negative gradient of f(X) with G(X) then we can write the equation of the mean shift vector as follows

$$M(X) = \frac{\sum_i X_i G(X)}{\sum_i G(X)}$$

The calculation of the term G(X) just involves taking the negative derivative of f(X): $G(X) = -f'(X)$. We take the negative gradient because otherwise the vector would point in the opposite direction!

(continued)

Expert Sect. 2.7 (continued)

The simplified MeanShift algorithm for assigning a point to a cluster then performs as follows:

- It initializes X randomly to one of the data points in the dataset and calculates M(X), the mean shift vector.
- It moves X by the M(X) mean shift vector.
- It repeats the above steps until convergence. At the end, X will be the centroid of the cluster the data point is assigned to!

In higher dimensions (several columns in the dataset) the algorithm works in the same way as explained above but instead of moving along a line the mean shift is undertaken in higher dimensions. A 2D example of the first three steps ('shifts') taken is shown below. Each step starts from where the last step ended (Fig. 2.23).

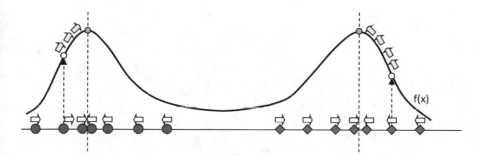

Fig. 2.19 Example data points

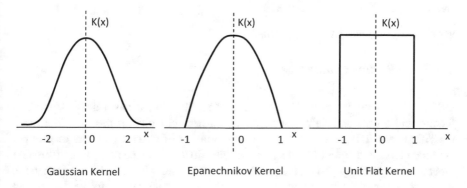

Gaussian Kernel   Epanechnikov Kernel   Unit Flat Kernel

Fig. 2.20 Distribution functions

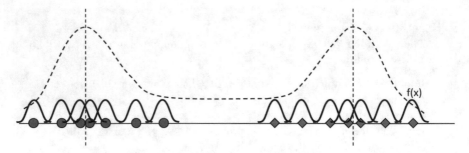

Fig. 2.21 Gaussian kernel applied to example data (1)

Fig. 2.22 Gaussian kernel applied to example data (2)

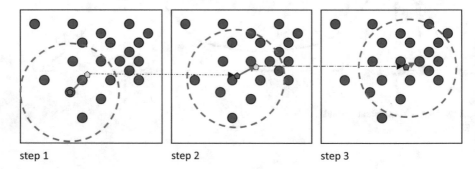

step 1   step 2   step 3

Fig. 2.23 The Simplified MeanShift Algorithm

2.3.3   DBScan Clustering

DBScan is a density based algorithm like MeanShift but it has a noise separation
feature and therefore it is not sensitive to outliers or noise in the data like other
algorithms are! DBScan also discovers the number of classes (clusters or groups)
automatically, and it can handle clusters with different shapes and sizes (this is a
great advantage compared to, e.g., k-Means which cannot handle all cluster shapes).
It can, for example, handle non-convex cluster shapes which cannot be discovered
with k-Means!

There are two main disadvantages of DBScan:

1. It does not work well with clusters which have varying density.
2. It does not work well with high dimensional data (many columns or features). This is an important disadvantage compared to, e.g., k-Means if you have a lot of features in your data.

DBScan is very similar to MeanShift and also works with a sliding window with a given radius. It has an extra parameter that defines the minimum number of data points in the sliding window, which is needed for considering the data points as one group or cluster (see below).

DBScan does not need the number of clusters as parameter but it does need two other parameters (the sliding window radius and the minimum number of points in the sliding window) which are difficult to estimate (the algorithm is sensitive to these parameters).

Remember that the basic idea behind density based clustering algorithms (like MeanShift and DBScan) is that clusters are dense areas of data points separated by regions of lower density. The maximum possible number of density-connected data points forms a cluster.

The algorithm is relatively simple and works as follows:

- It first selects a random data point and analyzes the surrounding small area in a sliding window (very similar to MeanShift). If the density of data points is sufficient (minimum number of points in epsilon radius) the region in the sliding window is considered as part of a new cluster.
- Next all of the neighboring points are analyzed with the same technique and the areas are merged if they have the same minimum density (these neighboring points are sometimes called 'density reachable' points = similar density). The areas (data points) are separated if they have a different density.
- The process above is repeated until all of the data points have been analyzed and added to a cluster or flagged as noise (outliers). An outlier is a point which does not have enough neighboring points in epsilon radius (i.e., there is < minimum number of points).

It may happen that a point flagged as noise later becomes part of a cluster because it is in the neighborhood of a cluster and it is a border point (on the outskirts of the cluster)! For an example, see Figs. 2.24 and 2.25. We call this an asymmetric behavior because if we start from a border point of a cluster (point A) the border point will not be added (low density region), but if we start from a point inside the cluster (point B) then border point A will be added because it is a neighbor (inside the same sliding window) of a point which is in the cluster already!

2.3.4   Hierarchical Clustering

Hierarchical clustering algorithms treat data points in a hierarchical manner. There are two types of hierarchical clustering algorithms—those that use a merge strategy and those that use a divide strategy while treating each data point.

Fig. 2.24 DBScan
algorithm

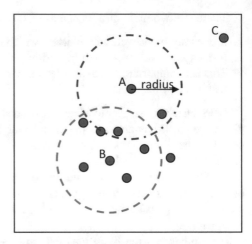

Minimum number of points: 5
Point A is a border point and part of the
cluster. The red dash-dot line indicates a low
density region (points < 5). Point C is an outlier.

The merge strategy first assigns each data point to a separate cluster (n data points = n clusters) and then merges (agglomerates) pairs of clusters (a similarity measure determines which clusters are merged) until all data points have been merged into a single cluster. The name of this type of algorithm is hierarchical agglomerative clustering (HAC). More about this algorithm later!

The divide strategy (the other type of hierarchical clustering) first assigns all data points to one cluster and then splits the cluster recursively into child clusters until all data points are in separate cluster.

In practice the merge strategy, the HAC method, is the most popular. The HAC algorithm is very simple and can be summarized as follows:

1. Assign each data point into a separate cluster.
2. Merge a pair of clusters into one cluster based on a chosen similarity measure (see later). For example, if the similarity measure is the centroid distance of two clusters then merge the two clusters which are closest to each other (centroid distance).
3. Repeat step 2 until all data points (clusters) are merged into one cluster.

The hierarchy of clusters can be represented as a tree (or dendogram) and each level of the tree corresponds to a step in the iteration above. Naturally each level corresponds to a different number of clusters the data is grouped into. This is one of

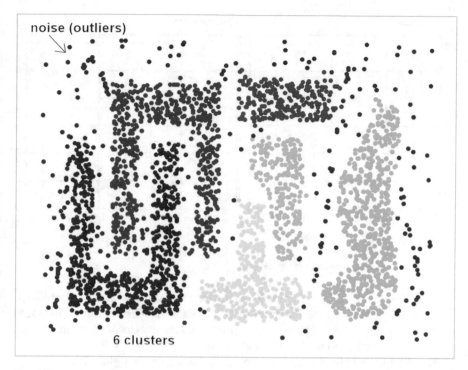

noise (outliers)

6 clusters

Fig. 2.25 DBScan example

the advantages of this algorithm because we can select from the hierarchical tree the best number of clusters after the algorithm is ready.

The HAC algorithm depends on the similarity measure chosen for the merging decision. The similarity measure of data points is first of all determined by a distance measure as in the other algorithms. The Euclidean distance is most commonly used for this purpose which can be written as follows:

$$d(x_i, x_j) = \frac{1}{n} \sum_{i=1}^{n} (x_i - x_j)^2$$

Depending on how we define the distance between two clusters (containing several data points) there can be several alternative similarity measures and algorithms:

- We can take the maximum distance from the collection of distances between all data points in the two clusters (pair wise).
- We can take the average of all distances instead of the maximum as explained here above.
- We can simply take the clusters' centroid distances. The centroids are calculated in the same way as we have seen for other algorithms.

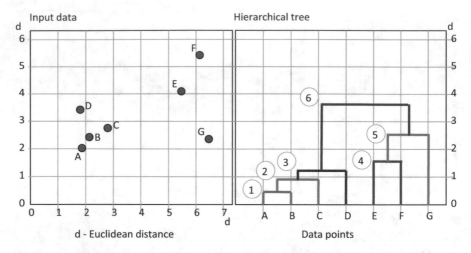

Fig. 2.26 Hierarchical clustering example

The 2D example on Fig. 2.26 explains how the algorithm works. For simplicity, both the input data and the hierarchical tree are shown with the same Euclidean distance scale. The drawing is not precise (the values are not calculated) and is meant only for showing the principles of the algorithm.

The steps to build the hierarchical tree are indicated on the right side of Fig. 2.26 by the numbers in circles. The first step is to merge points A and B because they have the minimum Euclidean distance. Then the centroid of the cluster (A, B) is closest to point C and therefore they are merged. Then D is added for the same reason. Points E and F are the next closest points and are therefore merged into a cluster (E, F). Then G is added because the centroid of (E, F) is the closest to G. And finally the tree is closed by merging the clusters (A, B, C and D) and (E, F and G) together. The height of the tree branches (horizontal lines) indicates the Euclidean distance between the clusters.

If we start to cut the tree on Fig. 2.26 vertically from the top then we get, for example, two clusters (A, B, C and D) and (E, F and G), at step/line 6 and three clusters, (A, B, C and D) and (E, F) and (G), at step/line 5, etc.

One of the disadvantages of this algorithm is that it is much slower than the other algorithms, but it may be useful if the input data contains hierarchical information which cannot be discovered by the other algorithms.

2.4   Reinforcement Learning Algorithms

Let us first summarize what we learnt in Chap. 1 about reinforcement learning.

We define reinforcement learning as a *general purpose decision making machine learning framework used for learning to control a system*. A reinforcement learning

Fig. 2.27 Reinforcement
learning system

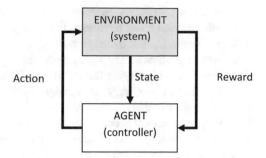

system can be symbolized by the interaction between a so-called *Environment* (*system*) and an *Agent* (*controller*) as indicated on Fig. 2.27. Many reinforcement learning applications train models in a virtual environment where the model plays a simulation over and over again and observes success and failure while trying different actions (trial and error).

The reinforcement learning system typically operates by initializing the *Environment* to a random state, but it could also start with a specific state. The *state* can mean many different things and depend on the problem. For example, it can be the speed of a car or it can also be several properties at a specific times step such as the speed, the direction, etc. The state is then passed to the *Agent* which calculates an appropriate action (for example, in the case of a car this could be increasing the speed or braking). Each time an action is taken and passed back to the *Environment* a reward is calculated (a numerical performance measure) for the last state transition and passed back to the *Agent* (reward signal). This is how the *Agent* knows if the action was right or wrong. This cycle is repeated until an end goal is reached e.g. the system reaches an expected state by reaching a destination, winning or losing a game, etc. We call this the end of an *Episode*. After an episode ends the system is reset to a new (random) initial state and a new episode begins. The reinforcement learning system cycles through many episodes during which it learns which actions are more likely to lead to the desired outcome or goal by optimizing the long term reward.

The three major components in a reinforcement learning system, the *Action*, the *State* and the *Reward*, are the glue of the whole system. How are the actions chosen? How are the states and the environment modeled? How is the reward determined? These are key questions and the answers explain how the whole system works; therefore, we will next look at these elements one by one in detail.

2.4.1 Action Selection Policy: How Does the Agent Select an Action?

Remember that the agent decides which actions are taken in which situations (states). In each reinforcement learning system there are a number of pre-defined possible actions from which the agent can choose. For example, in the case of a very simple car it could be two actions, accelerate and brake. When starting the learning process the agent does not know anything about the problem and thus must select actions by trial and error. This is exactly how, e.g., animals or humans approach unknown situations or problems. The agent must find out which actions are correct in which situations (states) by trying them out despite the risk of getting a negative reward (penalty). The agent simply takes random actions. This is the so-called *Exploration* action selection strategy.

During the learning process the agent stores (learns) which actions were successful (positive reward) in which states of the system, and later it can use this information to select actions which are more likely to lead to positive rewards in the long run. This is the so-called *Exploitation* action selection strategy.

The agent's actions affect the long term behavior of the environment. The best learning strategy on the long run is a well-balanced combination of the exploitation and exploration action selection strategies. The reason for this is not only because in the beginning of the learning the agent does not know anything about the problem and must take random actions but also because later, when the agent already knows a good set of actions, it should explore different sets of actions which may lead to even better results!

Which action the trained agent selects in which state defines the *Action Selection Policy*. The optimal action selection policy is the one that yields the highest expected future reward! We will see in the next section how this future reward can be calculated.

2.4.2   Reward Function: What Is the Cumulative Future Reward?

One of the most important and challenging tasks in reinforcement learning (RL) is the design of an effective reward strategy. If we fail to choose the right rewards the model will fail! We give a reward (or penalty) to the system after each action/step is taken. We often make the distinction between specific steps and give different rewards, for example, during an episode and at the end of an episode; there can also be some special steps specific to the problem which receive a different reward. Usually a higher reward (or penalty) is given for reaching a terminal state. This is the supervision (input) that we give to the system and this is how the agent knows if the action taken was right or wrong! More will be explained about this later when we discuss the examples.

But the RL system does not just simply employ the user-defined rewards! Each time an action is taken and passed back to the environment a so-called *discounted cumulative future reward* is estimated for the last state transition and passed back to the agent.

We often call this *discounted cumulative future reward* estimate the *Reward Function* or *Value Function*. The reinforcement learning system learns to maximize this reward function because maximum reward in the long run means that the best actions have been taken!

There are several keywords in the above definition which need some explanation. Why is it an estimate? Why do we have to take the future rewards into account and not only the immediate ones? Why do we have to discount future rewards and what does discounting exactly mean?

In order to perform well in the long run we need to estimate and take into account the future rewards we are going to get and not only the immediate ones! But future rewards are less certain than current rewards and therefore we discount (decrease the effect of) future rewards in order to take into account the uncertainty. This is very similar to discounted cash flow in finance!

It is an estimate because the agent does not know what the cumulative future reward will be after taking several actions but it can estimate it by using statistical theory! For more information see Expert Sect. 2.8.

Expert Sect. 2.8 How the Reward Function Is Estimated? Deep Q-Learning

Remember that an episode is several cycles of action-reward-state changes until the goal of the system is reached. The last state in an episode is called the *Terminal State*. One episode of the learning process forms thus a finite sequence of actions-rewards-states. If we denote the initial state with s_0, the first action with a_0 and the first reward after the first state change with r_1 then we can write the actions-rewards-states sequence as follows:

$$[s_0, a_0], [s_1, r_1, s_n], [s_2, r_2, a_2], \ldots [s_n, r_n]$$

The first element $[s_0, a_0]$ has no reward since we have just started the learning process and nothing has happened yet. The last element has no action because after reaching the terminal state s_n no more actions will be taken—r_n is the terminal reward.

In order to be able to calculate the discounted cumulative future reward (reward function) we must make the assumption that the probability of the next state s_{k+1} depends only on the current state s_k and action a_k and not on former states or actions, i.e., not on s_{k-1}, a_{k-1}. Thanks to this assumption we can use the theory of the so-called *Markov Decision Process* (MDP) to calculate the total reward of an episode as follows:

$$R = r_1 + r_2 + \ldots + r_n$$

This is the sum of all rewards. Since each reward is given at a specific time step we can also express the above equation as follows (t designates time):

$$R_t = r_t + r_{t+1} + r_{t+2} + \ldots + r_n$$

This gives us the cumulative future reward at any given time step.

(continued)

Expert Sect. 2.8 (continued)

We also know already that future rewards must be discounted because of the uncertainty regarding the results of future actions. The discounted cumulative future reward (reward function) of a Markov decision process is simply the discounted version of the above equation and can be written as follows:

$$R_t = \gamma^0 \cdot r_t + \gamma^1 \cdot r_{t+1} + \gamma^2 \cdot r_{t+2} + \cdots = \sum_{k=0}^{n} \gamma^k \cdot r_{t+k}$$

Here γ is the discount factor in the range $0 \le \gamma \le 1$. The greater the value of γ the more the future rewards are taken into account. A reward received k time steps in the future is worth only γ^k times (a lower value if $\gamma < 1$) what it would be worth if it was received immediately (the present value of a future reward).

With the same logic we can also express the discounted cumulative future reward (reward function) at time step t as follows:

$$R_t = r_t + \gamma.R_{t+1}$$

which says that the reward function at time step t is the sum of the reward function at times step t + 1 and the current reward r_t at time step t. This simply connects the reward function at different time steps!

We also know already that the aim of the agent is to maximize the reward function. A maximum reward will be received if the best actions are taken at any given state. For this reason we denote the maximum reward with $Q(s_t, a_t)$ which is the so-called Q-function. It is a function because it has different values for different state-action combinations. The letter Q comes from 'quality' which refers to the quality of the decisions/actions taken to reach the maximum reward. Because $Q(s_t, a_t)$ is the maximum reward we can write:

$$Q(s_t, a_t) = \max R_t$$

In a similar manner as with the reward function we can calculate the Q-function at any time step t as follows:

$$Q(s_t, a_t) = r_t + \gamma.Q(s_{t+1}, a_{t+1})$$

This equation is called the Bellman equation and it can be used to approximate the Q-function. In practice this is done with so-called *Deep Q-learning* which uses a neural network to learn the Q-function (neural networks are exceptionally good at learning arbitrary types of functions).

(continued)

Expert Sect. 2.8 (continued)

Figure 2.28 shows a possible neural network for this task. There are as many output nodes as the number of possible actions. Each output provides the Q-value for a given state (s_n) and action (a_n) combination with the condition that there is only one active state. This can be achieved by using an input vector which is set to 0 for all states except the active one which is set to 1.

After the neural network learns the Q-function it can give us the Q-value for any state-action combination. The agent can then select the action which results in the highest Q-value.

In practice the above $Q(s_t, a_t)$ equation is slightly modified and includes a so-called learning rate (α):

$$Q(s_t, a_t) = Q(s_t, a_t) + \alpha.[r_t + \gamma. \max{}_a Q(s_{t+1}, a) - Q(s_t, a_t)]$$

This equation may look complicated at first sight, but it is actually the same as before if we set $\alpha = 1$!

The meaning of this equation and of the α learning rate is important! In this equation, $Q(s_t, a_t)$ is the current or old Q-value, and the term '$r_t + \gamma. \max_a Q(s_{t+1}, a)$' is the learned or new Q-value. The term '$\max_a Q(s_{t+1}, a)$' is the Q-value calculated using the best action which provides the maximum Q-value.

If we look at the above equation with these remarks in mind, then we can see that the equation first determines the difference between the old and the learned new value and then multiplies this with the learning rate α, which is in the range $0 < \alpha \leq 1$. The learning rate will decrease this difference and will add this decreased difference to the old value. This means that if $\alpha < 1$ then we will slow down the learning process and let prior knowledge (contained in the old Q-value) be part of the new Q-value by a $1 - \alpha$ defined portion. If $\alpha = 1$ then the old value is replaced by the new one, and the learning is the fastest, but we ignore prior knowledge completely and use only the most recent information. In practice often a learning rate of $\alpha = 0.1$ is used which adjusts the Q-value slowly and lets prior knowledge dominate.

An improvement of Q-learning is the so-called *Double Deep Q-learning* in which two separate Q-value functions are trained, one for the selection of the next action and another one for the reward function (value function). The AI-TOOLKIT has built-in support for both algorithms.

2.4.3   State Model: How Does the Environment Behave?

We define the behavior of the environment in a model. The model influences how the agent's actions will transition the system into a new state. A model does not always exist in a reinforcement learning (RL) system or it may be something very simple or

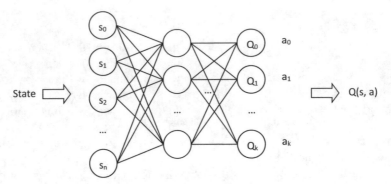

Fig. 2.28 Neural network example

very complex. A very simple example could be following the rules of physics and taking, e.g., gravity into account in an RL system; or we may decide to take into account the rolling resistance if we model something with wheels, etc. If the agent takes an action, such as accelerating a car, then the next state (e.g., location) will also depend on the environment model because of, e.g., the rolling resistance of the wheels. We can make our RL system as simple or as complex as we want! The environment model may of course contain many different rules and not only the rules of physics. In a business RL application these may be rules of decision making, or logistical routing, etc.

The examples in the next sections will help you to understand how to define such a model.

2.4.4   Example 1: Simple Business Process Automation

In this example we will design a reinforcement learning system (RLS) which is able to learn the best method for automating a very simple business process. The aim of the business process is to produce a specific product in a number of steps (tasks). There are 10 possible process steps A, B, C, D, E, F, G, H, I, J of which E is not available. What the steps (tasks) are and what product they produce are not important for this example. The simplified business process diagram can be seen on Fig. 2.29.

The outcome of the business process may be positive, delivering a good product, or negative, delivering a defective product. The aim is to find the best sequence of process steps which result in a good product.

Many real world problems can be represented in a grid layout where connected steps are next to each other. We call this kind of modeling of a problem a *Grid World*. It will be immediately obvious what this means if you look at the grid on Fig. 2.30 and compare it to the process diagram on Fig. 2.29. By transforming our simple business process diagram into a grid form we can use a standardized form of modeling.

Fig. 2.29 Example 1:
Simplified business process
diagram

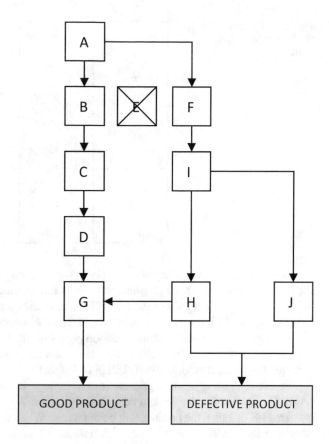

The process starts at grid cell A and ends at a good or defective product. A good product is always delivered through process step G. A defective product could be delivered through process step H or J. Process step E is not available; why it is not available is not important for this example but it could be, for example, because of the absence of an employee.

Next we will look at the possible actions at each process step or grid cell.

2.4.4.1   Actions

Each process step (or task) can be considered as a state in reinforcement learning terms, and at each state there are four possible actions which we represent with UP, DOWN, LEFT and RIGHT in the grid (the four way arrows in the grid +; possible moving directions). These four actions could be mapped to other problem related actions; for example, UP could mean acceleration, DOWN could mean braking, etc. In our example they just mean the transfer of the product to the next step. For example, at state A moving UP would transfer the product to state B, moving RIGHT would transfer it to state F, and moving LEFT or DOWN would do nothing

Fig. 2.30 Example 1:
Business process *Grid
World*

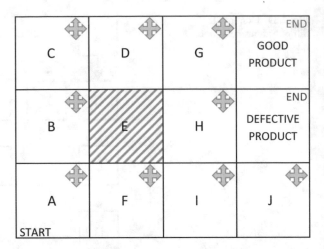

(remain at state A). When any state wants to transfer the product to E then nothing happens and the product stays where it is (E is not available). Finally, state G can deliver a good product by moving RIGHT, state H may deliver a defective product by moving RIGHT and state J may also deliver a defective product by moving UP. On Fig. 2.31 a possible action selection policy (or in short, policy) is shown with blue arrows. Remember that the policy is what each state (step) will do when it receives the product (UP, DOWN, LEFT or RIGHT).

We could think about this grid as, for example, a different person at each state (A, B, C . . .) who decides what to do with the product. When the product is delivered by person H or J then the product will be defective. When the product passes through person G the product is always good. A defective product could, e.g., be delivered because of an incorrect decision at a process step.

2.4.4.2   Design of a Reward Strategy

One of the most important and challenging tasks in reinforcement learning (RL) is the design of an effective reward strategy. If you fail to choose the right rewards the model will fail! In order to decide which rewards to use, we must first discover the outcomes of an episode.

Remember that an episode is several state changes (cycles of action-reward-state) until the goal of the system is reached. The goal is reached in the so-called terminal state, which is the last state in an episode.

In our simple business process there are two obvious outcomes and there is an additional outcome which is not so obvious but logical. The two obvious outcomes are delivering a good product and delivering a defective product. The less obvious outcome is that no product is delivered at all and instead it keeps moving around between the process steps, maybe not forever but for a long time. The intention of the business process is of course the delivery of the product as soon as possible.

Fig. 2.31 Example 1:
Action selection policy

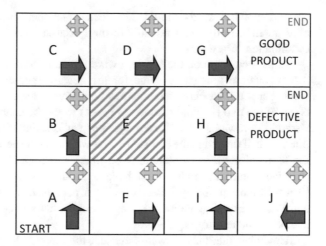

Each outcome should be rewarded according to its success (a good product) or failure (a defective product or no product at all). We need to assign an appropriate reward to all three outcomes!

The reward assigned to the delivery of the product is the *Terminal Reward*, which will be different in the case of a good product versus a defective product. In the case of a good product it will be a positive reward and in the case of a defective product it will be a negative reward (or penalty).

The reward assigned to the third outcome (no product or a delivery delay) should be a negative reward to discourage long processing times. Where and how should this negative reward be assigned? It is logical to assign this negative reward to each process step (whenever the product is transferred to another state or process step) in order to discourage moving around too long (unnecessary actions). We could call this the *Action Reward* (it is a penalty in this case).

The values of these rewards are very important, we cannot just choose any number! In our example we will use the following constant rewards for the three outcomes (for the reasons explained below):

1. Terminal reward—when a good product is delivered: +1
2. Terminal reward—when a defective product is delivered: −1
3. Action reward—when a step is taken: −0.04

The exact magnitude of these rewards is not important but the relationship between them is very important! You could, for example, choose (+10, −10, −0.4) and have the same result. They should of course have a reasonable magnitude—the reason for this will be clear after reading the explanation below.

Let us first concentrate on the terminal rewards. It is logical that we choose a positive reward for the delivery of a good product and a negative reward (penalty) for the delivery of a defective product. What would happen if we chose −0.5 instead of −1? The defective product would be penalized less than the good product is rewarded, which would result in more defective products. It is of course not always obvious which rewards should be selected and therefore it is often necessary to experiment with different rewards through trial and error in order to find the best combination of rewards.

Let us focus now on the action reward which is chosen to be −0.04. It is logical that we choose a negative value for this reward (penalty) in our example. What would happen if we chose a much lower negative value e.g. −0.8 or even −1.5? The effect of the two terminal rewards (+1, −1) would be canceled out! Remember that we add the rewards of each step together while calculating the cumulative reward (the terminal reward is the last step)! The weight of the action rewards (compared to the terminal rewards) would thus be much too high if we chose a much lower negative number, which would have a negative effect on the learning process (it would most probably fail or provide a poor result). The exact value of −0.04 comes from experimentation with this example, but you could of course also choose another value and try, e.g., −0.01.

Remember that these rewards are just the rewards for the current step (or state change) and the discounted cumulative future rewards (reward function) are calculated by our RL system! We provide these values as supervision to the RL system.

2.4.4.3   The Learning Process

Let us now look at how the learning takes place for this example but without going into too much detail, with focus on the main principles.

Remember that the reinforcement learning system typically starts to operate by initializing the environment to a random state but it could also be to a specific state. The state is then passed to the agent which calculates an appropriate action. Each time an action is taken and passed back to the environment a reward is calculated for the last state transition and passed back to the agent (reward function). This cycle is repeated until an end goal is reached and the episode ends. After the episode ends the system is reset to a new (random or specific) initial state and a new episode begins. The reinforcement learning system goes through many episodes during which it learns which actions are more likely to lead to the desired outcome or goal by optimizing the long term reward.

As a first step let us initialize our example environment to the policy shown on Fig. 2.32. Remember that policy means a set of actions for each state. The thick blue arrows show the initial actions for each state (process step). This is of course a very good initial policy, and a random policy could be much worse, but it is easier to understand the learning process in this way.

Figure 2.33 shows the rewards defined for this example. The action reward in each non-terminal-state cell means that we give '−0.04' reward (penalty) when leaving the cell in any direction (taking any step).

The agent takes several steps based on the selected policy until the terminal state is reached and the episode ends. Let us assume that the first three episodes are the ones shown on Figs. 2.34 and 2.36 (an exploration episode is also included).

Fig. 2.32 Example 1: Policy

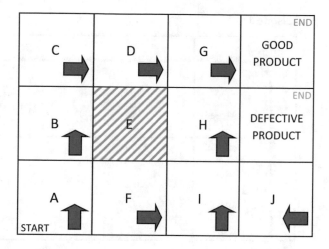

Fig. 2.33 Example 1: Rewards

The total reward for episode 1 with discount factor $\gamma = 1$ (no discount for simplicity) can be calculated as follows:

$$R_1 = -0.04 - 0.04 - 0.04 - 0.04 + 1 = 0.84$$

The total reward for episode 2 with discount factor $\gamma = 1$ can be calculated as follows:

$$R_2 = -0.04 - 0.04 - 0.04 - 0.04 + 1 = 0.84$$

The total reward for episode 3 with discount factor $\gamma = 1$ can be calculated as follows:

Episode 1

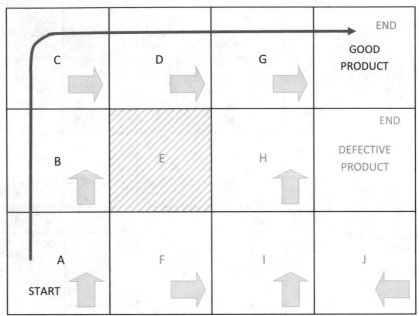

Episode 2

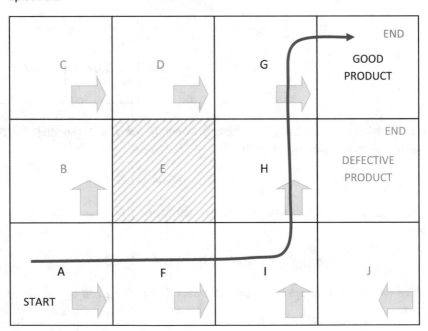

Fig. 2.34 Example 1: Several episodes—part I

$$R_3 = -0.04 - 0.04 - 0.04 - 1 = -1.12$$

The agent will try many episodes, learn the best policy and choose the best process routing in any step. What can we learn from the above three episodes and what would the agent learn? First of all, episodes 1 and 2 have the same total reward but the process steps in episode 2 pass through state H and the agent will learn (while running many episodes) that there is a big chance a defective product will be delivered if it moves to state H. For this reason the agent will prefer the process route of episode 1!

How does the agent select the policy in episode 1? It tries to minimize the total negative reward (and the risk of having a defective product with a large negative reward). For example, if the process is in state F, then the best route is $F \rightarrow I \rightarrow H \rightarrow G$! The policy is thus determined by taking the optimal route with the minimum total negative reward from any state. The agent learns this by trying out many combinations of process routes and calculating the reward function (discounted cumulative future reward). Do not be confused by the sign of the reward (negative or positive) and by the terms maximizing or minimizing! Just think about what we want to achieve based on the explanation in the previous sections.

The process route of episode 3 is a poor one because of the large negative reward (defective product). These 'wrong' episodes are as important for the learning as the better ones, e.g., of episodes 1 and 2.

The bottom right image on Fig. 2.35 is an exploration episode. Remember that exploration means that the agent must find out which actions are correct by trying them out despite the risk of obtaining a negative reward (penalty) or a lower total reward. Exploitation means that the agent leans towards actions which lead to positive results and avoid actions that do not (i.e., selected policies in episodes 1–3). The total reward for the exploration episode with discount factor $\gamma = 1$ can be calculated as follows:

$$R_e = -0.04 * 9 + 1 = 0.64$$

The effect of the action reward for transferring the product nine times is that the total reward decreases from the (previous) best value of 0.84 to 0.64. This is exactly what we wanted to achieve—unnecessary movements are penalized. The exploration episode on Fig. 2.35 shows, for example, that in state A when the agent wants to move left it stays in state A (there is no cell to the left; hence, the blue line turns back) but the -0.04 penalty is still added for the trial!

NOTE: Please note that this example is one of the demo projects in the AI-TOOLKIT and will be further explained in Sect. 9.2.3.10 while showing how to use the AI-TOOLKIT!

Episode 3

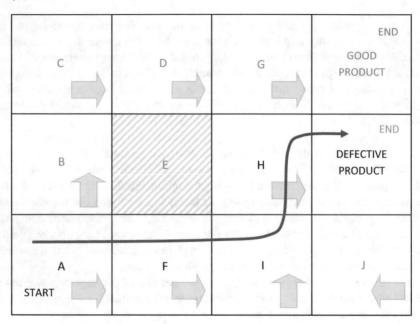

Exploration Episode

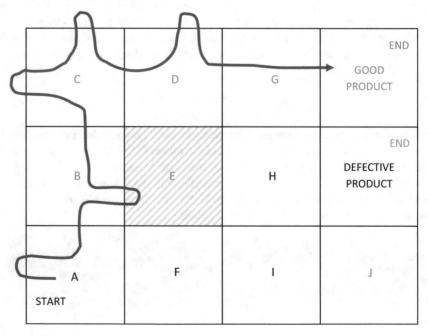

Fig. 2.35 Example 1: Several episodes—part II

2.4.5   *Example 2: Cart-Pole (Inverted Pendulum)*

In this example we will design a reinforcement learning (RL) system for one of the most studied models in control and robotics called cart-pole. The cart-pole is a simplified model of a real world physical system, having only the essential parts of the physics which are needed for studying the problem. This example demonstrates the second type of RL model which is built around a physical system.

The cart-pole is a cart which can move horizontally (left and right) in 1D, on a base, with the end of a pole attached to the cart in such a way that the pole can freely turn around the fixation point (i.e., like a pendulum). Figure 2.36 shows the cart-pole and its properties.

The cart-pole model takes into account the effect of gravity (g), the mass of the pole (m_p) and of the cart (m_c), and the length (l) of the pole. All other physical properties, such as friction, are neglected. If the cart moves left or right then the pole receives a force in the opposite direction because of its mass and turns around the fixation point and falls down. The aim of the cart-pole system is to move the cart horizontally (left or right), by applying a horizontal force (F), and balance the pole at the top of the cart so that it does not fall down. If the pole starts to fall down on one side then we move the cart in that direction to force the pole back into balance in the middle of the cart (i.e., to an upright vertical position). There is an additional requirement which limits the distance (x_{max}) from the center of the base because we do not want the cart to move too far in either direction.

The reinforcement learning agent will learn how to balance the pole vertically on top of the cart without human intervention.

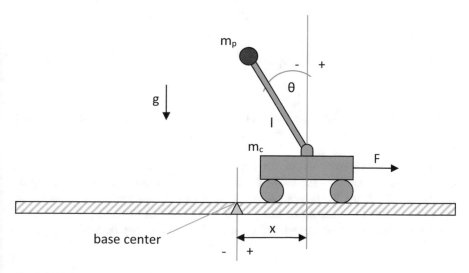

Fig. 2.36 The cart-pole simplified control system

2.4.5.1   Actions

There is only one action in this system, the horizontal force applied to the cart (F). When the force is positive the cart moves to the right and when it is negative the cart moves to the left. We use a constant F (always the same magnitude) force and only the sign changes.

2.4.5.2   Environment

The physical properties of the cart-pole system can be seen on Fig. 2.36. When applying a horizontal force (F) to the cart, the horizontal acceleration ($\ddot{x}$) of the cart and the angular acceleration ($\ddot{\theta}$) of the pole can be calculated (see below). Then the horizontal velocity (v) of the cart and the angular velocity ($\dot{\theta}$) of the pole are determined for a given time step (τ). The angle of the pole (θ) and the position of the cart (x) can be calculated from the angular velocity of the pole and the horizontal velocity of the cart, respectively, for any given time step. The equations have been derived from the laws of physics.

$$\ddot{\theta} = \frac{g \cdot \sin\theta - \cos\theta \cdot \left(\frac{F + m_p \cdot l \cdot \dot{\theta}^2 \cdot \sin\theta}{m_c + m_p} \right)}{l \cdot \left(\frac{4}{3} - \frac{m_p \cdot \cos^2\theta}{m_c + m_p} \right)}$$

$$\dot{\theta}_t = \dot{\theta}_{t-1} + \tau \cdot \ddot{\theta}_{t-1}$$

$$\theta_t = \theta_{t-1} + \tau \cdot \dot{\theta}_{t-1}$$

$$\ddot{x} = \frac{F + m_p \cdot l \cdot \left(\dot{\theta}^2 \cdot \sin\theta - \ddot{\theta} \cdot \cos\theta \right)}{m_c + m_p}$$

$$v_t = v_{t-1} + \tau \cdot \ddot{x}_{t-1}$$

$$x_t = x_{t-1} + \tau \cdot v_{t-1}$$

In the equations the letter t indicates the next state and t-1 the current state before applying the horizontal force to the cart. The interval between these two states is defined by the time step τ.

The horizontal movement of the cart and the angular movement of the pole depend on the time step—the larger the time step, the further the cart and the pole will move. Therefore, the magnitude of the time step is important and consideration must be given to the other properties (i.e., masses, pole length and applied force) to allow for several steps between terminal states (the pole falling down)!

2.4.5.3 Design of a Reward Strategy

Remember that one of the most important and challenging tasks in reinforcement learning (RL) is the design of an effective reward strategy. If you fail to choose the right rewards the model will fail! In order to decide which rewards to use, we must first discover the outcomes of an episode.

Remember that an episode is several state changes (cycles of action-reward-state) until the goal of the system is reached. In most of the RL problems the goal is reached in the so-called terminal state, which is the last state in an episode. But this example is a special case—there is no terminal state with a positive goal since we are trying to balance the pole for as long as possible. How should we then define an episode and what are the outcomes?

One of the possibilities is to define a minimum time period that we want the pole to be in balance for after starting from the vertical position. Another possibility is to choose a 'negative' goal and define an episode as the repeated attempts to balance the pole until failure. We will choose the second option in this example and will try to balance the pole for as long as possible.

We have chosen an episode as the repeated attempts to balance the pole until failure. But what is failure in this example? There are two possibilities for failure, or in other words, two possible negative outcomes. The first one is when the pole falls past a given θ angle. The second one is when the cart moves too far away from the center of the base. Thus, we have two terminal states which both represent a negative outcome or failure. The pole is reset to vertical after each failure!

We have now defined an episode and we know all of the outcomes. There are two negative outcomes and there is one continuous positive 'outcome' while the pole is in balance (the opposite of failure as explained above, with θ and x within limit!). It is time now to design our rewards.

We should give a positive reward to the agent when the pole is in balance and we should give a penalty in the case of failure. Because our goal is to balance the pole for as long as possible, we can only give a positive reward at each time step while the pole is in balance—let us make this a reward of +1. In the case of failure let us give a reward of 0. The value of 0 has been determined by experimenting with the problem. A value of -1 would also be logical but then the weight of the negative cases would be too high. Remember that the discounted cumulative future reward will be maximized!

The reward strategy in this example is very simple, with just two values (+1, 0), but it works well. We could also design a more complicated reward strategy by checking the horizontal position of the cart and the angular position of the pole and, for example, giving a higher reward to positions closer to the center of the base (position) and to the vertical axis (pole angle). In this way the agent would have much more information during the balancing trials. A 100% perfect reward strategy does not exist and we always introduce some sort of error with our design of the rewards. The aim is, of course, to minimize this error as much as possible!

2.4.5.4   The Learning Process

Remember that the reinforcement learning (RL) system typically operates by initializing the environment to a random state but it could also be to a specific state. The state is then passed to the agent which calculates an appropriate action. Each time an action is taken and passed back to the environment a reward is calculated for the last state transition and passed back to the agent (reward function). This cycle is repeated until an end goal is reached and the episode ends. After the episode ends the system is reset to a new (random or specific) initial state and a new episode begins. The reinforcement learning system cycles through many episodes during which it learns which actions are more likely to lead to the desired outcome or goal by optimizing the long term reward.

The first question arising is, what are the states in our example? In the previous example the states were the different process steps (grid cells). States are the properties of the model (or environment) which we want to use to model the dynamics of the system, in this case the cart-pole. Choosing which available properties to use is important for the success of learning! The states we choose will be the input data for the RL model. We will choose the following four properties:

- **Position** of the cart.
- **Velocity** of the cart.
- **Angle** of the pole.
- **Angular velocity** of the pole.

The learning process or simulation goes as follows:

- The pole is reset to vertical position. All states (position, velocity, angle, and angular velocity) are reset to 0.
- A random but constant horizontal force (F) is applied to the cart. It is random because it can be positive or negative (right or left direction) but we always use the same value!
- The resulting states are calculated with the equations presented in Sect. 2.4.5.2.
- The next step is to check if the cart-pole has reached the terminal state. Remember that the terminal state is when the pole is out of balance (falls past a given angle) or when the cart is too far from the center of the base. We could formulate this as $|x| > \Delta x_{threshold}$ and $|\theta| > \Delta \theta_{threshold}$ (both mean failure). Absolute values are used because we treat the two sides of the center and the vertical axis as positive and negative.
- If we have reached the terminal state then no reward is given (0) and the episode ends. The process restarts from the beginning (first point). If we have not reached the terminal state then a + 1 reward is given to the agent. The last action and reward are stored (learning).
- According to the current exploration/exploitation policy the agent chooses a random action (exploration) (constant horizontal force F) or it selects the best action (exploitation) for the current state.

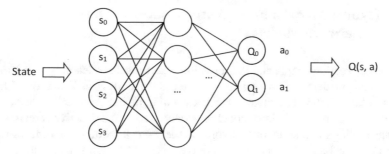

Fig. 2.37 Deep Q-learning neural network

- The resulting states are calculated with the equations presented in Sect. 2.4.5.2.
- The previous steps are then repeated, etc.

We can of course not run the simulation endlessly if the pole is in balance; therefore, we set the maximum number of steps in an episode to 200. The cart-pole RL system is considered to have learned well if the average reward over the last 100 episodes is equal to or greater than 195 (this corresponds to 97.5% accuracy (195/200)) with the selected (!) reward strategy (+1, 0). An average reward of 200 per episode is the best possible result (100%).

But how does the agent select the action in the case of exploitation (using past experience) and how does it learn (store the experience)? We have answered this question already—deep Q-learning. Remember that deep Q-learning (a neural network based algorithm) is used to approximate the (Q) reward (or value) function, which is the discounted cumulative future reward (see Sect. 2.4.2)! For this task we have to design a neural network structure. It is never easy (without former experience with similar problems) to define the right neural network and we must often use trial and error. The four states (position s_0, velocity s_1, angle s_2, and angular velocity s_3) used to model the dynamics of the cart-pole system are fed to the neural network as input data (see Fig. 2.37). The outputs of the network are the Q-values for each action. Each state combination (position, velocity, angle, and angular velocity) and action ($\rightarrow$positive force a0, $\leftarrow$ negative force a1) will be associated with Q-values and the optimal action is the one with the highest Q-value (highest cumulative future reward)!

EXPERT NOTE: In the case of double deep Q-learning (an improved form of Q-learning) the same (a duplicate) neural network is used to estimate the probability of each action, which is then used to select the next action, the action with the highest probability. The outputs of the network are then the probabilities of each action.

NOTE: Please note that this example is one of the demo projects in the AI-TOOLKIT and will be explained in more detail later when we look at how to use the AI-TOOLKIT!

2.5   Hybrid Models: From Autoencoders to Deep Generative Models

The former sections explained the most important machine learning (ML) models (algorithms) in detail. This section will give an overview of a family of hybrid models which contain two linked machine learning models, a so called *encoder* and a *decoder*. This type of hybrid model is widely used in many active research today and the findings thanks to these hybrid models are important building blocks of several practical applications as for example face recognition where feature extraction is being done with an encoder (for more information about this read Chap. 6).

We will first look at Autoencoders which were the starting point of a lot of similar hybrid models as for example the Variational Autoencoder (VAE) and the Generative Adversarial Network (GAN). These hybrid models are not very useful on their own (although there are some interesting applications as for example denoising, generating creative images, etc.) but the insides obtained during their research are successfully used in other types of models as for example face recognition, machine translation, etc.

2.5.1   The Autoencoder (AE)

An autoencoder contains two linked machine learning models (usually neural networks or convolutional neural networks) which are trained together. The output of the first machine learning model is fed to the second machine learning model. An autoencoder first simply converts its input to a reduced representation (less data– encoder), we could call this reduced representation *feature space*, and then converts this feature space back to its original form (decoder). The encoder has exactly the same functionality as a PCA model has (see Sect. 4.3.2.2), dimensionality reduction and feature extraction, it extracts the most important features of the input data (with the most variation/information) and by this it also reduces the size of the data (dimensionality reduction). The second ML model, the decoder, then learns how to convert this reduced representation of the input data back to its original form.

We can simply train an autoencoder with the same techniques as we use for training single neural networks (backpropagation, stochastic gradient descent, etc.— see Sect. 2.2.2.3) by minimizing the difference between the input and the output of the autoencoder (loss function).

Autoencoders are mostly used for dimensionality reduction (like PCA), for feature extraction (see face recognition in Chap. 6) and for data visualization (lower dimensional data which contains more distinct information can be visualized better than the original data).

Since autoencoders may take a huge amount of data as input and reduce it, the question rises whether we could use autoencoders for data compression, reducing the size of the data then sending it through an IT network and then restoring the data back to its original form in order to reduce network traffic? The answer is no because the reduction is lossy (the autoencoder does not reconstruct the data completely) and because the autoencoder learns only to reduce (compress) a specific type of data and cannot be used for other types of data. This is also the reason why we should not use the term *data compression* in case of autoencoders.

2.5.2   The Variational Autoencoder (VAE) and Generative ML Models

The aim of the so called generative machine learning models is to train a hybrid ML model which can be used to create (generate) new data similar to the input data used during the training of the model. For example, if a generative model is trained on images of dogs then the model will be able to create new images of dogs which will be similar to the input but somewhat different. It is also possible to inject extra information into such a generative model and by this modify the newly created images (for example change the orientation, color, style, etc.), more about this later.

Variational Autoencoder (VAE) is a modified version of an autoencoder, the modification is simply learning the feature space (conditional) probability distribution (usually a complex gaussian distribution) instead of learning to describe the reduced representation (feature space) of the data itself. This is similar to density based unsupervised learning (see Sect. 2.3.2). If you are new to the subject then the words probability distribution will probably be a bit confusing but it simply means that we try to estimate (probabilistic, uncertainty) the complex distribution of the reduced representation of the input data, for example the distribution of the colored pixels in an image with a complex function which is modeled with a machine learning model (neural network). Neural networks are very good in learning complex probability distribution functions.

If we know the approximate distribution of the feature space data, where knowing means that we can model it, the we can sample from this distribution in order to create the input data for the second machine learning model (decoder) which then learns how to convert this data back to the original input data (as we have seen in case of the autoencoder in the former section).

How can we generate new data with this hybrid model? After training the model we simply throw away the encoder and sample from the learned feature space probability distribution and run the data through the decoder (second ML model) which will produce the new data similar to the input data originally used during the training of the hybrid model. We could call this setup the *Generator*.

The training of the VAE is very similar to the training of an autoencoder but there are some extra steps (for example the so called reparameterization trick between the two ML models) necessary which allow us to simply use backpropagation through all layers of both ML models (jointly learning the two models by stochastic gradient descent).

Generative models are not very useful on their own (more about this later) but form the basis for a lot of active research. One of the interesting applications of generative models is the learning of the feature space (distribution), analyzing this feature space and then transforming/modifying it in some way and then generating a modified (hopefully improved) version of the data. If we are able to find more systematic information in the feature space (e.g. the reason for specific phenomena) then we can make use of this knowledge to create new improved versions of the data, for example improved chemical substances, improved drugs, etc. In this context generative models are a kind of simplification and feature extraction techniques but the question remains whether the direction of this simplification is relevant and whether it can be directed.

Generative models are often used for learning to generate images, also slightly modified ones, but it takes a considerable amount of resources and time to train such models (especially in case of bigger images) and the quality of the output is often not very good. VAE's are known to generate blurry images, because of the way how they learn the input data but can handle more diverse samples; another type of generative model the Generative Adversarial Network (see next section) can generate sharper but less diverse images. Researchers are currently trying to combine these two models.

If the input data contains some kind of systematic information as for example the orientation of a face by including a lot of slightly rotated face images then the trained VAE contains this information in the learned feature space (embedding). If we can extract this information and can inject it into another model then we can generate a different rotated face. The practical business use of such techniques is questionable but it helps researchers to better understand how this feature space is constructed.

2.5.3   Generative Adversarial Network (GAN)

The GAN is similar to the Variational Autoencoder explained in the former section, it is also a generative hybrid model which contains two ML models, but there are some important differences. The GAN is usually used with images (we will focus on this application) but may also be applied to other kinds of data. The name 'adversarial' comes from the way how the two ML models work together or better to say against each other. The first model learns to generate synthetic (artificial or fake) images and the second model learns to distinguish between real and fake images. During the training of the hybrid model the first generator ML model must improve in generating images which look like real images and the second ML model must improve in discriminating between real and fake images. Both models are getting better and better in what they are doing.

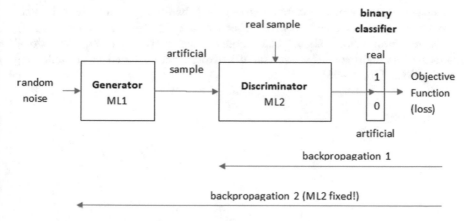

Fig. 2.38 Generative adversarial network (GAN)

The architecture of a GAN can be seen on Fig. 2.38. The input on the left side is random noise (gaussian distribution) which is converted by the generator neural network to artificial sample (artificial or fake images), then passed to the discriminator which also receives real samples (real images). The discriminator then decides if an image is real or artificial. Finally the objective function is evaluated and the error is backpropagated through the two networks. The training of the GAN happens in several stages, first the discriminator is trained with real data, then it is trained with artificial data from the generator (in both cases backpropagating for updating the weights of the discriminator neural network), then the generator is trained by using random noise as input and by backpropagating through the whole dual network (without changing the weights of the discriminator!) in order to update the weights of the generator neural network. In this way the discriminator gives feedback to the generator how it can generate better (more real) data (images).

It is very time consuming to train a GAN and it often does not converge to the optimal solution (equilibrium between how well the generator works and how well the discriminator works) or the convergence is very slow. There is a lot of ongoing research in this subject.

Instead of training a binary classifier, real (1) or artificial (0), we can also supervise the hybrid model with labels, for example by using the type of the object (dog, cat, horse, etc.).

We can also inject into this hybrid model extra information which captures the main attributes of the data, for example orientation, color, etc. See Sect. 2.5.2.

GAN's has been used for generating new creative images, for simulating face aging, image-to-image translation (for example changing the style of the image), etc. The amount of resources and time needed for training these kinds of models is considerably high especially in useful dimensions, and the models often do not converge to the desired solution or collapse after several days of training. There is a lot of active research in this area also. Even if direct useful practical business applications are not very probable the work being done helps researchers to better understand hybrid machine learning models and the findings can be used in other types of models (for example as we have seen in case of face recognition in Chap. 6).

More information about autoencoders, generative models, variational autoencoders and generative adversarial networks can be found in the following publications: [16, 94–96].

Chapter 3
Performance Evaluation of Machine Learning Models

Abstract After learning about the most important machine learning models in the previous chapter, this chapter explains how to evaluate the performance of machine learning models. Evaluating the performance means that we determine how well the model learns the given subject (data) and how well it applies this learning in practice. Each major form of machine learning has its own performance measures and they are often application dependent. This chapter explains the different types of performance evaluations and how and where to apply them. After reading this chapter you will be able to understand the performance measures reported by machine learning algorithms.

3.1 Introduction

This chapter introduces the performance evaluation of machine learning models. The performance of a machine learning model can be defined in different ways and it is usually application dependent. We have seen, in the first chapter, the importance of using different training and test datasets and the importance of the generalization performance of the machine learning model. We have also seen one of the performance metrics, the mean squared error, in the case of the simplest machine learning model—linear regression. This chapter will introduce several other performance metrics which may be used together or separately depending on the application. Each performance metric has its advantages and disadvantages, which is one of the reasons for having many different types of performance measures. Some metrics underestimate the accuracy or overestimate it (depending on the data), others are not always representative and some are better used together to provide a more complete picture of the model's performance.

We have seen in the previous chapters that there are three major categories of machine learning models (supervised, unsupervised and reinforcement learning) and we will see later that there are also some hybrid machine learning models, such as recommendation models (Sect. 8.4). Each of these types of models has its own performance measures.

© Springer Nature Switzerland AG 2021
Z. Somogyi, *The Application of Artificial Intelligence*,
https://doi.org/10.1007/978-3-030-60032-7_3

Supervised learning can be further divided into classification and regression which both have different performance measures. The most important supervised learning performance measures will be introduced in the next section.

Unsupervised learning may have so-called external and internal performance measures depending on whether reference labels are available or not. The aim of unsupervised learning is to classify (group or cluster) the input data based on an internal measure of cluster compactness which is, e.g., the sum of squared error in case of k-means clustering (see Sect. 2.3.1). This measure of compactness is then used in the optimization of the clusters and as a stopping criterion. There are several internal performance measures available to check how well the data is distributed in the different groups or clusters. In case we want to check how an unsupervised model performs and we have labeled data (reference labels) then after clustering the input data we can compare the class assignments to the original class assignments and calculate the performance measures in a similar way as in supervised learning. We call these external performance measures. In general we can say that calculating the performance of an unsupervised learning model is more difficult than calculating the performance of a supervised learning model. Some of the important unsupervised learning performance measures will be introduced in Sect. 3.3.

Reinforcement learning has a special performance measurement system called the reward system. Usually a so-called cumulative discounted future reward is calculated for several episodes which are then used as performance measure (see Sect. 2.4).

There are several special hybrid machine learning models which all have their own performance measures (e.g. recommendation systems) and which will be discussed in the appropriate sections about these models.

3.2   Performance Measures of Supervised Learning

Figure 3.1 summarizes all supervised learning performance measures. Each of these will be explained in the following sections. All classification measures are based on the so-called *Confusion Matrix* which will also be explained. General measures are measures which are calculated for the model's overall performance and single class measures are calculated for the model's performance concerning a specific class. Single class measures are useful because we can evaluate the model's performance focusing on only one class.

3.2.1   RMSE

The root mean squared error (RMSE) is a very simple and effective measure of model performance in the case of regression models. RMSE was discussed in Sect. 1.2.1 in Chap. 1.

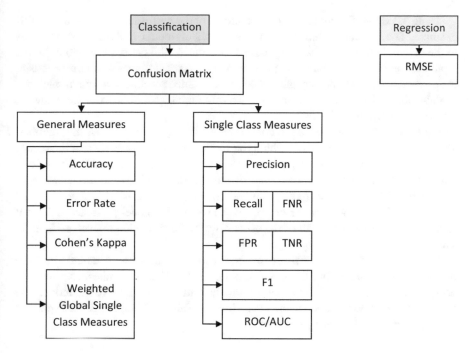

Fig. 3.1 Supervised learning performance measures

Table 3.1 Input data example		f_1	f_2	...	f_m	decision
	1	...	...	...	...	A
	2	...	...	...	...	C
	...	...	...	...	...	...
	N	...	...	...	...	B

3.2.2 The Confusion Matrix

The confusion matrix was invented in order to show in one simple table which kind of mistakes a machine learning model makes. Based on this table it is also possible to derive all of the performance measures of the machine learning model.

In order to make the introduction of the confusion matrix and all performance measures as clear as possible let us define a simple example. Let us assume that we have a machine learning model which was trained with an input (training) dataset. The test dataset contains N number of records which contain m number of features (columns) and one decision variable (column). The decision values are three classes A, B and C (the principle is the same in the case of any other number of classes). Each record is labeled with these classes. For an example of what the input data could look like see Table 3.1 (without real data filled in).

Table 3.2 shows the confusion matrix and all of the performance measures for this example. The horizontal dimension of the confusion matrix (columns) corresponds to the original classes, where 'original' classes means the classes (decisions) in the input dataset which we know are all correct. The vertical dimension corresponds to the predicted classes which are generated by the trained machine learning model. We evaluate the performance of the machine learning model by comparing the predicted classes to the original classes.

Let us look at the values in the confusion matrix in Table 3.2. The first number, 45, is the count of the records in which the model predicted correctly that the record belongs to class 'A' (the model predicted 'A' and the original is also 'A'). The second number, 5, in the first row is the count of the records in which the model wrongly predicted 'A' but it should have predicted 'B' (the original is 'B'). And the final number, 0, shows that there are no records in which the model predicted 'A' but it should have predicted 'C'. All of the other rows in the confusion matrix are filled in similarly.

We can see immediately from the confusion matrix without calculating any performance measures that the trained model makes the most mistakes when the original class is 'B' (5 + 2). We can use this information to improve the model! Do not worry if you do not understand everything in Table 3.2 yet, it will be explained step by step in the following sections!

3.2.3 Accuracy

The accuracy of a classification model is simply the number of correctly identified records divided by the total number of records. We can express this also with the values in the confusion matrix as follows:

$$Accuracy = \frac{\sum T}{\sum (T + F)} \cdot 100$$

where ΣT is the sum of cases (records) where the class was correctly identified, which is the diagonal of the confusion matrix ($45 + 44 + 48 = 137$), and $\Sigma(T + F)$ is the total number of cases, which is the sum of all values in the confusion matrix (150). 'T' stands for 'true' and 'F' stands for 'false'. The accuracy is then $137/150 = 0.913$ or 91.3%.

We can also express the error rate of the model instead of the accuracy. The error rate is just 1-accuracy. In this example it is $1-0.913 = 0.087$ or 8.7%. The model makes a mistake in 8.7% of the cases.

Accuracy is a very simple and powerful measure but it may not always be adequate for all types of problems. For example, if there is a significant difference between the number of records with class A, B and C (we call this imbalance of the classes), then the accuracy may not be precise (it may be higher than it is in reality).

Table 3.2 Example performance measures

Confusion Matrix

predicted	original A	B	C	Σ
A	45	5	0	50
B	6	44	0	50
C	0	2	48	50
Σ	51	51	48	

Multi-class measures:

		95% c.i.	Equation
Accuracy	91.3%	4.5%	$(TP+TN)/(TP+TN+FP+FN)$
Accuracy$_{chance}$	0.3333		$(P_{e-original-A}*P_{e-predicted-A})+(P_{e-original-B}*P_{e-predicted-B})+(P_{e-origina-iC}*P_{e-predicted-c})$
Cohen's Kappa	87.0%	5.4%	$(Accuracy-Pe)/(1-Pe)$
Weighted Global Precision	91.2%	4.5%	$((TP+FN)_A*P_A + (TP+FN)_B*P_B + (TP+FN)_C*P_C)/(TP+FN)_{ABC}$
Weighted Global Recall	91.3%	4.5%	$((TP+FN)_A*R_A + (TP+FN)_B*R_B + (TP+FN)_C*R_C)/(TP+FN)_{ABC}$
Weighted Global F1	91.3%	4.5%	$((TP+FN)_A*F1_A + (TP+FN)_B*F1_B + (TP+FN)_C*F1_C)/(TP+FN)_{ABC}$

95% confidence interval (c.i.): $\pm1.96*sqrt(accuracy*(1-accuracy)/N)$

Single-class measures:

		Equation	
Precision	90.0%	$TP/(TP+FP)$	*also called PPV (positive prediction value)
Recall	88.2%	$TP/(TP+FN)$	*also called TPR (TP rate), Sensitivity
FNR	11.8%	$FN/(TP+FN)$	*also called Miss rate
F1	89.1%	$2*Precision*Recall/(Precision+Recall)$	
TNR	94.9%	$TN/(FP+TN)$	*also called Specificity
FPR	5.1%	$FP/(FP+TN)$	

Precision	88.0%
Recall	86.3%
FNR	13.7%
F1	87.1%
TNR	93.9%
FPR	6.1%

Precision	96.0%
Recall	100.0%
FNR	0.0%
F1	98.0%
TNR	98.0%
FPR	2.0%

A

predicted	original A	B	C
A	TP	FP	FP
B	FN	TN	TN
C	FN	TN	TN

B

predicted	original A	B	C
A	TN	FN	TN
B	FP	TP	FP
C	TN	FN	TN

C

predicted	original A	B	C
A	TN	TN	FN
B	TN	TN	FN
C	FP	FP	TP

A second disadvantage of accuracy is that it does not explain how the different classes are misclassified and this could be important in some applications. For example, let us assume that in a medical hospital class 'A' means a serious operation, class 'B' means medication and 'C' means the exit from the therapy. Let us concentrate on class 'A' and call the cases where 'A' is wrongly predicted, because in reality it should be 'B' or 'C', false positive (FP) cases. Let us call the cases where 'B' or 'C' is wrongly predicted, because in reality it should be 'A', the false negative (FN) cases. It is obvious that if we wrongly predict 'B' or 'C' instead of 'A' (a serious operation) the patient may die because of the lack of an operation. However, if we wrongly predict 'A' instead of 'B' or 'C' the patient may have an unnecessary operation but there isn't a high risk of dying. The type of mistake the machine learning model makes may thus be very important in some cases! If we know which mistakes are more costly than others, then we can use this information to improve the machine learning model for the specific application.

3.2.4 Cohen's Kappa

The kappa statistic was invented in order to quantify the reliability of two raters (two people who rate or evaluate the same thing in parallel to each other). This experiment often happens, for example, in healthcare. It is sometimes also called inter-rater reliability. Kappa takes into account the possibility that some agreement (i.e., that both vote in the same way) occurs by chance (coincidence) and not because they are both confident of their vote. The kappa statistic is therefore thought to be a more robust measure than percentage agreement (i.e., the percentage of votes which are the same). For example, if there are two people who are evaluating a product as part of an experiment, then we can make a global statistical conclusion about that product by combining their votes. The kappa statistic is only valid in the case of two raters and not if there are more than two!

We can use the kappa statistic in the case of a machine learning model because the machine learning model can be thought of as one of the raters who predicts (votes) on an item (record), and the original known decision values (votes) can be thought of as if they come from the second rater. In this way we get a kappa statistic for our predictions which can be viewed as a kind of robust accuracy which has been corrected for chance.

The kappa statistic is especially useful if there is a significant imbalance between the different classes in the dataset (if there are many more records for one of the classes than for the others) because then it is more reliable than accuracy. Kappa is not sensitive to class imbalance in the data!

Cohen's kappa can be calculated as follows:

$$\kappa = \frac{Accuracy - Accuracy_{chance}}{100 - Accuracy_{chance}}$$

Accuracy in this equation is the accuracy introduced in the previous section as a percentage. Chance accuracy is the accuracy which can be achieved if we predict by chance (random classifier; just guessing). The chance accuracy can be calculated as follows:

$$Accuracy_{chance} = \frac{1}{N^2} \sum_k (n_{ko} \cdot n_{kp})$$

where 'N' is the total number of records, 'k' is the number of classes (decision values; there are three classes in this example: A, B and C), n_{ko} is the number of times rater 'o' (original) choses class 'k' and n_{kp} is the number of times rater 'p' (predicted by the machine learning model) choses class 'k'.

The above formula has been derived by using statistical probability theory and it is simply the sum of the probabilities of the raters choosing randomly in all cases.

Let us calculate the chance accuracy first. $N = 150$. The probability that class 'A' is chosen by chance (P_{e-A}) is the product of the original chance probability and the predicted chance probability for class 'A' ($P_{e-A} = P_{e-original-A} * P_{e-predicted-A}$), where $P_{e-original-A} = (45 + 6 + 0)/150$ and $P_{e-\ predicted\ -A} = (45 + 5 + 0)/150$. In the above equation $n_{ko} = 45 + 6 + 0$ and $n_{kp} = 45 + 5 + 0$. So $P_{e-A} = (45 + 6 + 0) \cdot (45 + 5 + 0)/ (150 \cdot 150) = 0.1133$. We can calculate P_{e-B} and P_{e-C} in the same way giving $P_{e-B} = 0.1133$ and $P_{e-C} = 0.1067$. The chance accuracy is then $0.1133 + 0.1133 + 0.1067 = 0.33$.

Kappa can then be calculated as follows: $(91.3 - 33.3)/(100 - 33.3) = 86.95\%$. Cohen's kappa tells us that the accuracy most probably overestimated the performance of the model. However, keep in mind that all measures are statistical estimates!

3.2.5 Single Class Performance Measures

Before introducing the single class performance measures one by one we first need to categorize the values in the confusion matrix. The values in the confusion matrix can be grouped into four categories:

- True positive value (TP).
- False positive value (FP).
- True negative value (TN).
- False negative value (FN).

The categorization of the values in the confusion matrix is class dependent. In Table 3.2, under the confusion matrix, the categorization of each class (A, B and C)

Table 3.3 Categorization of classes

A		original		
		A	B	C
predicted	A	TP	FP	FP
	B	FN	TN	TN
	C	FN	TN	TN

is shown. Why the categorization is different for each class will become clear after we define what these four categories mean.

True Positive (TP)

A true positive value is obtained when the model correctly predicts that an item (data record) belongs to the selected class. Where 'selected class' means simply that we calculate the single class performance measures for that class. For example, if the selected class is 'A', then it means correctly predicting 'A' when the original class indicates 'A'. Table 3.3 shows that in case of class 'A' the first cell belongs to the category TP.

False Positive (FP)

A false positive value is obtained when the model predicts that an item (data record) belongs to the selected class when it should not have. For example, if the selected class is 'A', then it means wrongly predicting 'A' when the original class indicates 'B' or 'C'. The second and the third cells in the first row of Table 3.3 are FP values.

True Negative (TN)

A true negative value is obtained when the model correctly does not predict that an item (data record) belongs to the selected class. For example, if the selected class is 'A', then it means correctly predicting that the item belongs to any class other than 'A'. A prediction of 'B' or 'C' is also a true negative since it does not matter what is chosen as long as it is not 'A'! This is the reason for the extra blue TN values shown in the matrix (Table 3.3) if class 'A' is selected.

False Negative (FN)

A false negative value is obtained when the model does not predict that an item (data record) belongs to the selected class but should have. For example, if the selected class is 'A', then it means wrongly predicting 'B' or 'C' when the original class indicates 'A'.

By using the values belonging to these four categories in the confusion matrix the single class performance measures can be calculated.

3.2.5.1   Precision

$$precision = \frac{TP}{TP + FP}$$

Precision is the proportion of correctly identified positive cases among all positives. A positive case means that the class which is selected was predicted by the model. Precision is sometimes also called the positive prediction value (PPV). In this example the precision for class 'A' can be calculated as follows: the TP value = 45, the FP value = 5 + 0 (there are two FP's!) and so the precision is then = 45/50 = 90%.

The precision is sensitive to imbalanced data which means that it will report a different performance (value) if the proportion of the cases with one of the classes changes, even if the real performance of the model does not change. Let us demonstrate this with a simple example by increasing the number of cases with class 'B' and 'C' by a factor of 10 while keeping the same classification performance!

Figure 3.2 shows the original confusion matrix on the left and the modified confusion matrix on the right with the accuracy, kappa and precision below each matrix. All values in the original columns 'B' and 'C' (Confusion Matrix 1) have been multiplied by 10 to generate the modified values (Confusion Matrix 2).

Cohen's kappa shows that the performance of the modified model is basically the same but the precision has decreased dramatically for class A! The accuracy has increased slightly but not dramatically. This proves that kappa is not sensitive to class imbalance in the data.

	Confusion Matrix 1				Confusion Matrix 2		
	original				original		
	A	B	C		A	B	C
predicted A	45	5	0	predicted A	45	50	0
predicted B	6	44	0	predicted B	6	440	0
predicted C	0	2	48	predicted C	0	20	480

Accuracy	91.3%	92.7%
Cohen's Kappa	87.0%	87.1%
Precision A	90.0%	47.4%

Fig. 3.2 Confusion matrix example

3.2.5.2   Recall and FNR

$$recall = \frac{TP}{TP + FN}$$

$$FNR = 1 - recall = \frac{FN}{TP + FN}$$

Recall is sometimes also called the true positive rate (TPR) or sensitivity. The false negative rate (FNR) is complementary to recall. Recall is the correct positive classification rate (the proportion of positive records that were correctly classified), while FNR is the incorrect positive classification rate (the proportion of positive records that were incorrectly classified). Both recall and FNR are not sensitive to imbalanced data.

The recall for class 'A' in this example is calculated as follows: $45/(45 + 6 + 0) = 45/51 = 88.2\%$. Please note that there are two FN values which must be summed up to obtain the overall FN value for class 'A'!

3.2.5.3   TNR and FPR

$$TNR = \frac{TN}{FP + TN}$$

$$FPR = 1 - TNR = \frac{FP}{FP + TN}$$

The true negative rate (TNR) is sometimes also called specificity. The false positive rate (FPR) is complementary to TNR and it is sometimes also called fallout. TNR is the correct negative classification rate (the proportion of negative records that were correctly classified) and FPR is the incorrect negative classification rate (the proportion of negative records that were incorrectly classified). Do not be confused by the fact that FPR is the false positive rate but we say that it is the proportion of negative records that were incorrectly classified! The false positives are the incorrect negatives because the model incorrectly assigns some records to class 'A' which actually belong to class 'B', and therefore the records which are known to be negative records (because they belong to classes other than the selected class) are incorrectly classified!

Both TNR and FPR are not sensitive to imbalanced data.

3.2.5.4   F1 Score

$$F1 = \frac{2 \cdot precision \cdot recall}{precision + recall}$$

F1 is the harmonic mean of precision and recall and thus combines both measures into one. F1 can also be thought of as a kind of accuracy. F1 is sensitive to imbalanced data.

3.2.5.5   Weighted Global Performance Measures

It is possible to calculate the weighted average of all single class performance measures in order to get one weighted global precision, recall, F1, etc. The weights are the number of cases in each class. The weighted global precision, for example, can be calculated as follows:

$$precision_G = \frac{N_A \cdot precision_A + N_B \cdot precision_B + N_C \cdot precision_C}{N_A + N_B + N_C}$$

where N_A is the number of records with class 'A', N_B is the number of records with class 'B' and N_C is the number of records with class 'C' which can be calculated from the confusion matrix by adding together the TP and FN values per class (the sum of one column).

3.2.5.6   The ROC Curve and the AUC Performance Measure

The ROC curve is used to visualize the balance between the benefits (the positively identified records—TPR or recall) and the costs (the mistakes made—FPR) of the machine learning model. The aim is of course to reduce the number of mistakes and increase the correctly identified cases. Some machine learning models even use the area under the ROC curve (AUC) for optimization purposes. A bigger area under the ROC curve means a better performance.

The ROC curve can be drawn by plotting several FPR and recall (TPR) values, where FPR is on the horizontal axis and recall is on the vertical axis. Since the confusion matrix provides only one recall/TPR point we need to find a way to calculate several points. This can be done if we know the probability assigned to each prediction, because if we vary the threshold of the probability for assigning a record to a specific class, then we can get several confusion matrixes and recall/TPR value pairs. A simple example of this procedure can be seen in Table 3.4. The output of the machine learning model (prediction) is the three probability values per record which indicate the assignment probability of each class. We have chosen five probability threshold values. For example, 0.9 means that if the probability of a

Table 3.4 ROC curve example

	original	probability			threshold				
		p_A	p_B	p_C	0.9	0.8	0.7	0.6	0.5
1	A	0.7	0.1	0.2	C	C	A	A	A
2	B	0.8	0.05	0.15	C	A	A	A	A
3	A	0.4	0.5	0.1	C	C	C	C	B
4	C	0.1	0.3	0.6	C	C	C	C	C
5	A	0.9	0.1	0	A	A	A	A	A
6	B	0.7	0.1	0.2	C	C	C	C	A
7	C	0.5	0.3	0.2	C	C	C	C	A
8	C	0.6	0.3	0.1	C	C	C	C	A
9	A	0.5	0.2	0.3	C	C	A	A	A
10	C	0.7	0.2	0.1	C	C	A	A	A

ROC A

Confusion Matrix

threshold 0.9

		original		
		A	B	C
predicted	A	1	0	0
	B	0	0	0
	C	3	2	4

FPR   A   0.0%
Recall A   25.0%

threshold 0.8

		original		
		A	B	C
predicted	A	1	1	0
	B	0	0	0
	C	3	1	4

16.7%
25.0%

threshold 0.7

		original		
		A	B	C
predicted	A	2	2	1
	B	0	0	0
	C	2	0	3

50.0%
50.0%

threshold 0.6

		original		
		A	B	C
predicted	A	2	2	2
	B	0	0	0
	C	2	0	2

66.7%
50.0%

threshold 0.5

		original		
		A	B	C
predicted	A	3	2	3
	B	1	0	0
	C	0	0	1

83.3%
75.0%

class is greater or equal to 0.9 the record will be assigned to this class. Depending on the threshold there is a different set of class predictions and a different confusion matrix. Based on each confusion matrix the FPR/recall pairs can be calculated and the ROC curve can be drawn for each class. A global ROC curve could also be drawn by using the weighted global values of FPR and recall.

3.3   Performance Measures of Unsupervised Learning (Clustering)

Clustering an unknown dataset is often difficult because of the dimensionality of the data, the distribution of the data points (records), the noise in the data, etc. This is the reason there are so many types of clustering algorithms, as we have seen in Sect. 2.3. Some researchers even think that it is not possible to design a single clustering algorithm which is able to cluster all types of datasets because of the required differences in similarity measures and the tradeoff each algorithm must make. We have seen in Chap. 2 some of the different types of similarity measures and their advantages and disadvantages (tradeoffs). Because of these differences it is also often difficult to assess the performance of such clustering models.

The most important characteristics of a good clustering model are as follows:

- It should be able to find the correct number of clusters in which the data points are grouped. This is often difficult in case of an unknown dataset.
- It should group similar items together and put dissimilar items in different groups. This depends entirely on the similarity metric used by the model.

There are two main groups of methods to evaluate the performance of a clustering model:

- Internal criterion based and
- External criterion based.

Internal criterion based performance evaluation methods assess the internal cohesion of data points in each cluster (togetherness or unity) and how good the separation is between clusters (clarity of distinction) or the combination of cohesion and separation.

External criterion based performance measures are available if the classification of the input data points is known and the predicted clustering (classification) can be compared to the original (reference) clustering. These external measures are therefore similar to supervised learning performance measures, but as we will see in the following sections they are not the same because of the comparison between two possibly different labeling systems (reference and original). For example, if the clustering algorithm assigns a label 'A' to a group of points it is unlikely that the reference set has the same label 'A' for the same group of points. This additional uncertainty must be incorporated in the evaluation metrics!

3.3.1   Internal Criterion Based Performance Measures

In order to assess the performance of clustering algorithms two internal criteria are chosen:

- The cohesion (togetherness or unity) of the data points (records) in each cluster.
- The separation (clarity of distinction) of the clusters.

In practice a hybrid criterion combining cohesion and separation is used most often because it combines the advantages of both criteria.

The cohesion of data points in a cluster can be measured in different ways. Most often the sum of squared errors (SSE) is used in combination with the Euclidean distance of the points from the centroid of the cluster. The smaller the sum of squared errors, the closer the data points are to their centroid and the compacter the cluster is, or in other words, the higher the cohesion in the cluster (group).

$$SSE(C_i) = \sum_{X \in C_i, c_i \in C_i} EuclideanDist(c_i, X)^2$$

The SSE can be calculated using the equation above where C_i is cluster i, X is the data in cluster i and c_i is the centroid of cluster i. *EuclideanDist* is the function to calculate the Euclidean distance with the two indicated parameters.

The cohesion criterion based on SSE, and the internal performance measures based on the cohesion criterion, may be less representative in the case of clustering methods which use a non-distance based similarity metric such as, for example, a density based similarity metric (DBSCAN, MeanShift, etc.). In the case of density based clustering methods, the SSE based internal performance measure may have a lower value than distance based models (e.g., k-Means) have but the clustering may be as good or even better.

The separation of clusters can also be measured in different ways. Most often the cluster sizes weighted sum of squared distances of the centroids from the overall centroid is used (SSC). *EuclideanDist* is the function to calculate the Euclidean distance with the two indicated parameters.

$$SSC = \sum_{i=1}^{n_k} n_i \cdot EuclideanDist(c_i, c)^2$$

In this case n_k is the number of clusters, n_i is the number of data points in cluster i (cluster size), c_i is the centroid of cluster i and c is the overall centroid (the average of all centroids).

The SSC based internal criteria and performance measures may also be less representative in the case of, for example, density based clustering methods because of the different similarity metric. Lower values do not always mean worse results when comparing density based and distance based models! The usefulness of SSC depends on the distribution of the data points.

Because we usually want a high cohesion and high separation most of the internal performance measures combine cohesion and separation in their algorithms.

Let us now look at some of the commonly used internal performance measures which are also available in the AI-TOOLKIT.

3.3.1.1   The Silhouette Coefficient

The Silhouette coefficient combines cohesion and separation into one general measure [48]. It is calculated for each data point (record) and then these values are averaged to get the global Silhouette coefficient for the clustering model. The name Silhouette comes from the fact that a higher value means that the boundary (silhouette) of the clusters is more distinct ('visible').

$$S_i = \frac{min\,(b_i) - a_i}{max\,(a_i, b_i)}$$

S_i is the Silhouette coefficient for point (record) i, where a_i is the average distance of point i to all the other points in its own cluster and b_i is the average distance of point i to all the points in other clusters; $min\,(b_i)$ means that we take the closest cluster.

The value of the Silhouette coefficient is between -1 and $+1$. Higher values mean a better clustering result, or in other words, that the point is well matched to its own cluster and weakly matched to neighboring clusters. When a_i decreases (smaller distances within clusters) then cluster cohesion (compactness) increases, and when b_i increases (greater distances between clusters) then the separation of clusters increases. Negative values mean overlapping clusters.

The global Silhouette coefficient is then calculated as follows:

$$Silhouette = \frac{1}{n} \sum_{i=1}^{n} S_i$$

where n is the number of data points in the dataset.

3.3.1.2   The Calinski-Harabasz Index

The Calinski-Harabasz index (CHI) combines cohesion and separation into one general measure, and it is based directly on the previously introduced SSE and SSC measures [47]:

$$CHI = \frac{SSC}{SSE} \cdot \frac{n - n_k}{n_k - 1}$$

where n is the number of data points, n_k is the number of clusters, SSE is the within cluster sum of squared errors and SSC is the between cluster weighted sum of squared errors (see Sect. 3.3.1).

Remember that SSE is a measure of cohesion and SSC is a measure of separation. Higher score thus means higher cohesion (lower SSE) and better separation (higher SSC). The CHI measure may be less representative for non-distance based models (e.g., density based models) for the same reason as explained earlier in the case of SSE and SSC. Its usefulness depends on the distribution of the data points.

3.3.1.3   The Xu-Index

The Xu-index is another variation on these similar types of internal performance metrics and can be calculated as follows [49]:

$$Xu = D \cdot \log\left(\sqrt{\frac{SSE}{D \cdot n^2}}\right) + \log(n_k)$$

where D is the number of features (columns) in the dataset (dimension), n is the number of data points and n_k is the number of clusters.

The Xu-index is usually negative and lower values mean better clustering results.

All of the internal performance measures we have reviewed may also be used for determining the correct number of clusters when this is not known (as is usually the case). More information about this will be provided in the next section!

3.3.1.4   Determining the Optimal Number of Clusters

If there is no labeled reference data available, then it is possible to use one or more of the internal performance measures to find the optimal number of clusters. Finding the optimal number of clusters is not an exact science but rather a rule of thumb. Depending on how far the clustering model is from the optimal solution (the optimal number of clusters) there are two methods which can be used to estimate the optimal number of clusters:

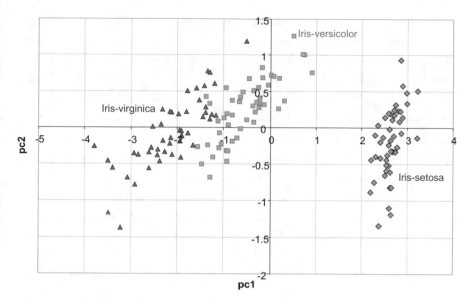

Fig. 3.3 The reduced Iris dataset with two principal components

- If the model is far from the optimal solution, then the so-called *elbow method* should be used. The elbow method is based on the fact that the cohesion and separation (see previous sections) of the clusters (and points) do not change significantly after a specific level is reached, and this signals that we are in the neighborhood of the optimal cluster size. Since all internal performance measures are based on the cohesion and separation criteria we can use them with the elbow method.
- If the model is close to the optimal solution, then the minimum or maximum (depending on the performance measure) value of one or more of the internal performance measures indicates the optimal cluster size.

Since we usually do not know if the model is close to the optimal solution or not, we first need to use the second method and if that does not give a clear answer then we turn to the first method. It is also possible that we may need to follow an iterative procedure and go back to the second method at the end for the final estimate of the cluster size.

In order to make the process outlined above clearer, let us determine the optimal number of clusters in the case of a simple example. We will use the Iris dataset which is explained in Sect. 8.3. The dataset contains three clusters (classes) which are known (a labeled supervised dataset). This will come in very handy while checking the validity of the method. The dimension of the Iris dataset has been reduced using principal component analysis (PCA) which helps to visualize the dataset as you can see on Fig. 3.3. The reduced dataset is just used for the visualization below and not for clustering.

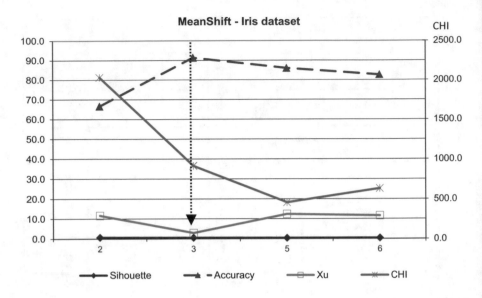

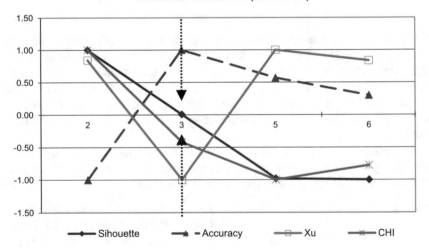

Fig. 3.4 MeanShift clustering results (AI-TOOLKIT)

The Iris dataset has been clustered with the AI-TOOLKIT using two different methods: the MeanShift (density based) and k-Means (distance based) clustering methods. This will also help to show the validity of the internal performance measures for these methods, which use two different similarity metrics.

In the case of MeanShift the kernel bandwidth is varied, and in the case of k-Means the number of clusters is varied directly. Since in the case of MeanShift the number of clusters cannot be easily set, cluster sizes 2, 3, 5 and 6 are recorded. In

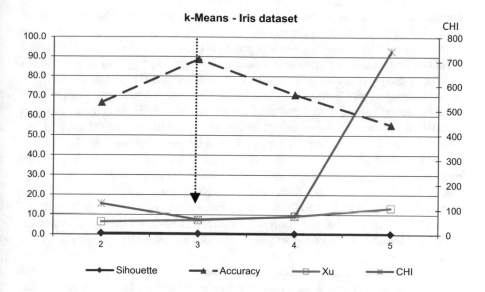

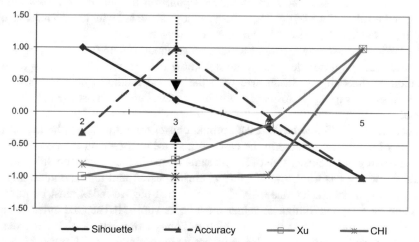

Fig. 3.5 k-Means clustering results (AI-TOOLKIT)

the case of k-Means cluster sizes 2, 3, 4, 5 and 6 are used. All internal performance metrics and the accuracy (the reference labels are known!) provided by the AI-TOOLKIT are recorded for each setting and the results are shown on Figs. 3.4 and 3.5. For both charts there is a normal and a normalized version for better understanding.

The horizontal axis on Figs. 3.4 and on Fig. 3.5 shows the cluster sizes and the (left) vertical axis shows the performance measures. The CHI values are drawn on the right axis when not normalized because of the large values. This is the reason why it is useful to normalize the values.

The MeanShift clustering model gives a good global clustering accuracy (91.33%) despite the fact that the Iris dataset is complicated by two clusters which are overlapping (versicolor and virginica). We can see this clearly on Fig. 3.3.

Let us imagine just for a minute that we do not know the accuracy since in most cases clustering is used with datasets which are not labeled (there is no reference to calculate the accuracy). As explained before we start with the second method and check if one or more of the internal performance measures indicate the optimal cluster size. The Xu-index must be minimized and the CHI and Silhouette values must be maximized! The minimum Xu-index value indicates clearly that a cluster size of 3 is the optimal value. Both the CHI and Silhouette values indicate a cluster size of 2 if we just look at their maximum value. We know at this point that the optimal cluster size is 2 or 3. Because of the sudden high jump in the CHI value ($>$ 2000) and because not all measures show the same cluster size we should try the elbow method.

According to the elbow method we must take the *next* value after the optimal value (minimum in case of the Silhouette coefficient and maximum in case of the CHI value), which in both cases (CHI and Silhouette) indicates a cluster size of 3 (the optimal value is in both cases the first point on the left if we look at the normalized results). If we do not know the accuracy calculated with known reference labels then we should choose 3 as the optimal cluster size.

The k-Means clustering model provides less accuracy (88.66%). Let us imagine once more, just for a minute, that we do not know the accuracy. There is no clear indication of the optimal cluster size with the second method (the minimum or maximum of the various performance measures); therefore, we must use the elbow method.

According to the elbow method the optimal cluster size is indicated by the *next* value after the minimum value in the case of the Xu-index (-1 on the normalized chart) and that gives a cluster size of 3. The same is true in the case of the Silhouette coefficient, which has a maximum value at cluster size 2 and thus the next value indicates an optimal cluster size of 3. Remember that the Xu-index must be minimized and the CHI and Silhouette values must be maximized! The CHI index has a strange sudden jump at cluster size 5 which in combination with a high Xu-index value and a very low Silhouette value indicates a non-optimal point. For this reason we must also choose a cluster size of 2 as the starting point for the CHI index, and then the next value indicates the optimal cluster size, which is 3.

As you can see if, we are further from the optimal solution, the estimation of the optimal cluster size is more difficult, which is of course not illogical. In general this 'rule of thumb' method can be trusted if enough points have been calculated and if several performance measures are used. If there is no clear indication of the optimal cluster size according to the above procedure, then this most probably means that the model is far from the optimal solution and you should try another model or change the parameters significantly.

Many more internal performance measures exist than the three introduced in the previous sections (Silhouette, Xu and CHI) and this is also the subject of ongoing research, but the three measures presented here are the three most useful and most used today. Most of the other existing internal performance measures operate according to similar principles.

3.3.2   External Criterion Based Performance Measures

External criterion based performance measures are available if the classification of the input data points is known and then these measures compare the predicted clustering (classification) to the original (reference) clustering. These external measures are therefore similar to supervised learning performance measures, but they are not the same because of the comparison between two possibly different labeling systems (reference and original). For example, if the clustering algorithm assigns a label 'A' to a group of points it is unlikely that the reference set has the same label 'A' for the same group of points. This additional uncertainty must be incorporated in the evaluation metrics!

This section will introduce the most important external criterion based performance measures and demonstrate how to use them in the case of a simple example.

3.3.2.1   Purity Measure

If we assign to each cluster the label (class) which is most frequent in that cluster then we can calculate the accuracy (purity) of the clustering with the following formula:

$$Purity = \frac{\sum_{i=1}^{K} n_{max}^i}{n}$$

where K is the number of clusters, n^i_{max} is the count of the most frequent class in clusters i and n is the total number of points in the dataset. For example, if we have a dataset with two known classes A and B, and the clustering result shows two clusters containing the following:

- Cluster 1: A A A B A A B A A
- Cluster 2: A A B B B B B B A

then the majority class for cluster 1 is A and for cluster 2 is B. The purity of this clustering result is then $(7 + 6)/18 = 72.2\%$.

The value of purity is between 0 and 1 (or 0 and 100%) and higher values indicate better clustering. One of the disadvantages of purity is that with increasing numbers of clusters its value gets closer to 1.

3.3.2.2 Contingency Table

Several external performance measures can be calculated with the help of the so-called contingency table, which is very similar to the confusion matrix used in the performance evaluation of supervised learning models.

Remember that we cannot just compare the reference labels with the labels from the clustering because the labeling systems are not the same!

Manning et al. [50] suggest viewing clustering as a series of decisions, one for each of the N (N-1)/2 pairs of points (records) in the dataset. We want to assign two points to the same cluster if and only if they are similar. A true positive (TP) decision assigns two similar points to the same cluster; a true negative (TN) decision assigns two dissimilar points to different clusters. There are two types of possible errors, a false positive (FP) decision assigns two dissimilar points to the same cluster and a false negative (FN) decision assigns two similar points to different clusters. This is similar to what we have seen in case of supervised learning.

The contingency table for the clustering model can be constructed with these four types of decisions as follows:

	Same cluster	Different clusters
Same class	TP	FN
Different classes	FP	TN

The external performance measures discussed in the following sections are based on the contingency table.

3.3.2.3 Rand Index

The Rand index measures the percentage of correct decisions and therefore it is a corrected global accuracy of the clustering model.

$$RandIndex = \frac{TP + TN}{TP + TN + FP + FN}$$

3.3.2.4 Precision, Recall and F1-measure

$$Precision : P = \frac{TP}{TP + FP}$$

$$Recall : R = \frac{TP}{TP + FN}$$

$$F1 = \frac{2 \cdot P \cdot R}{P + R}$$

Precision counts the well classified points (records) per cluster. It is sometimes also called positive prediction value (PPV). The precision is sensitive to imbalance in the data.

Recall evaluates each cluster with regard to the inclusion of a specified class. It is sometimes also called true positive rate (TP rate). Recall is not sensitive to imbalance in the data.

F1 combines precision and recall into one measure. F1 is sensitive to imbalance in the data.

There is also a modified version of the F1-measure which gives more weight to recall. This penalizes false negatives more strongly than false positives by selecting $\beta > 1$ in the following equation [50]:

$$F_\beta = \frac{(\beta^2 + 1) \cdot P \cdot R}{\beta^2 \cdot P + R}$$

3.3.2.5   External Criterion Based Performance Measures: Practical Example

In this section we will invoke a practical example which shows how to calculate the performance measures introduced in the previous section. This example is aimed at advanced readers because of the combinatorics (an area of mathematics) involved, and it was inspired by an example in [50]. The AI-TOOLKIT calculates these measures automatically; thus, there is no need for manual calculations.

Let us assume that a dataset was clustered and the results are presented near the top left corner of Table 3.5. There are three classes designated with 'triangle', 'circle' and 'diamond' which are distributed by the clustering model into three clusters c1, c2 and c3. For example, there are 6 triangles, 1 circle and 0 diamonds in cluster 1, etc.

The first step is to calculate the values in the contingency table—TP, TN, FN and FN.

The total number of positives (TP + FP), or pairs of points, in the same cluster, can be calculated with the total number of points in each cluster, which is 7 in the case of c1, 7 in the case of c2 and 6 in the case c3. Remember that we view clustering as a series of decisions, one for each of the N (N-1)/2 **pairs of points** (records) in the dataset. For this reason we cannot just sum up the individual elements but must take into account all possible groupings (arrangement of configurations).

Combinatorics is an area of mathematics which can be used to calculate the number of possible combinations. For example, if we have two objects, A and B, then we can calculate how many times these two objects can be distributed into one

Table 3.5 External criterion based performance measures: practical example

Input data cluster distribution

	c1	c2	c3	Σ
△	6	1	2	9
○	1	5	0	6
◇	0	1	4	5
Σ	7	7	6	

		FN	TN
c1	△	18	60
	○	5	8
	◇	0	0
c2	△	2	4
	○	0	30
	◇	4	2
c3	△	0	0
	○	0	0
	◇	0	0
	Σ	29	104

Contingency table

	Same cluster	Different clusters
Same class	32	29
Different classes	25	104

TP + FP	57
TP	32
FP	25
TN+FN	133
TN	104
FN	29

Purity	75.0%
Rand Index	71.6%
Precision	56.1%
Recall	52.5%
F1	54.2%

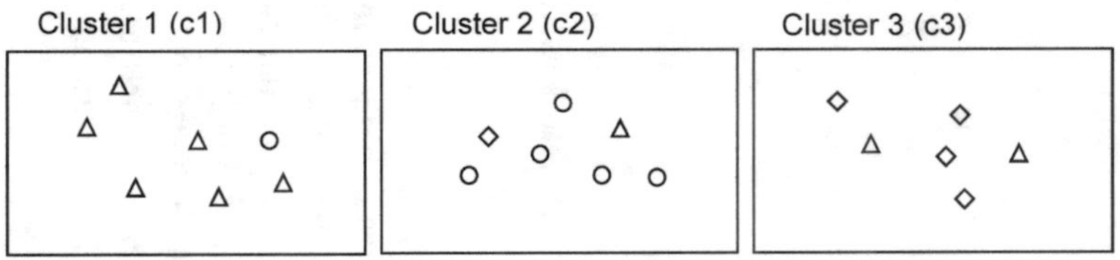

Cluster 1 (c1)

Cluster 2 (c2)

Cluster 3 (c3)

group. The answer is 1, the combination $\binom{2}{2} = 1$, there is only one grouping possible, A and B in one group (AB). But if we have three objects, A, B and C, then the combination $\binom{3}{2}$ results in 3 because there are three possible groupings (BC|A, AC|B and AB|C). The order of the elements is not important. The bottom number in the combination is always 2 in this case because we are working with a pair of points. We distribute a pair of points to a cluster.

According to the above we can calculate the total number of positives as follows:

$$TP + FP = \binom{7}{2} + \binom{7}{2} + \binom{6}{2} = 57$$

With the same logic we can also calculate the true positives (TP). Remember that a true positive (TP) decision assigns two similar points to the same cluster; we have 6 triangles in c1, 5 circles in c2, 4 diamonds in c3 and two triangles in c3. Adding the two last triangles in c3 is necessary because they are in the same cluster (TP) (it does not matter in which cluster).

$$TP = \binom{6}{2} + \binom{5}{2} + \binom{4}{2} + \binom{2}{2} = 32$$

And finally FP can be easily calculated as follows:

$$FP = (TP + FP) - TP = 25$$

The next step is to calculate the negatives. There is no combinatorics involved in this step because negatives break our rule of distributing a pair of points to the same cluster, instead single points are distributed ($\binom{n}{1} = n$). Remember that a true negative (TN) decision assigns two dissimilar points to different clusters, and a false negative (FN) decision assigns two similar points to different clusters.

The true negatives for the triangles, circles and diamonds can be calculated as follows:

$$TN_\Delta = 6 \cdot (5 + 0 + 1 + 4) + 1 \cdot (4 + 0) = 64$$
$$TN_O = 1 \cdot (1 + 2 + 1 + 4) + 5 \cdot (2 + 4) = 38$$
$$TN_\diamond = 0 \cdot (1 + 2 + 5 + 0) + 1 \cdot (2 + 0) = 2$$
$$TN = TN_\Delta + TN_O + TN_\diamond = 104$$

The calculation above simply counts how many times a point can be replaced by another class in another cluster. For example, if there are two points of class A in one cluster and three points of class B in another cluster then there are six possible ways to exchange the two classes in the two clusters without exchanging two at the same time.

The false negatives can be calculated in a similar way (here we replace the same classes):

$$FN_\Delta = 6 \cdot (1 + 2) + 1 \cdot 2 = 20$$

$$FN_O = 1 \cdot (5 + 0) + 5 \cdot 0 = 5$$

$$FN_. = 0 \cdot (1 + 4) + 1 \cdot 4 = 4$$

$$FN = FN_\Delta + FN_O + FN_. = 29$$

With the above results we can calculate the performance measures easily by using the equations introduced in the previous sections. The results can be seen in Table 3.5.

Chapter 4
Machine Learning Data

Abstract Machine learning models (algorithms) learn from data. This chapter explains how to collect, store and preprocess the data for machine learning. We will also look at the data strategy and the connection to machine learning data strategy. All data related subjects are very important because the quality of the machine learning model depends greatly on the data quality! We will look at data pre-processing (cleaning, transformation, sampling, resampling, feature selection, normalization, etc.) in more detail because of its importance.

4.1 Introduction

Machine learning data may come from several sources such as business documents and data, sensors (IoT devices), social media, web stores, video cameras, mobile devices, internal and external data stores, microphones, etc. The data may have different types and formats, for example, numerical values, text, images, audio recordings, etc. It is crucial to the success of machine learning that all of these data are collected, pre-processed and stored in a standardized and efficient manner! We must create an organization wide data strategy which includes a machine learning data strategy.

Data processing and delivery for machine learning have some specific needs compared to other types of business data which must be addressed!

The following sections will explain the specific requirements of machine learning with regard to data (collection, pre-processing, storage, etc.) and how to integrate these into an organization wide data strategy.

4.2 Data Strategy

First of all we need to define what an organization wide data strategy is, in order to be able to integrate our machine learning data strategy. The SAS White Paper [5] summarizes the organization wide data strategy very well (Fig. 4.1):

© Springer Nature Switzerland AG 2021
Z. Somogyi, *The Application of Artificial Intelligence*,
https://doi.org/10.1007/978-3-030-60032-7_4

Fig. 4.1 Data strategy

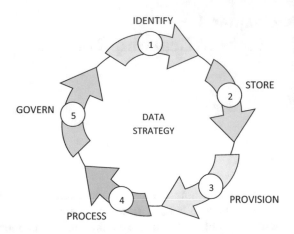

Organizations need to create data strategies that match today's realities. To build such a comprehensive data strategy, they need to account for current business and technology commitments while also addressing new goals and objectives. The idea behind developing a data strategy is to make sure all data resources are positioned in such a way that they can be used, shared and moved easily and efficiently. Data is no longer a byproduct of business processing—it's a critical asset that enables processing and decision making. A data strategy helps by ensuring that data is managed and used like an asset. It provides a common set of goals and objectives across projects to ensure data is used both effectively and efficiently. A data strategy establishes common methods, practices and processes to manage, manipulate and share data across the enterprise in a repeatable manner.

A data strategy must address data storage, but it must also take into account the way data is identified, accessed, shared, understood and used. To be successful, a data strategy has to include each of the different disciplines within data management. There are five core components of a data strategy that work together as building blocks to comprehensively support data management across an organization: identify, store, provision, process and govern [5].

1. **IDENTIFY:** *"Identify data and understand its meaning regardless of structure, origin or location.*

   Establishing consistent data element naming and value conventions is core to using and sharing data. These details should be independent of how the data is stored (in a database, file, etc.) or the physical system where it resides. It's also important to have a means of referencing and accessing metadata associated with your data (definition, origin, location, domain values, etc.). In much the same way that having an accurate card catalog supports an individual's success in using a library to retrieve a book, successful data usage depends on the existence of metadata (to help retrieve specific data elements). Consolidating business termi-nology and meaning into a business data glossary is a common means to addressing part of the challenge.

2. **STORE:** *Persist data in a structure and location that supports easy, shared access and processing.* The key is to make sure there's a practical means of storing all the data that's created in a way that allows it to be easily accessed and

shared. You don't have to store all the data in one place; you need to store the data once and provide a way for people to find and access it.

3. **PROVISION:** *Package data so it can be reused and shared, and provide rules and access guidelines for the data.*

4. **PROCESS:** *Move and combine data residing in disparate systems, and provide a unified, consistent data view.* At most companies, data originates from both internal and external sources. Internal data is generated from dozens (if not hundreds) of application systems. External data may be delivered from a variety of different sources (cloud applications, business partners, data providers, government agencies, etc.). While this data is often rich with information, it wasn't packaged in a manner to be integrated with the unique combination of sources that exist within each individual company. To make the data ready to use, a series of steps are necessary to transform, correct and format the data.

5. **GOVERN:** *Establish, manage and communicate information policies and mechanisms for effective data usage.* The role governance plays within an overall data strategy is to ensure that data is managed consistently across the company. Whether it is for determining security details, data correction logic, data naming standards or even establishing new data rules, effective data governance makes sure data is consistently managed, manipulated and accessed. Decisions about how data is processed, manipulated or shared aren't made by an individual developer; they're established by the rules and policies of data governance." [5]

4.3   Machine Learning Data Strategy & Tasks

As an extension to the data strategy defined in the previous section we will define a machine learning data strategy framework. Figure 4.2 shows a summary of the framework which can be used to design a machine learning data strategy in your organization.

Different machine learning (ML) applications and algorithms need different types and amounts of data and different data delivery methods. Because of this the development of a machine learning data strategy must start with the study of how the ML workflow influences data acquisition and storage, data delivery and data pre-processing steps.

The ML workflow is often served from a dedicated ML database where the pre-processed data is stored. But it is also possible that the pre-processed data is directly fed to the ML workflow, or that no pre-processing is needed because of the built-in automatic data pre-processing in the ML workflow (see Fig. 4.2). For example, the AI-TOOLKIT has several automatic data pre-processing steps and the data features (columns) can be chosen as part of the design process which significantly simplifies the work to be undertaken.

The output of the ML application is often saved into a database before further processing (e.g. before feeding it to a dashboard, to another application, etc.).

Each part of the framework will be explained in the following sections.

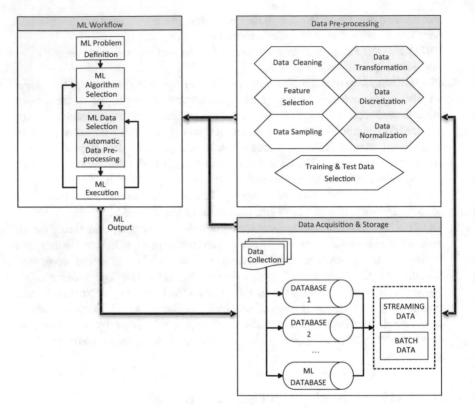

Fig. 4.2 Machine learning data strategy framework

TIP: Machine learning data strategy development is a continuous process! When new ML applications are needed the strategy may need to be extended and/or modified in order to accommodate all requirements.

4.3.1   Machine Learning (ML) Workflow

The simplified ML workflow can be seen on Fig. 4.3.

Each ML application must start with the definition of the problem the application will solve. Naturally the necessary data will depend on this. Next the type of ML algorithm is selected which also influences the necessary input data. Remember that there are three main types of ML algorithms: supervised learning, unsupervised learning and reinforcement learning. Remember also that each of these types of ML algorithms have several possible implementations, for example, supervised learning

Fig. 4.3 Simplified ML
workflow

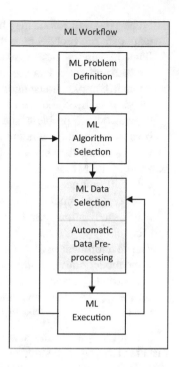

may utilize a support vector machine, neural network, random forest, etc. Further-more, supervised learning may involve classification or regression.

How do all of these options influence the type of data, the amount of data and the data delivery method?

The answer to this question depends on several factors:

1. The problem definition will provide information about what kind of data is needed for the ML application, and if this data is immediately available in internal or external databases, or if some kind of data collection is needed. Data collection is a huge subject in itself (measurements, questionnaires, digital search...) and must also be standardized in order to provide the necessary quality and efficiency! The collected data must then be stored according to the previously defined data strategy (identify, store, provision, process and govern)! It is always better to overestimate the necessary amount of data, and the number of features (columns or variables) in the data, because during ML model development machine learning experts will need to search for the best combination of features and optimal amount of the data! See 'Data Acquisition and Storage' on Fig. 4.2.
2. For supervised learning we need separate datasets for training, testing and inference (prediction). For unsupervised learning we only need a training and an inference (prediction) dataset. Remember from Sect. 1.2.1 that the training and test datasets must be independent from each other and must be identically distributed! All of these will influence our machine learning data strategy. See 'Data Pre-processing' on Fig. 4.2. More about this later!

3. Some ML applications may have an automatic data pre-processing capability and extra data pre-processing will not be needed or just partially needed.
4. Some ML applications may allow the dynamic selection of features (columns) in the data and extra data pre-processing may not be needed.
5. Some ML applications/algorithms will need data to be delivered in one batch and others may need the data to be streamed. The data delivery method will thus depend on the problem and the selected algorithm. Or it is also possible that because of the huge amount of data needed we decide to use an algorithm which uses streamed data instead of a fixed size dataset.
6. Most of the ML algorithms require numerical input data and thus non-numerical data will first need to be converted to numbers. Some other algorithms will use other types of input data, e.g., text or images. Data cleaning, data transformation and discretization may be needed. We will talk about these tasks in the next section.
7. The development of an ML model is an iterative process in two ways. First of all, if the model does not provide satisfactory results, then another type of algorithm may need to be selected which may influence the necessary input data. Secondly, for the same reason, the amount of data or the number of features in the data may need to be modified (see Fig. 4.3).

It is up to you to find all of the ML workflow parameters which influence the machine learning data strategy in your organization and design a strategy which will provide high quality and efficient machine learning. The previous chapters and sections about ML algorithms will help you to understand all of the requirements.

4.3.2   Data Pre-Processing

After the necessary data is identified it must often be pre-processed before it is fed to our machine learning (ML) work flow. The most commonly used pre-processing steps will be explained hereunder.

A lot of these steps can often be automated in data pre-processing, which also ensures that all of the data is handled in a standardized and efficient way (Fig. 4.4).

4.3.2.1   Data Cleaning

One of the most commonly used pre-processing steps is the cleaning of the data. There can be empty records, empty values, incorrect values, non-valid values (e.g. text instead of numbers), etc. All of these irregularities must be resolved before any further steps can be undertaken!

We call the list of values in several columns (features), or in other words one row of data, a *data record*. A value is the data in a column of a specific row. If you

Fig. 4.4 Data
pre-processing

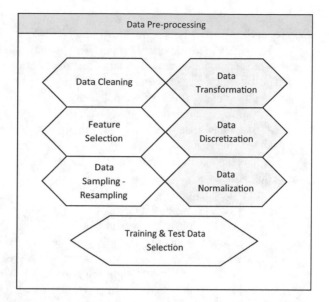

imagine the data as a table, then the value is the data in a table cell and a record is a table row.

Empty records must be removed. Empty (missing) values may be **removed** together with the whole record, or they may also be **replaced** with, for example, the mean of the whole column, the mean of the neighboring values, the most frequent value in the column, or with 0, etc. It is up to you to decide what the best solution is for your specific problem. The same applies in the case of wrong or non-valid values. The process of finding and replacing values can be automated!

It is important to make clear and well communicated rules about how the data cleaning should be performed in order to make sure that the data is always cleaned in the same way and that this does not confuse the results of machine learning! Imagine that one dataset is cleaned by replacing empty values with 0 and another one is cleaned by replacing empty values with 1, this will unintentionally influence the results of machine learning!

Data cleaning can be automated well, but the visual inspection of the data and manual data analyses are often very useful for detecting all kinds of possible problems in the data. Therefore, a data analysis platform may be indispensable!

A little known technique, so-called variation analysis, may be used to detect problems in the data. Variation analysis will pinpoint possible erroneous data points where unexplained variation occurs. More information about this technique can be found in Expert Section 4.1. The AI-TOOLKIT has a built-in automatic variation analysis tool.

Expert Section 4.1 Variation Analysis

Variability is a very important property of the data. Variability measures how the data is spread around the center of the distribution.

The two most common measures of variability are the variance and the standard deviation. The variance is the average squared distance of each data point from the mean and it is has no unit because it is just a measure of the spread of the data. The standard deviation is the square root of the variance and it has the same unit as the data.

Outliers have a huge effect on the variance (and on the standard deviation) and this is one of the reasons why you will want to find the root cause of outliers in your data!

There is always variation in the data but the data can be *in control* or *out of control* depending on the type and level of variation. When data is out of control it means that there is something wrong in the data. With some simple statistical techniques it is possible to determine if the variation in a specific feature variable makes the feature out of control or if it is just normal variation which causes no harm. **This is important in order to remove variation which would confuse the machine learning model.**

The tool which we will introduce in this section to study variation is the so-called X&R control chart. The AI-TOOLKIT data analyses module has a built-in X&R control chart tool!

We can categorize most of data into three groups:

1. Discrete count (Poisson distribution). E.g., a count of events.
2. Discrete classification (binomial distribution). E.g., 'yes/no', 'good/ bad', etc.
3. Continuous (normal distribution).

Because it is much more reliable to base our decisions on continuous time series data it is always preferable to convert other types of data (discrete count or classification) into continuous time series data. This can be done by:

• In the case of discrete classification ('yes/no', 'good/bad', etc.), using the time period between the occurrences of the values (e.g. between 'yes' happening and 'no' happening, etc.) to form a continuous data series.
• In the case of discrete count data measuring the elapsed time between the specific events that are being counted.

The X&R chart is the most frequently used control chart because it employs the central limit theorem to 'normalize data', and therefore it is not very important what the underlying distribution is. But there are two other less frequently used control charts: for classification (binomial) data, the p-chart, and for count (Poisson) data, the u-chart. These two charts are not necessary for our purposes and thus will not be discussed in more detail.

(continued)

Expert Section 4.1 (continued)

You can see an example of the X-chart on Fig. 4.5 and of the R-chart on Fig. 4.6. On Fig. 4.5 you can also see the important zones and limits of control charts. The control limits are three standard deviations away from the mean ($\mu \pm 3\sigma$): the upper control limit (UCL) in a positive direction (+) and the lower control limit (LCL) in a negative direction (−).

Between the control limits the chart is divided into six zones, three on each side of the mean (below and above the mean/centerline): zone C which is up one standard deviation away from the mean, zone B which is from one to two standard deviations away from the mean, and zone A which is from two to three standard deviations away from the mean (i.e. until the control limit). In the case of the R-chart, if any of the lower limits (LCL, zone B and zone C) are lower than 0, then that control limit is not valid and is not taken into account.

The labels of the different lines and signs used on the control charts are shown on Fig. 4.7.

How to use Control charts

When using the control charts we have to pay attention to two problem areas:

1. *The control limits*—check if any point is outside the control limits which means that the data is out of control (non-explainable variation).
2. *The spread of data points between zones A, B and C*—the spread of the data points between the three zones, A, B, and C, can signal specific problems in the data. An overview of all possible problems with the data is given on Figs. 4.8 and 4.9.

If we detect a problem then we have two options, we can try to find the reason for the problem in the data or we can just remove the data point. If we are able to find the reason for the problem, then we can try to eliminate the problem (e.g., in the data collection, or studied process, etc.) and repeat the process of collecting the data. If it is a systematic problem then many erroneous data points may exist and it could be very helpful to eliminate the problem instead of just removing the bad points.

TIP: Always look at the R-chart first and check if there are out of control points (points outside the control limits) or other problems. The control limits on the X-chart are derived from the average range so if the R-chart is out of control then the control limits on the X-chart are meaningless. Once the effect of the out of control points from the R-chart is removed look at the X-Chart.

On Figs. 4.8 and 4.9 you can see the illustration and explanation of the different problems the control chart can detect (based on, but modified from, [7, 8, 10]) (the AI-TOOLKIT automatically identifies all of these cases):

- 1 point beyond zone A
- 9 points in a row on the same side of the centerline

(continued)

Expert Section 4.1 (continued)
- 6 points in a row steadily increasing or decreasing
- 2 out of 3 points in zone A or beyond
- 4 out of 5 points in zone B or beyond
- 14 points in a row alternating up and down
- 15 points in a row in zone C, above and below the centerline
- 8 points in a row on both sides of the centerline with none in zone C

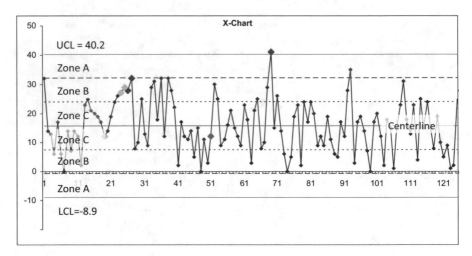

Fig. 4.5 X-chart

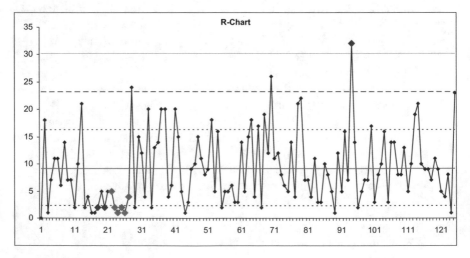

Fig. 4.6 R-Chart

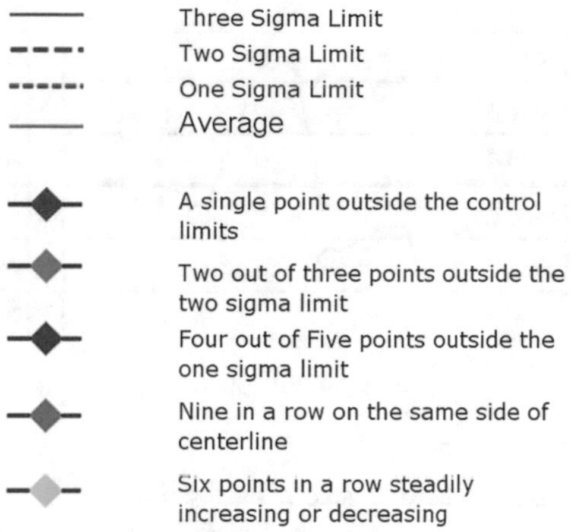

Fig. 4.7 Control chart signs

————	Three Sigma Limit
— — —·	Two Sigma Limit
- - - - - -·	One Sigma Limit
————	Average
—◆—	A single point outside the control limits
—◆—	Two out of three points outside the two sigma limit
—◆—	Four out of Five points outside the one sigma limit
—◆—	Nine in a row on the same side of centerline
—◆—	Six points in a row steadily increasing or decreasing

4.3.2.2   Data Transformation

Data transformation may be needed in several circumstances. For example, if we want to reduce the data by combining several features together, or if we want to modify specific features, or if we want to convert textual categorical data to numerical values, etc.

Data Reduction with Principal Component Analyses (PCA)
Why would we want to reduce the data and combine several features (columns) together? Because more data does not always mean better machine learning! A limited amount of high quality data, where high quality means not only that the data is clean but that it contains just enough information for the purpose of machine learning, is often much more useful! Furthermore, less data makes machine learning faster and it may also solve the problem of not enough memory! In order to get to this high quality data, one must invest enough time into researching the problem and the data instead of just using any data which is available! The data has the most important influence on the effectiveness of machine learning!

We can reduce a large dataset with many features (columns) significantly without losing the underlying relationships in the data with so-called principal component analyses (PCA). This reduction is achieved by transforming the dataset into a new set of features, the principal components, which are uncorrelated and ordered so that the first few retain most of the information (variation) present in the original dataset. The computation of the principal components can be reduced to the calculation of the so called covariance (or correlation) matrix of the features and the derivation of the eigenvalues and eigenvectors [11].

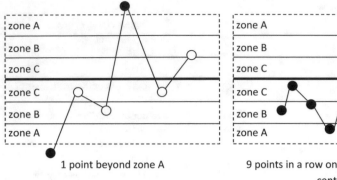

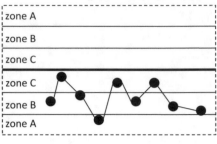

1 point beyond zone A

9 points in a row on the same side of the centerline

1 point beyond zone A means that the point is outside the control limits. Statistically this situation can mean one of three things:

- A shift in the mean of the data
- An increase in the standard deviation
- A single problem (effects only 1 point)

This situation occurs when there is a shift in the mean of the data.

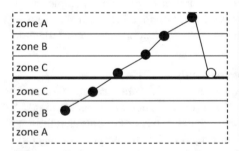

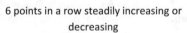

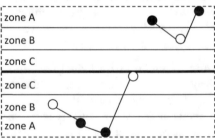

6 points in a row steadily increasing or decreasing

2 out of 3 points in zone A or beyond

This situation signals a trend (systematic change) in the mean of the data.

This situation signals a shift in the mean or an increase in the standard deviation of the data.

Fig. 4.8 Control chart problem cases: part I

The AI-TOOLKIT has built-in support for principal component analyses (see templates). A summary of the PCA algorithm and more details are explained in Expert Section 4.2.

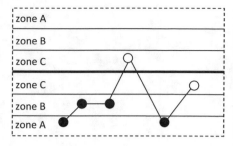

4 out of 5 points in zone B or beyond

This situation signals a shift in the mean of the data.

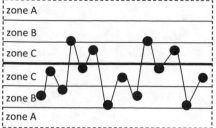

14 points in a row alternating up and down

This arrangement of the points signals a systematic change in the data.

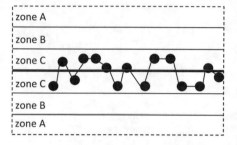

15 points in a row in zone C, above and below the centerline

This situation signals that there is a problem with the sampling/sub-grouping of the data, more specifically it means that the observations in a subgroup come from sources with different means.

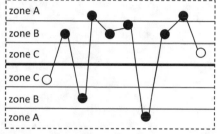

8 points in a row on both sides of the centerline with none in zone C

This situation signals that there is a problem with the sampling/sub-grouping of the data, more specifically it means that the data in the different subgroups come from different sources with different means.

Fig. 4.9   Control chart problem cases: part II

Expert Section 4.2: Principal Component Analyses [11]

Suppose that we have p number of features in a dataset (p columns of data). Let us call these features $x_1, x_2 \ldots x_p$, where $x_1, x_2 \ldots x_p$ are vectors. PCA looks for a few m derived features ($m \ll p$) that preserve most of the information in the dataset, defined by the variances of the original features and by the correlation between the original features. Let x denote the whole dataset $\{x_1, x_2 \ldots x_p\}$.

(continued)

Expert Section 4.2 (continued)

PCA can reduce x, a large dataset with many features (columns), significantly without losing the underlying relationships in the data. How does this work?

As a first step PCA searches for a linear function $\alpha_1^T.x$ having maximum variance, where α_1 is a vector of p number of constants $\alpha_{11}, \alpha_{12}, \ldots \alpha_{1p}$, and 'T' is the transpose of the vector α, so that

$$\alpha_1^T x = \alpha_{11}x_1 + \alpha_{12}x_2 + \ldots + \alpha_{1p}x_p = \sum_{j=1}^{p} \alpha_{1j}x_j$$

The next step is to search for a linear function $\alpha_2^T.x$, uncorrelated with α_1^T. x and having maximum variance. And so on, so that at the kth step a linear function is found, $\alpha_k^T.x$, which has maximum variance and is uncorrelated with the previous functions $\alpha_1^T.x, \alpha_2^T.x, \ldots \alpha_{k-1}^T.x$. We call $\alpha_k^T.x$ the kth principal component.

There are p number of principal components (equal to the number of features in the dataset) but often most of the information in the dataset can be explained with only m number of principal components where m < < p. And this reduction in features or complexity reduction is what we want to achieve!

In order to make this clearer, let us look at an example with two features (two columns of data).

The upper image of Fig. 4.10 shows the plot of the data. The lower image shows the plot of the principal components calculated according to the method explained above. The exact values are not important and therefore not indicated. Only the proportions of both plots are important.

The features x_1 and x_2 are highly correlated and there is significant variation in both of them. There is, however, more variation in the x_2 direction (the vertical spread of the points is bigger than the horizontal spread). From the principal components plot it is clear that there is more variation in the direction of pc_1 than in x_1 and x_2. There is nearly no variation in the direction of pc_2 and thus we can say that pc_2 does not contain significant information.

We can say that if there is substantial correlation between p features, then the first few principal components will account for most of the information in the original data.

The calculation of the principal components can be reduced to a number of relatively simple tasks. The first step is to calculate the so-called covariance (or correlation) matrix of the p features (p data columns). The covariance matrix contains the covariance between features i and j if i ≠ j or the variance of feature i if i = j. The kth principal component can be calculated with $pc_k = \alpha_k^T$.

(continued)

Expert Section 4.2 (continued)
x, where α_k is the so-called **eigenvector** of the covariance matrix corresponding to its kth largest so-called **eigenvalue** λ_k. If α_k has a unit length (requirement) then the variance of pc_k is equal to λ_k.

In practice the correlation matrix is often used instead of the covariance matrix because it provides better results in many cases.

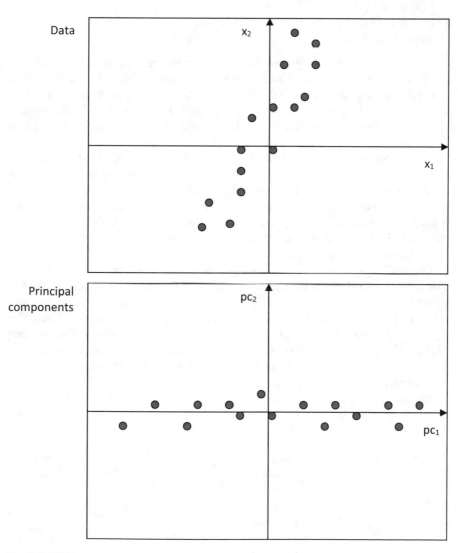

Fig. 4.10 PCA example

Converting Categorical Values to Numbers

There are two main types of categorical data:

- Nominal—these are categorical values without order; e.g., [red, green, blue, yellow]
- Ordinal—these are categorical values with order; e.g., [less good, good, very good]

The three most used techniques to convert categorical data to numerical values are the following:

- Integer encoding
- One-hot encoding
- Binary encoding

The integer encoder converts each categorical value to a sequential integer value. For example, [less good, good, and very good] is replaced with [1, 2, and 3]. Each categorical feature (column) will then have a similar encoding, for example, a second column could be [small circle, large circle] which is converted to [1, 2]! Ordinal categorical values may also be converted in this way. Table 4.1 shows an example of integer encoding.

In the case of nominal categorical values, some machine learning models may introduce a bias (error in the learning) during integer encoding if they assume that the sequential integer values are on a meaningful scale. For example, in the case of [red, blue, violet], if red = 1, blue = 2 and violet = 3, then violet > red, blue < violet, etc. introduces a special relationship in the data between the colors which in reality does not exist! In this case the integer encoded categorical values could be further converted with one-hot encoding or binary encoding.

One-hot encoding replaces each integer encoded value with a series of values. The number of values in the series is equal to the number of unique categorical values in the column. The numbers in the series are all 0 except for one value which is set to 1 (a different one in each category). For example, [red, blue, violet, green] is replaced with [1, 0, 0, 0, 0, 1, 0, 0, 0, 0, 0, 1], where, e.g., red is converted to the series {1, 0, 0, 0}, blue to the series {0, 1, 0, 0}, etc. This will increase the number of columns (features) in the dataset by the number of values in each category! Table 4.2 shows an example of one-hot encoding.

Binary encoding replaces each integer encoded value with its binary value and the binary values are split into a series of 0's and 1's. The longest binary value

Table 4.1 Integer encoding example	Original data		Integer encoded data	
	Column 1	Column 2	Column 1	Column 2
	Good	Small circle	1	1
	Good	Large circle	1	2
	Less good	Normal circle	2	3
	Very good	Large circle	3	2
	Less good	Normal circle	2	3

Table 4.2   One-hot encoding example

Original Data		One-hot Encoded Data					
c1	c2	c1	c2	c3	c4	c5	c6
good	small circle	1	0	0	1	0	0
good	large circle	1	0	0	0	1	0
less good	normal circle	0	1	0	0	0	1
very good	large circle	0	0	1	0	1	0
less good	normal circle	0	1	0	0	0	1

Table 4.3   Binary encoding example

Original Data		Binary Encoded Data							
c1	c2	c1	c2	c3	c4	c5	c6	c7	c8
good	small circle	0	0	0	1	0	0	0	1
good	large circle	0	0	0	1	0	0	1	0
less good	normal circle	0	0	1	0	0	0	1	1
very good	large circle	0	0	1	1	0	0	1	0
less good	normal circle	0	0	1	0	0	0	1	1

determines the number of added columns. Shorter binary vales are extended with 0's. Table 4.3 shows an example of binary encoding. The binary value of 1 is 0001, the binary value of 2 is 0010, and 3 is 0011.

If there are more than 15 categories in a column then, e.g., the integer value 16 is converted to 00010000 in binary form. This will add eight columns per converted column (eight digits)! Values below 16 which have a digit length of four in binary form will be extended with 0's to have a length of eight digits. If we used one-hot encoding then we would have to add 16 columns instead of just eight!

Binary encoding may be useful when there are many categories because it adds a much smaller number of extra columns than one-hot encoding! Fewer columns mean less memory consumption and faster processing. When there are only a few categorical values then use one-hot encoding instead of binary encoding, because binary encoding is somewhere in the middle between integer encoding and one-hot encoding concerning the extra relationship added by the encoding, which is not in the original data.

4.3.2.3   Data Discretization

Data discretization means that we convert continuous values in a column into categories or groups. For example, if we have the temperature in a column then we could convert temperature ranges into the categories 'very cold', 'cold', 'normal', 'hot' and 'very hot', and then convert these categories with integer encoding to a sequence of numbers. Or we could also replace temperature values in a range with the integer value of the average temperature of the range. Discretization may sometimes help to improve machine learning performance by simplifying the data.

4.3.2.4   Data Sampling

Sampling means that we choose which data to use from all of the available data. We normally use all of the data that we have collected or have in storage. Sometimes we collect data randomly (e.g., not every client fills in a questionnaire, only some chosen ones—random sampling) or we collect all of the data over a period of time (i.e., every client fills in the questionnaire), or we collect data only from a specific group, etc. The method we choose will depend on the specific business case and the goal we want to achieve!

There are two important assumptions which should be met in order to comply with statistical theory while sampling data:

- The sample is a **random** sample (i.e., we have not systematically selected some specific types of values and neglected all other types of values) and it must be **representative** of the whole dataset (sometimes called the population).
- The population is assumed to be **normally distributed** (see Expert Section 4.3).

Before collecting the data we need to decide the amount of data we will collect, which is called the sample size. It is a sample because we only collect a subset of the data, and not all of the data, for obvious reasons—we cannot collect data endlessly.

The sample size depends on several factors but the most important one is the sampling error. The sampling error is the error we make by collecting only a subset of the data, and thus by increasing the sample size we decrease the sampling error. On the other hand, a sample size which is too large may also cause some problems; for example, it takes much more time and cost to collect the data.

The appropriate sample size can be estimated by the simplified statistical formula in Eq. 4.1 [6]:

$$n = \frac{4\sigma^2}{B^2} \tag{4.1}$$

where n is the sample size, σ is the standard deviation of the sample and B is the sampling error on the *mean* (half of the 95% confidence interval) that we agree to accept. It is an estimate because we do not know beforehand the standard deviation of the sample we will collect and we need to estimate it, for example, from historical data (data collected in the past) or by just making a reasonable guess. An example of the sample size calculation can be seen in Sect. 4.3.2.5.

NOTE: Most of the data you will collect will have a normal distribution but there are, of course, also populations, and samples of these populations, which are not normally distributed, and instead have a binomial, Poisson, or exponential distribution. More information about this subject can be found in Expert Section 4.3.

4.3.2.5 Example: The Introduction of a New Bank Service: Sample Size

A bank decided to introduce a new kind of service to its clients. In order to review the effectiveness of the new service, they decided to collect data on the use of the new service by the clients and to feed this to a machine learning model.

After careful consideration the bank decided to collect data on customer arrival times and on satisfaction with the service using a scale between 1 and 10 (1 being not satisfied, 10 being very satisfied). In this way they also hoped to be able to see whether long waiting lines/queues were forming. They knew that by collecting customer arrival times they also have automatically the number of clients using the services in a specific period.

In order to collect the data for studying the arrival times, the bank had to decide how many data points they needed (the sample size).

The necessary sample size can be calculated from Eq. 4.1, where B is the acceptable sampling error on the mean of the arrival times and σ is the estimated standard deviation of the arrival times. From historical data the standard deviation was chosen to be 7.12 min and B was chosen to be 0.45 min.

$$n = \frac{4\sigma^2}{B^2} = \frac{4 \times 7.12^2}{0.45^2} = 1001$$

Expert Section 4.3: Distribution of the Data

Most of the data you will collect will have a normal distribution but there are, of course, also populations, and samples of these populations, which are not normally distributed and instead have the binomial, Poisson, or the exponential distribution. We will discuss all of these distributions hereunder in more detail.

The Normal Distribution

The most important distribution is the normal distribution. The normal distribution is characterized by its symmetric bell shaped curve and by its mean and standard deviation as shown on Fig. 4.11. In many cases the

(continued)

Expert Section 4.3 (continued)

distribution is not completely symmetric but skewed to the left or to the right. However, most of these skewed distributions can be approximated by a normal distribution and the empirical rules apply.

There is an important rule in the case of a symmetric or nearly symmetric bell shaped distribution which is approximately a normal distribution (many datasets which you will encounter are approximately normally distributed!): approximately 95% of the data are within two standard deviations of the mean, which means the range of [Mean $\pm$ 2σ], and approximately 99.7% of the data are within three standard deviations of the mean, which means the range of [Mean $\pm$ 3σ].

There is a very useful MS Excel function which can be used to generate a normally distributed dataset if you know the mean and the standard deviation: '=NORMINV (p, μ, σ)'.

where p are values in the range of [0, 1] (related to the standardized normal probability distribution), μ is the mean of the data and σ is the standard deviation of the data. You could use this function to generate a dataset if you know the mean and standard deviation from a historical experiment but there is no data available or if you want to replace missing or incorrect data!

Most data analysis software (e.g., MS Excel) has built-in support for testing whether a data is normally distributed!

The binomial distribution

The binomial distribution is the distribution of a discrete variable which has only two possible values in the form of 'yes/no' decisions (e.g., 0/1, male/female, present/missing—as in the number of missing VAT numbers in 100 invoices every month, etc.) and where each experiment to get/measure the values is identical. This implies that the probability of each value being 'yes' equals p and the probability of it being 'no' is 1-p (obviously, because there are only two possible values).

If we know the number of experiments performed (n) and the probability (p) of the 'yes' event, then we can calculate the mean and standard deviation of the binomial distribution:

$$\mu = n \cdot p$$

$$\sigma = \sqrt{n \cdot p(1 - p)}$$

For example, if we check in a bank whether 1000 clients (n = 1000) are male or female, and we know that the probability of being a male is 70% (p = 0.7) and that the data (gender) has the binomial distribution, we then know from the above equations that the mean of the data is 700 (1000*0.7) and

(continued)

Expert Section 4.3 (continued)

the standard deviation is 14.5 ($\sqrt{1000 \cdot 0.7(1 - 0.7)}$). This also gives us the expected number of 700 male clients!

And because the empirical rules which apply to the normal distribution also apply to the binomial distribution, if n·p > 5 and n·(1-p) > 5, we can also say that we are 95% certain that the number of male clients will be 700 ± 2(14.5), or between 671 and 729 [6]! Please note that this is the rule we saw before, which states that approximately 95% of the data are within two standard deviations of the mean (Mean ± 2σ).

The Poisson distribution

If the data concerns the count of events occurring within a specified period of time, then the data has a special distribution, which is called the Poisson distribution (see Fig. 4.12). The data values are integer values. The Poisson distribution is characterized by a single parameter λ, which is equal to the Mean and also to the variance of the distribution. The standard deviation is thus the square root of λ. A commonly occurring example of a Poisson distribution is, e.g., the number of customer arrivals in an hour (λ — number of arrivals per hour), or the number of units of a product sold in a week (λ = number of products sold per week).

The exponential distribution

If the data are, e.g., the time periods between events, such as the times between customer arrivals, then the data has a special distribution which is called the exponential distribution, defined with parameter λ (see Fig. 4.13). Both the mean and the standard deviation of the data are equal to the reciprocal of λ (μ = σ = 1/λ).

There is a very close relationship between the Poisson and the exponential distributions, which is also shown by the shared λ parameter.

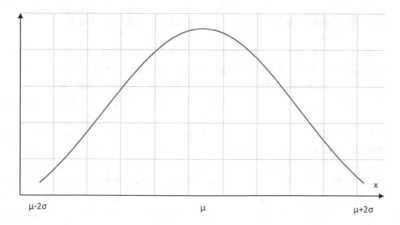

Fig. 4.11   The Normal distribution

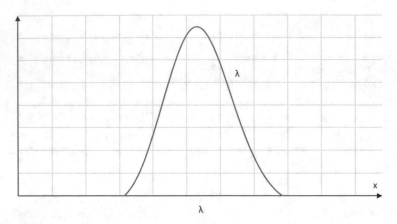

Fig. 4.12 The Poisson distribution

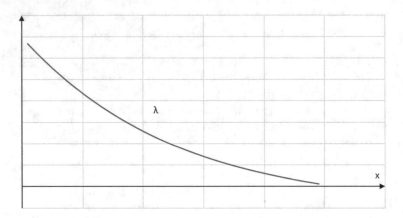

Fig. 4.13 The exponential distribution

4.3.2.6   Data Resampling to Remove Class Imbalance in the Data

In supervised classification problems, imbalance in the data means that there are many more data records belonging to a specific class than to other classes (type I), or that a subset of a specific class has many more records than another subset of the same class (type II). A simple 2D illustration of these problems can be seen on Fig. 4.14.

Why is imbalance in the data a problem? Because it reduces (per class) the prediction performance. If some classes are overrepresented, the ML model will 'focus' more on these classes and will be less precise concerning the minority classes. Minority classes are the classes which are underrepresented and majority classes are the classes which are overrepresented.

In many real world applications there is a significant imbalance in the data. For example, in the case of anomaly or problem detection, there are usually many more

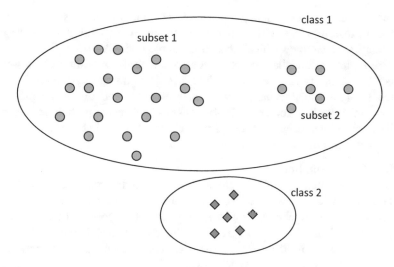

Fig. 4.14 Class imbalance in the data (type I and type II)

data records with normal (non-anomalous) cases than anomalous cases. Intrusion detection, fraud detection, preventive maintenance, root cause analysis… are all examples which involve imbalance in the data.

There are two important considerations when using imbalanced datasets:

- The ML model performance evaluation must take into account the imbalance in the data and must provide appropriate metrics which are not sensitive to this imbalance.
- The training of the ML model must take this imbalance into account.

More information about the performance evaluation and about which metrics are not sensitive to imbalance in the data can be found in Chap. 3.

In the training of the ML model we can take imbalance in the input data into account in two ways:

- By applying special weights in the ML model for each class, based on the proportions of the classes—minority classes will then have a higher weight.
- By pre-processing the input data and correcting the imbalance.

It is much easier and more efficient, and in many cases results in a better ML performance, if we pre-process the data instead of using special imbalance modified models. Therefore, we will focus on correcting imbalance by pre-processing the data.

It is also important to note that not all imbalances in the data will cause major ML performance problems! And some ML models may even be less sensitive to imbalance than others. It is important to use the right performance metrics and take action by pre-processing the data if needed.

There are many types of data pre-processing techniques and it is also the subject of ongoing research. We will introduce hereunder two useful oversampling and undersampling techniques which are also available in the AI-TOOLKIT.

Oversampling means that we create new data records with minority classes and undersampling means that we remove data records with majority classes.

Each pre-processing technique has its advantages and disadvantages, and it depends on the specific problem and dataset as to which technique should be applied. Usually a selection of different resampling techniques is applied in order to combine their advantages. For an example, see Sect. 8.5.7.

Random Undersampling

Random undersampling removes random data records from the majority class (or classes) and in this way reduces the imbalance problem. One of the major disadvantages of this technique is that it may remove important information (variation) from the data. Therefore, be careful while using this technique and check the performance metrics of the model and per-class performance. The advantage of this technique is that it is fast and easy to apply.

This technique alone will not remove all imbalances in the data and usually it should be applied in combination with other resampling techniques.

Random Oversampling

Random oversampling duplicates random data records from the minority class (or classes) and in this way reduces the imbalance problem. One of the disadvantages of this technique is that it may cause over-fitting or degraded generalization performance because it artificially decreases the variation in the data. It is easy and fast to apply.

Borderline Synthetic Minority Oversampling Technique (BSMOTE)

The borderline synthetic minority oversampling technique (BSMOTE) [36, 37] is one of the best advanced oversampling techniques today. Not only does it reduce the imbalance problem but it also often results in a higher ML performance by increasing the clarity of class separation. BSMOTE generates new synthetic samples between minority samples (data records in a minority class) and some of their closest neighbors, along the borderline of the class samples. In order to clarify this, let us look at a very simple 2D example on Fig. 4.15.

The reason why the algorithm concentrates on the samples (points) at the borderline, is that these points are more at risk of being misclassified than the internal points. There is also a built-in check for noise detection in order to prevent the addition of extra noise points (if all nearest neighbors are from another class).

In some situations it may be useful to add extra points even if a border cannot be detected (e.g., if the other classes are far away), and therefore the AI-TOOLKIT automatically detects these cases and adds extra internal points.

Undersampling with TOMEK Links Removal

A TOMEK link [35] is formed by two data records (nodes) if they belong to different classes and if they are each other's nearest neighbors. This can only happen if one of the records is noise or if they are located on the boundary between the two classes

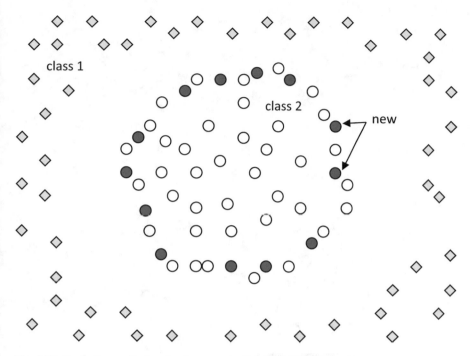

Fig. 4.15 Borderline synthetic minority oversampling (BSMOTE) Example

(borderline). See Fig. 4.16 for a simple example. If class 2 is selected for undersampling then the dark blue points will be removed.

This method may be applied to all classes in order to remove noise and to clean up the boundary between the classes if there are enough samples. In some applications it is only applied to the majority class.

There are many other types of resampling algorithms, and there are also many variations of the above resampling techniques, but they all operate according to similar principles. The combination of random undersampling, random oversampling, borderline synthetic minority oversampling (BSMOTE) and TOMEK links removal is a very effective solution to the imbalance problem and an extended version of this is available in the AI-TOOLKIT!

4.3.2.7   Feature Selection

Feature selection is the process in which we select which features (columns) to feed to the machine learning model. It is often an iterative process because if the performance of the machine learning model is not satisfactory, then we need to add or remove specific features (columns) from the data and retrain the model. It is important to have domain expertise in order to be able to select the best collection of features instead of just using trial and error! For example, if we want to train a

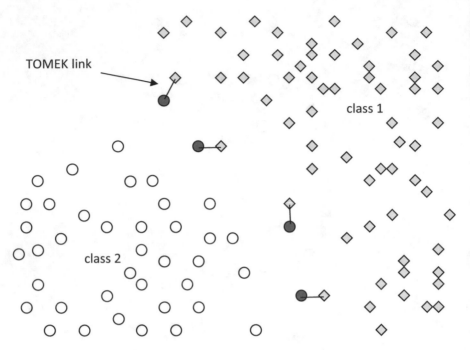

Fig. 4.16 TOMEK links for undersampling

machine learning model to detect cancer, then we must have experts in cancer research who tell us which features are most probably the best ones to use.

Sometimes variation analysis is used to remove features which do not contain enough information for the machine learning model. Features (columns) with very low variation are then removed. One must, however, be careful with this because sometimes non-normalized features may have very different amounts of variation but may all contribute to the learning of the problem to some extent. See also Expert Section 4.1 about variation analysis.

4.3.2.8 Data Normalization: Scaling and Shifting

A dataset may have many features (columns of data) with different scales. For example, a financial dataset may contain a feature with the age of a client and another feature with the amount of their yearly salary. The age may be in the range [18, 100] and the salary may be in the range [0, 1000000]. Many machine learning algorithms will have problems with the scale difference between these two features and the search for the optimal solution fails. Another problem may arise if the distribution of a feature (the data in a column) is not zero-centered, because the solution search (e.g., gradient descent) will be very inefficient (zigzagging). More information about this problem can be found in Expert Section 2.1.

The solution to all of the above problems is data normalization. Normalization scales and shifts the data in all columns into a pre-defined zero-centered range, for example $[-1, 1]$. The variation (knowledge) in the data will stay, but all of the features will have the same scale and will be zero-centered.

In some circumstances we may have to apply normalization several times during a machine learning model training process because, for example, activation functions in a neural network undo the effect of normalization. More information about normalization can be found in Expert Section 2.3 (batch normalization).

The AI-TOOLKIT performs automatic normalization in all circumstances; we do not have to manually pre-process the data.

4.3.2.9   Training/Test Data Selection

We make a significant distinction between how well the machine learning model performs (success of learning) on a training (learning) dataset and on a test dataset that is not seen during the learning phase! The accuracy of the machine learning model on these two datasets is an important indication of how good the trained model is! More information about this can be found in Sect. 1.2.1. We first divide the input dataset into two parts: one part will be used for learning (training) and one part will be used for testing. How this division happens is very important for the validity of the model and its accuracy.

The rules of training/test data division are as follows:

- The original dataset (before division) must be representative for the problem it helps to describe! We must have enough data points in the dataset! (Related information can be found in Sects. 4.3.2.4 and 4.3.2.7).
- The training and test datasets must be independent from each other! This means that the same data record must not be present in both datasets.
- The datasets must be identically distributed! This is an important statistical requirement because then both datasets are representative for the problem they describe.

One of the techniques used to achieve the two last rules is randomly shuffling the data before it is divided (e.g., 80% for training and 20% for testing).

4.3.3   Data Acquisition & Storage

Different machine learning (ML) applications and algorithms need different types and amounts of data, and different data delivery methods. Data acquisition and storage is the process by which the necessary data are collected, stored and transmitted for further processing (Fig. 4.17).

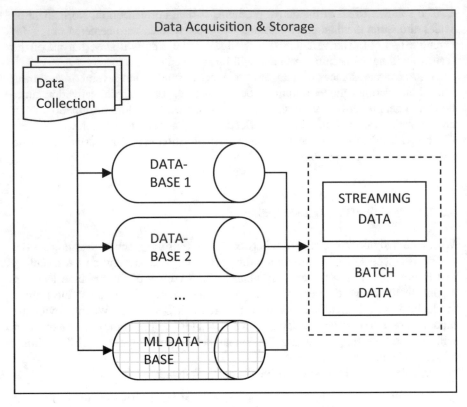

Fig. 4.17 Data acquisition & storage

We have seen before that there are five core components of the data strategy that work together as building blocks to support data management across an organization: IDENTIFY, STORE, PROVISION, PROCESS and GOVERN (see Sect. 4.2).

We must incorporate these five building blocks into our data acquisition and storage strategy! Data must be well identified and cataloged, and must be stored and packaged in such a way that it is easy to share, access and process, manage and communicate information policies and mechanisms for effective data usage, etc.

Machine learning data must often be pre-processed in several different ways (see Sect. 4.3.2). Therefore, it is often very useful to have a dedicated database for pre-processed ML data. Data storage is very cheap, and it becomes cheaper and cheaper every day; therefore, some duplication of the data, if it is done with the aim of more efficiency and clarity, is useful.

4.3.3.1   Data Collection

Machine learning (ML) data may come from many different sources such as business documents and data, sensors (IoT devices), social media, web stores, video cameras, mobile devices, internal and external data stores, microphones, etc. The data may have different types and formats, for example, numerical values, text, images, audio recordings, etc. In a data driven business culture all kinds of data are collected and stored for future use. Often data is also collected on purpose because it is needed for a specific task such as ML.

The subject of data collection is a huge subject in itself—measurements, questionnaires, capturing IoT signals, extracting data from business processes, etc. which must be standardized and well documented! We must make sure that the right things are collected and in the right way! It is not the aim of this book to explain this subject in detail but you may find a lot of information in other sources.

4.3.3.2   Data Transfer

How the data is transferred to data storage and to machine learning is important. There are two main data transfer mechanisms:

- Discrete data is simply forwarded in one or more big chunks or batches. This is often called batch data transfer.
- Continuous streaming data must be handled with some kind of stream processing platform. This is often called streaming data transfer.

Batch data transfer is the simplest of the two and does not need special care. Streaming data transfer is a much more complex issue because unless we capture and store the streaming data for further batch transfer, we need special stream data pre-processing and machine learning algorithms which can handle streaming data. Most machines learning training algorithms use data in one chunk which must fit into the memory; there are, however, some ML algorithms which can use streaming data, but most of them have a lower accuracy then batch ML models. There is also a difference in training and inference. Inference (prediction) with a trained machine learning model can use streaming data which is fed, record by record, to the model. There is no need for a batch of data for inference.

The question that then arises is whether we actually need a huge amount of streaming data to train a machine learning model. There is some disagreement about this among machine learning experts. I personally think that, in most cases, a small amount (where small can be from several MBs to several GBs!) of high quality data is best for training a machine learning model! Streaming data is often full of errors and very difficult to clean, and may contain too much unnecessary information which may even confuse the machine learning model. This, of course, may be different in the case of inference (prediction) because it may be useful to have a streaming inference system which makes continuous predictions!

Part III
Automatic Speech Recognition

Chapter 5
Automatic Speech Recognition

Abstract This chapter is entirely dedicated to automatic speech recognition (ASR) which is one of the most complex fields of machine learning. Topics from signal processing and the properties of the acoustic signal to acoustic and language modeling, pronunciation modeling and performance analysis will all be explained in an easily comprehensible manner. After reading this chapter you will also understand how the open source software package in the AI-TOOLKIT, called VoiceBridge, works.

5.1 Introduction

Because of the complexity and usefulness of automatic speech recognition (ASR) a whole chapter will be dedicated to this subject. ASR contains a combination of several of the machine learning models and techniques we have already seen, such as supervised learning and unsupervised learning (clustering).

One of the most complex fields of machine learning is natural language processing (NLP). Automatic speech recognition is a sub-category of NLP. The aim of ASR is to provide a word sequence (often called a transcription) which corresponds to an input speech. The input speech is captured in the form of a speech waveform (acoustic signal), which is the combination of sound waves representing the change in air pressure over time (the creation and spread of sound—theoretical physics of sound waves).

There are many useful applications of ASR, for example, voice dialing, call routing (e.g., in order to automate call centers), data entry by speech, command and control (computers, vehicles, appliances...), etc. A GPS in a car or a smart phone which can be commanded by human speech is a good example of ASR.

We can think of ASR as a supervised classification problem, despite the fact that it is the combination of several machine learning and statistical models. Remember (Sect. 1.3) that we speak about *supervised learning* when the input to the machine learning model contains extra knowledge (supervision) about the task modeled in the form of a kind of label (identification). For example, in the case of an e-mail spam filter the extra knowledge could be labeling whether each email is spam or not. The

© Springer Nature Switzerland AG 2021

Z. Somogyi, *The Application of Artificial Intelligence*,

https://doi.org/10.1007/978-3-030-60032-7_5

machine learning algorithm then receives a collection of emails which are labeled spam or not spam, and by this we supervise the learning algorithm with the extra knowledge. The label (or class) can be a simple integer number, for example, 0 for not spam and 1 for spam.

In ASR the classes are not single labels but structured objects consisting of a word sequence according to some grammatical rules. The inputs are the acoustic signals and the classes are the associated sentences in text form (transcription). This is much more complex than standard supervised classification of single labels because there is a large number of possible classes and because of the number of possible word combinations in a natural language.

Automatic speech recognition (ASR) can also be thought of as a process in which we search for every possible sentence in a given language which could match the input acoustic signal and then choose the best match. Because the acoustic signal of speech is very variable (may contain different pronunciations, noise, etc.), the best match will never be an exact match but a probable best match, and therefore a probabilistic model is used. Because there are a lot of possible word combinations we need an efficient way of searching for sentences which have a high probability of matching the input sentence (or word sequence). This efficient search mechanism is one of the most important parts of NLP and ASR, and it is based on the extended form of so-called *Regular Expressions*.

A text in any language may contain a sequence of letters, numbers, spaces, punctuation, etc. We call one segment of such text a *string*. Regular expressions define how we select (search) one or more strings of a text. The mechanism of regular expressions was extended to so-called *Finite State Automata*, which are just one step away from the (weighted) *Finite State Transducer* on which the *Hidden Markov Model* (HMM) is based. The HMM is widely used in automatic speech recognition (and also in NLP) as part of an efficient search mechanism in acoustic models which convert an input signal to a sequence of words or phones. More information about this can be found in Appendix! This is an advanced and complex subject, and not very important for understanding how ASR works, but it may be interesting to some readers.

In order to perform ASR we first need to train and assemble an ASR model. We also have to test the model by comparing the input transcriptions to the predicted output transcriptions. When we are satisfied with the trained ASR model we can use it to transcribe an input speech. In supervised classification terms this is called prediction or inference. Both the training and inference of an ASR model have three similar stages which can be summarized as follows (Fig. 5.1):

1. The first is the so-called *feature extraction* stage in which the input acoustic signal is converted to numerical features that best represent the input speech for our purposes. There is a reason why we do not just use the acoustic signal itself, we will see this later.
2. The second, in the case of training the ASR model, is the so-called *Acoustic Modeling* stage in which we calculate the 'likelihood' (occurrence probability) of phones or sub-phones based on the numerical features extracted in the first stage

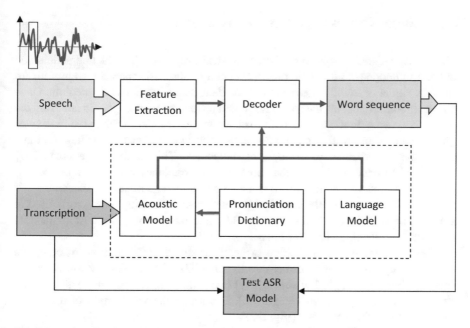

Fig. 5.1 Automatic speech recognition (training and inference)

and on the input text transcriptions. Sentences are decomposed into words, phones or sub-phones (more about this later). The acoustic model learns which phones or sub-phones are most likely to be generated by the input features (or vice-versa). In the case of inference the acoustic model is used to predict the most likely phone or sub-phone sequences.

3. The third is the so-called *Decoding* stage in which the acoustic model (i.e., the acoustic likelihoods), together with a dictionary of word pronunciations and a language model (N-gram, more about this later), is used to produce the most probable sequence of words (the transcription). In the case of training the accuracy of the model is tested.

In the following sections the most important elements of ASR will be introduced. But first a bit of signal processing will be explained in order to understand the input acoustic signal, which is composed of sound waves. Each stage of the speech recognition process (feature extraction, acoustic modeling and decoding) will be explained in detail. The pronunciation and the language model will also be introduced. As usual (like in the other parts of the book) we will focus on the understanding of the principles and not on the details of the mathematical algorithms!

5.2   Signal Processing: The Acoustic Signal

The acoustic signal of speech has some physical properties that must be understood in order to understand the basic principles behind automatic speech recognition. An acoustic signal or sound is generated by the disturbance of an elastic medium. Once this disturbance happens, the sound wave will propagate with a specific rate from the source into some direction. The disturbance compresses and expands the molecules (pressure variation) in the immediate neighborhood of the elastic medium and this passes the disturbance on to adjacent molecules. This sound wave created by pressure disturbance may reach the human ear and produce the sensation of hearing. Many types of music, sound, noise and also speech are generated in this way.

Figure 5.2 shows how the sound propagates from a vibrating tuning fork (used in music). The sound wave is a simple sinusoid and has two important properties: its *frequency* and its *magnitude (or intensity)*.

The *frequency* of a sound wave or acoustic signal is the number of complete vibrations (cycles) per second measure in hertz (Hz). The frequency can be calculated by dividing the speed of sound [m/s] (a constant depending on the air pressure and temperature) by the wavelength [m]. Increasing the wavelength of the sound wave decreases the frequency (Fig. 5.2)!

The tuning fork (Fig. 5.2) generates sound at a single frequency. All other real life sounds, such as musical instruments or human speech, are much more complex, containing several sound waves with different frequencies as can be seen on Fig. 5.3.

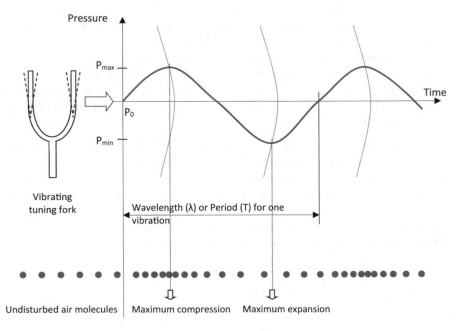

Fig. 5.2 Sound wave propagation from a vibrating tuning fork

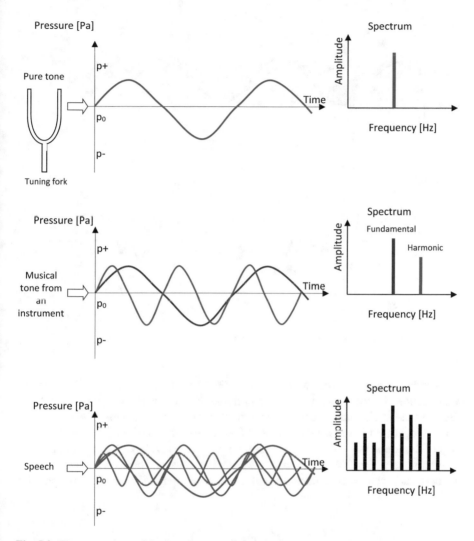

Fig. 5.3 The comparison of the sound waves of speech, musical tone and pure tone

The complex sound wave generated by speech can be decomposed into several sine waves with different frequencies, and vice-versa; the sine waves can be combined to get the complex sound wave of speech. The combination of several sine waves is just the addition of the frequencies at each time step (see Fig. 5.4). The *fundamental frequency* of a sound wave is the frequency of the lowest frequency (biggest wave length) decomposed sine wave. All other frequencies are called harmonics. The so-called *Spectrum* shows the different frequency components of a sound wave.

The second important property of an acoustic signal is the magnitude of the acoustic energy it contains, which is often called sound intensity or acoustic power.

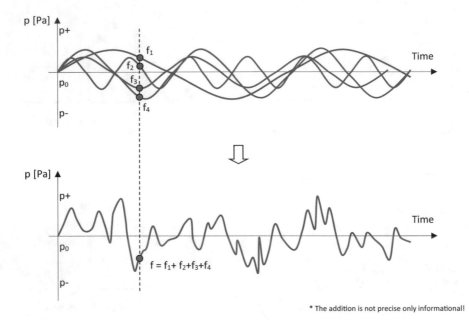

Fig. 5.4 The combination of several sine waves into one complex sound wave

The intensity of the sound wave is proportional to the *Amplitude* of the pressure disturbance described above. The amplitude is the maximum pressure (P_{max}). Because the human ear perceives sound intensities in an approximately logarithmic way, a logarithm-based measurement unit—the decibel (dB)—is used for the intensity of the sound (see Eq. 5.1). For example, the noise level in a quite living room is approximately 30 dB (faint), in an active business office and conversational speech is 60 dB (loud), at a music concert is 90 dB (very loud) and the human pain threshold is 140 dB.

$$L_p = 20 \cdot \log_{10}\left(\frac{p}{p_{ref}}\right) \text{ [dB]} \tag{5.1}$$

The acoustic power can be calculated with Eq. (5.1). As we have stated above the unit of acoustic power is the decibel [dB], and it is calculated from the sound pressure p and from a reference sound pressure p_{ref} which is chosen to be 20 µPa for sound in air ($20*10^{-6}$ Pa = 0.00002 Pa) [21].

The frequency range of a female voice is approximately 260–4000 Hz. The male voice is between 130–2100 Hz. The C tone on a piano is 250 Hz.

5.2.1   The Pitch

While frequency is a physical property of the sound wave, pitch is a subjective property which depends on how humans perceive a sound with a given frequency. The pitch may be used in combination with other features (e.g., MFCC) in acoustic models (we will see this later when we talk about feature extraction), and therefore it is important to have a basic understanding of it.

"Pitch is generally described as the psychological correlate of frequency, such that high-frequency tones are heard as being "high" in pitch and low frequencies are associated with "low" pitches (ANSI 2004)" [23]. According to the same ANSI standard: "Pitch depends mainly on the frequency content of the sound stimulus, but it also depends on the sound pressure and the waveform of the stimulus."

It is possible to express the pitch as a function of the frequency where the unit measure of pitch is the so-called *mel* (created by Stevens and Volkmann's (1940)) [23]. We call this the *mel scale*, see Fig. 5.5. "The reference point on this scale is 1000 mels, which is defined as the pitch of a 1000-Hz tone" at 40 dB. "Doubling the pitch of a tone doubles the number of mels, and halving the pitch halves the number of mels. Thus, a tone that sounds twice as high as the 1000-mel reference tone would have a pitch of 2000 mels, while a tone that is half as high as the reference would have the pitch of 500 mels." [23].

From Fig. 5.5 it is clear that the relationship between frequency and pitch is not linear and that, for example, between 1000 and 3000 Hz the pitch only changes by 1000 mels (i.e., from 1000 to 2000 mels) instead of 2000 (see the indicative arrows). This is how humans perceive the different frequencies.

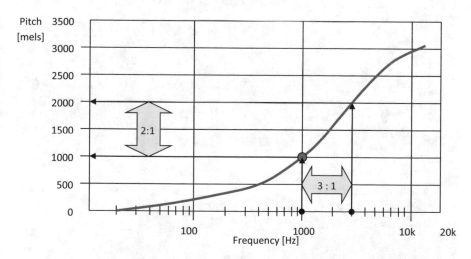

Fig. 5.5 The relationship between frequency in hertz and pitch in mels based on the findings of Stevens and Volkmann (1940); modified from [23]

5.2.2   The Spectrogram

The energy in a speech signal is highly variable and depends on the time, frequency and sound intensity (or sound level). We could say that it is three dimensional. In order to show these three dimensions, the time, frequency and sound intensity, the so-called *Spectrogram* was created.

Figure 5.6 shows the audio wave and spectrogram of the sentence "BUT NOW NOTHING COULD HOLD ME BACK". The words are marked below the audio wave. The horizontal axis of the spectrogram is the time in seconds, the vertical axis is the frequency in kHz and the sound intensity is shown with colors. Deeper red colors mean a higher intensity or sound level.

The characteristics of the speech shown on the spectrogram, the frequency and intensity combinations, are very speaker and language dependent. The sentence in Fig. 5.6 is spoken by a young native English-speaking woman. You can see on the spectrogram that the highest intensity in a wide frequency band is on the words "NOW" and "BACK" and at the beginning of "NOTHING". This is very pronunciation dependent behavior. If a non-native speaker were to speak these words, then

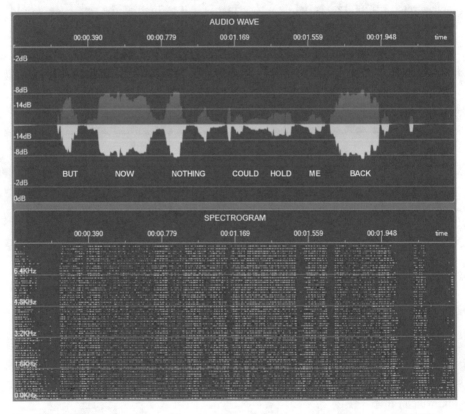

Fig. 5.6 The audio wave and spectrogram of the speech of a sentence (source: AI-TOOLKIT)

the energy distribution in the different frequency bands could be totally different. The same is true if a man were to speak the sentence instead of a woman. We could say that each person has their own spectrographic signature which can be viewed with a spectrogram.

The AI-TOOLKIT contains an audio editor (see Sect. 9.2.9) which generated the images on Fig. 5.6. You can use the audio editor to view, analyze, modify and listen to your sound recordings.

5.3 Feature Extraction and Transformation

Feature extraction is one of the most important parts of speech recognition because if we do not succeed in extracting the right features, then the speech recognition will fail or will not have sufficient accuracy!

Remember that we interpret automatic speech recognition (ASR) as a supervised classification problem. The input observations are the acoustic signals and the supervised labels (classes) are the sentences in text form (transcription). We have many sentences coupled to as many acoustic signals. But the acoustic signal has a very transient (variable) nature and cannot directly be used as an input to a machine learning model. Because of the transient nature of the acoustic signal the machine learning model would not be able to find a trend in the signal which can be coupled to the text transcription (words, phones, sub-phones). We need to find a way to extract some kind of information or 'signature' from these waves which is representative of the acoustic signal and can be used as input for our machine learning model.

Our aim is to develop a method which extracts features from the acoustic signal in such a way that if there are two different acoustic signals, then for the same linguistic units (word, phone or sub-phone) a similar feature set will be extracted and for different linguistic units a significantly different feature set will be extracted. The bigger the difference between features representing different linguistic units the better the speech recognition will be!

Why would two acoustic signals be different if they belong to the same sequence of words? There are three types of reasons:

1. Speaker dependent reasons—pronunciation differences, voice differences because of gender or age, differences because of dialect, etc.
2. Environment dependent reasons—noise, acoustics of a room, external sound sources, etc.
3. Signal processing reasons—how the speech is registered (microphone) and transferred for processing, etc.

From the previously defined aim of feature extraction it is clear that we need to filter out, as much as possible, these negative effects from the extracted features, so that we can model linguistic unit variation and not the variation because of the three reasons listed above.

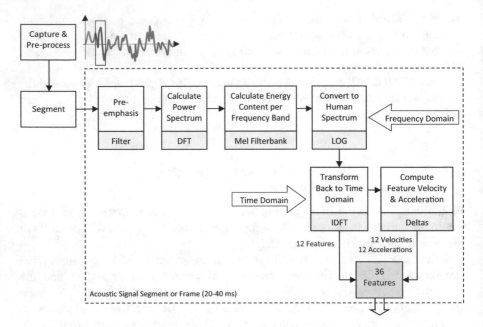

Fig. 5.7 Feature extraction (AI-TOOLKIT)

In short, feature extraction works as follows: we cut the acoustic signal into small 20–40 ms segments. The 20–40 ms size is determined by ASR research. This range provides the best ASR performance (a much smaller window would not contain enough information and a much bigger window would contain too much variation). Each segment is analyzed in pre-determined frequency bands and from each frequency band a representative property is calculated. This is somewhat similar to the spectrogram (see Sect. 5.2.2). These properties per frequency band form the feature vector per audio segment, which is a kind of 'signature' of that acoustic signal segment. The same phone sequences will have very similar signatures depending, of course, on the audio signal and speaker. If you look at the spectrogram on Fig. 5.6 you will get a feeling for this signature by noticing how the high energy regions (red) are formed per segment. After feature extraction there is often a feature transformation step in which we transform the extracted features in order to improve speech recognition (see Sect. 5.3.1).

Let us look at feature extraction in more detailed steps (see also Fig. 5.7):

1. First the audio signal is captured and may be pre-processed. Noise, echo… may be removed or the audio signal may be improved in some other way. It is important to consider how the ASR model will later be used. If we train the model with a transformed acoustic signal, then the input to the trained ASR model should be similarly transformed. It is also possible that we do not pre-process (transform) the audio signal at all.

2. The audio signal is segmented into 20–40 ms frames with a frame shift smaller than or equal to the frame width (e.g., 10 ms). The frame shift is the shift in time by which we move the segmentation window on the signal. If the frame shift is smaller than the frame width, then overlapping regions are extracted from the signal. Each frame is fed into the following steps.

3. A pre-emphasizing filter (Hamming window) is often applied, as a first step, to the acoustic signal frame in order to smooth the energy spectrum of the signal through all frequencies. This can be useful because of the big difference between the energy content in low frequencies and the energy content in high frequencies of human speech. Low frequencies have much more energy. By smoothing out this difference we improve phone or sub-phone recognition. Smoothing here means the amplification of the signal, especially in the high frequency regions.

4. The power spectrum of the audio frame is calculated by applying the discrete Fourier transform (DFT).

5. The energy content for a number of selected frequency bands is calculated by using the so-called *Mel Filterbank*. Remember the Mel scale from the discussion about the pitch (Sect. 5.2.1)! The Mel filterbank is a collection of overlapping triangular filters (usually 26) which collect the energy content for each triangular filter placed and spaced on the power spectrum (step 3) of the audio frame according to the Mel scale. See Fig. 5.8 (note that only a small part of the audio frame is shown).

6. The energy values from step 4 are converted into the human sensing spectrum by taking the logarithm of each value—remember that the human ear perceives sound intensities in approximately a logarithmic way (see Sect. 5.2)!

7. Each value is transformed back to the time domain by applying the inverse discrete Fourier transform (IDFT) (in some implementations the inverse discrete cosine transform (IDCT) is used instead of IDFT). This will also de-correlate the values. The correlation is caused by the overlapping Mel filters in step 4. The resulting 12 features are often called the MFCC features. The number 12 comes from the number of Mel filters in step 4, usually 26 filters, but several filter values are removed in the high frequency regions because they do not contain relevant information for speech recognition. Depending on the implementation this number may be slightly different.

8. Finally the changes in the MFCC feature values, MFCC velocity and the change in MFCC velocity (acceleration), between two time frames are calculated, which results in $2 \times 12 = 24$ additional features. These are often called delta (velocity) and delta-delta (acceleration) features. At the end we have 36 features per audio frame which form the input feature vector to the acoustic model (see Sect. 5.4).

After feature extraction there is often a further step in which we transform the extracted features in order to improve speech recognition (see Sect. 5.3.1).

It is still possible to extend the above steps by adding more features (energy, pitch, etc.). For example, the pitch (see Sect. 5.2.1) is sometimes added as an extra

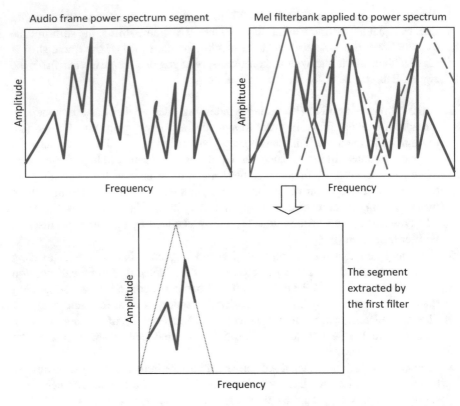

Fig. 5.8 Applying the Mel filterbank (only a small part of the audio frame and the Mel filterbank is shown)

feature because it is also representative of the pronunciation of the words, especially in Asian languages.

The above steps were developed and improved during many years of (still ongoing) speech recognition research. More information about these steps can be found in the references and further reading section of listed publications at the end of the book.

5.3.1 Feature Transformation

After feature extraction there is often an additional step in which we transform the extracted features in order to improve speech recognition. This is sometimes called feature space transformation. There are many types of feature transformation methods (and more are still being developed), but the following three are used the most often:

- Linear discriminant analysis (LDA) transform.
- Maximum likelihood linear transform (MLLT).
- Speaker adaptive training (SAT).

5.3.1.1 Linear Discriminant Analysis (LDA) Transform

"Linear discriminant analysis is a standard technique in statistical pattern classification for dimensionality reduction with a minimal loss in discrimination." [28]. In simple terms, LDA attempts to perform dimensionality reduction in order to reduce the number of features and, at the same time, it tries to make sure that the difference between the representations (features) of different classes (different linguistic units—phones, sub-phones) does not increase. A reduction in the difference between the features of different linguistic units would result in poorer speech recognition accuracy, which is why we want to prevent this from happening.

The LDA algorithm takes the feature vectors and creates two matrices: one representing the within-class scatter (W), calculated with the covariances of each vector, and the other one the between-class scatter (B), calculated with the means of all vectors. Next LDA tries to find a linear transformation (θ) which maximizes the ratio between B, modified by the linear transformation (θ), and W, also modified by the linear transformation (the ratio $= \theta B\theta^T/\theta W\theta^T$). Both matrices are multiplied with the transformation matrix and with its transpose. The maximization of this ratio is an optimization problem (using the so-called eigenvectors and eigenvalues), and it normally results in dimensionality reduction by applying the resultant linear transformation (θ) matrix to the feature vectors. Please note that the word "class" in this instance means the feature vector representing a linguistic unit (phone, sub-phone). For more information on the algorithm please refer to [26, 28].

LDA is most often applied in speech recognition together with the MLLT transform (see the following section) and on top of the MFCC + delta + delta-delta features (see Sect. 5.3). The reason for this is because MLLT improves class discrimination.

5.3.1.2 Maximum Likelihood Linear Transform (MLLT)

MLLT is applied after applying the LDA transform and it aims to improve class discrimination between the classes of feature vectors. In other words, it tries to improve the difference between the representations (features) of different classes (different linguistic units—phones, sub-phones). The greater the difference between the features of different linguistic units, the better the speech recognition accuracy. This is a small first step towards speaker adaptive training (SAT) and feature normalization (removing the effects of the speaker and the environment—discussed in the beginning of Sect. 5.3), which will be examined in more detail below.

The MLLT algorithm applies a diagonalizing linear transformation to the output of the LDA transformation. For more information on the algorithm please refer to [28].

Applying LDA + MLLT does not always result in a better accuracy and it also makes the training process significantly slower. Some authors report improved accuracy when using sub-phones as linguistic units [26]. The combination (MFCC + delta + delta-delta) + SAT provides, in most cases, much better performance and accuracy (see below).

5.3.1.3 Speaker Adaptive Training (SAT)

The aim of SAT is full speaker adaptive training and feature normalization (removing the effects of the speaker and the environment, e.g., noise. For SAT the training data must be divided into groups per speaker (and/or per environmental effect). SAT is applied on top of the MFCC + delta + delta-delta features (see Sect. 5.3) and it transforms the feature vectors by removing the variation caused by different speakers or environmental effects.

The SAT algorithm applies several statistical techniques to identify the variation caused by different speakers (or caused by an environmental effect, e.g., noise). When this variation is identified it can be removed from the feature vectors, or in other words, the feature space can be normalized. For extended information on the algorithm please refer to [29].

Separating the training data (acoustic signals) per speaker and applying SAT results in a significant WER (word error rate) reduction (i.e., accuracy improvement).

5.4 Acoustic Modeling

Acoustic modeling is the second step in automatic speech recognition (ASR). Very simply put, we train a machine learning model to learn which feature vectors (extracted in the second step) produce the phones or sub-phones in the input transcription. This is also a probabilistic model, and thus the acoustic model will provide the most probable sequence of phones depending on the input feature vectors. The model has no prior knowledge of words or language structure (grammatical rules); this will be added in the subsequent decoding step.

The sentences in the transcription are decomposed into words. Words are decomposed into phones and sub-phones. The pronunciation dictionary is used to split words into phones (see Sect. 5.6). Phones are modeled with the so-called *Hidden Markov Models* (HMM; refer to Appendix) (see Fig. 5.9). We use an HMM because it is the kind of probabilistic model we need, and because there are well known and efficient algorithms to train the parameters of the HMM model (we will see later how).

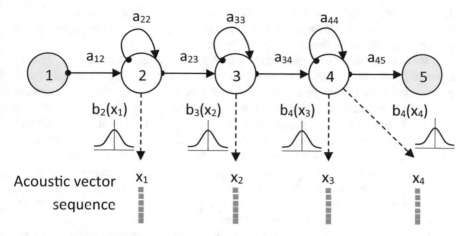

Fig. 5.9 Five-state HMM phone model; modified from [19]

The most important reason for using phones instead of words is that the number of phones is much lower than the number of words in a natural language, which makes training a phone model much more feasible than training a word model with a large vocabulary. The reason we do not just use letters is that phones and sub-phones are very much related to how words are pronounced (to the sound) and thus to the acoustic signal. If we pronounce several letters of a word in sequence, then it doesn't sound the same as pronouncing a word with the same letters.

There are two main types of HMMs for phone modeling: mono-phone and tri-phone. The mono-phone HMM uses only one phone. The tri-phone HMM uses three phones: the current phone, the previous phone and the next phone in the sequence. The most important reason for using a tri-phone model instead of a mono-phone model is that the articulation of a phone strongly depends on the surrounding phones and this brings context dependency into the model instead of just looking at one phone at a time. Tri-phone models provide significantly better accuracy!

Figure 5.9 shows the graphical representation of a five-state HMM model which could be a tri-phone model. The model has one input and one output state (1 and 5) and three internal states (2, 3 and 4). Each internal state corresponds to an observable symbol (a phone). There is a transition probability between each state which is designated with $A = \{a_{ij}\}$. The model decides at each time step according to this transition probability whether to move to the next state or stay in the current state. Each transition to state j produces an observable symbol (phone or sub-phone) and a hidden (non-observable) output probability distribution designated with $B = \{b_j(x_t)\}$ (i.e., the 'hidden' part of the Hidden Markov Model). Note that B is the probability of generating symbol s_t while in state j and having x_t as a feature vector. From this output probability distribution the most probable feature vector x_t at that time step (state) is deduced. You have probably noticed that there is an extra output distribution in state 4 shown with $b_4(x_4)$. This may happen when the HMM stays in the same state because of the transition probability a_{44}. Even if there is no state change but there is a self-transition a feature vector is generated. This may happen if a phone is pronounced two times, one after the other.

If we concatenate (connect) all of the HMM phone models for a sentence, then the acoustic likelihood of the sentence produced by the input acoustic signal can be estimated and this is how the decoder in the next stage will work. The most common method of computing acoustic likelihoods is by using *Gaussian Mixture Models* (GMM). Gaussian refers to the normal distribution which is used to approximate $B = \{b_j(x_t)\}$ (see above). We train the acoustic model with the mean and variance of the normal distribution as extra parameters (the mean and variance or standard deviation define a normal distribution completely). Because in reality the B distribution may not be normal (Gaussian) we often model B with a weighted mixture of normal distributions (which can approximate any distribution), and this is where the name of the Gaussian mixture model comes from.

Training an HMM model involves choosing the number of states (5 in the example above), the number of symbols (given by the transcription and the pronunciation dictionary) and the estimation (learning) of the two probability densities A and B. These parameters are estimated from the training data, consisting of the acoustic signal of speech, and from the associated input transcriptions (input text). During the training of the acoustic model there is also segmentation and alignment of the acoustic phone models, the acoustic signal and the transcription. The reason for this is that we have the input transcription per sentence and the appropriate acoustic signal per sentence and we need to make sure that the right phones are scheduled at the right time step of the acoustic signal. The pronunciation of some words (and their phones) may take longer than the pronunciation of other words and thus the signal cannot just be divided.

Do not worry if you do not understand this completely! HMM modeling is a complex topic and it is not very important to fully understand it if you just want to use ASR, but you should understand the big picture. An HMM is a kind of statistical probabilistic model which helps us to efficiently search for the right words (phones) which correspond to the acoustic signal. More information can be found about this subject in Appendix.

5.5   The Language Model (N-gram)

The language model (LM) is one of the most important machine learning models in natural language processing (NLP) and thus also in automatic speech recognition (ASR). As you will see, it is very simple but has far reaching advantages.

The LM is a statistical framework which is used to model the most probable sequence of words in a natural language. In most NLP tasks, and thus also in ASR, an LM is a crucial piece of our methodology in order to reduce the search space of possible word combinations. There are a lot of words in a natural language and theoretically all combinations of these words are possible, but there are of course specific rules in each language which limit these combinations. The aim of the LM is to learn and find these limited numbers of combinations more easily.

One of the most important advantages of the LM is that it also captures the grammatical rules existing in a natural language and thus it avoids the need to create and use formal, and often complex, grammatical rules.

The statistical probability theory behind the LM is based on the probability of word sequences. The probability that the sentence "I am reading this book" exists in English is much higher than the probability of the existence of the sentence "book am this I reading" (the words are just mixed up). The sequence of words is important and we will see later that we only need to look at a few neighboring words instead of all words. A well trained LM (where well trained means that the LM is trained with a large number of sentences) can predict the next word in a sequence from only the previous N-1 words. And from this comes the name 'N-gram language model'! From the conditional probabilities of the possible next words, the joint probability of a possible whole sentence can be estimated. We call it conditional probability because of the condition of having a specific neighboring word (more about this later!).

Why is this important? It is important because in speech recognition the sounds of words are often not clear, or different words may have a similar sound, and we need a mechanism to decide which word is spoken. And we can decide this with a good accuracy (depending on the accuracy of the LM) by looking at a few preceding words and using a language model.

A logical way of making an LM would be to calculate the probability of each subsequent word based on all previous words in a sequence of words. But this is an impossible task because of the huger number of word sequence combinations in any natural language. Thanks to statistical theory we can approximate this conditional probability with only the preceding N-1 words.

We designate the conditional probability that the word w_i will occur given the previous word sequence $w_1, w_2, \ldots, w_{i-1}$ with $P(w_i \mid w_1, w_2, \ldots, w_{i-1})$. If we only use one preceding word, then we call this LM a 'bigram' and designate the conditional probability that the word w_i will occur with $P(w_i \mid w_{i-1})$. If we use the two preceding words, then we call the LM a 'trigram' and designate the conditional probability that the word w_i will occur with $P(w_i \mid w_{i-2}, w_{i-1})$. The range of N used most often, in practice, is 2–4 but it may also go higher. Higher N means higher accuracy but significantly higher processing time and complexity.

Training an LM model is very simple but may be time consuming because we need many sentences. The conditional probabilities for individual words, $P(w_i \mid w_1, w_2, \ldots, w_{i-1})$, can be estimated by simply counting the occurrences of words given the chosen N (e.g., bigram or trigram) preceding words $w_{i-N+1}, w_{i-N+2}, \ldots, w_{i-1}$ (see the example on LM training hereunder). From these word conditional probabilities we can calculate a word sequence (or sentence) probability by simply combining several word conditional probabilities. For example, the probability of the word sequence "I am reading this book" can be calculated with a bigram (a single preceding word) LM based on the word conditional probabilities as follows:

$$P(\text{I am reading this book}) = P(\text{I}|< s >) \cdot P(\text{am}| \text{I}) \cdot P(\text{reading}|\text{am})$$
$$\cdot P(\text{this}|\text{reading}) \cdot P(\text{book}|\text{this}) \cdot P(< /s >|\text{book}).$$

where $<s >$ designates the start of the sentence, and $</s >$ the end of the sentence. We need these in order to be able to apply our bigram model to the whole sentence (i.e., including the first and last words of the sentence)!

Expert Sect. 5.1 Language Model Training Example
This is a simple example to demonstrate how the word conditional probabilities and the sentence probabilities are calculated.
The training data are as follows:

"Peter is reading a book"
"Reading this book is interesting"
"I am reading"

Let us calculate the conditional probability of P(I am reading this book) with a bigram language model trained on the above training data.

We do not care about capital letters and we assume that all of the words are converted to lowercase (this is what we usually do). We also do not care about punctuation and we assume that all punctuation marks are removed in a pre-processing normalization step.

The word conditional probabilities can be calculated from the above training data as follows:

$$P(\text{I}|< s >) = count(< s >, \text{I})/count(< s >) = 1/3$$

$$P(\text{am}| \text{I}) = count(\text{I}, \text{am})/count(\text{am}) = 1/1 = 1$$

$$P(\text{reading}|\text{am}) = count(\text{am}, \text{reading})/count(\text{reading}) = 1/2$$

$$P(\text{this}|\text{reading}) = count(\text{reading}, \text{this})/count(\text{this}) = 1/1 = 1$$

$$P(\text{book}|\text{this}) = count(\text{this}, \text{book})/count(\text{book}) = 1/2$$

$$P(< /s >|\text{book}) = count(\text{book}, < /s >)/count(< /s >) = 1/3$$

Remember that, for example, P(I| < s>) means the conditional probability of having an 'I' after the start sentence sign <s>, which means having an 'I' at the beginning of a sentence. The 1 / 3 comes from the single occurrence of 'I' at the beginning of the sentences in the training data (i.e., in "I am reading") and from three occurrences of the end of a sentence (i.e., there are three sentences in the training data).

(continued)

Expert Sect. 5.1 (continued)

Finally, the conditional probability of the sentence we are looking for is as follows:

$$P(\text{I am reading this book}) = P(\text{I}|< s >) \cdot P(\text{am}| \text{I}) \cdot P(\text{reading}|\text{am})$$
$$\cdot P(\text{this}|\text{reading}) \cdot P(\text{book}|\text{this})$$
$$\cdot P(< /s >|\text{book})$$
$$= 1/3 \cdot 1 \cdot 1/2 \cdot 1 \cdot 1/2 \cdot 1/3 = 1/36 = 0.0278.$$

This probability value is of course far too low for a valid sentence but do not forget that we only have three sentences in our training data! We would need much more training data to get a valid and high probability. We need several millions of words to train a good LM model!

For training a language model we use the same methodology as we use for training other machine learning models. We create training and test datasets (see Sect. 1.2), we train the LM on the training dataset and then we test it on the test dataset. We evaluate an LM by simply calculating the sentence probabilities in both the training and the test datasets by using the trained LM. The accuracy of the LM is the probability of the test dataset (or training dataset) divided by the number of words in the dataset.

In practice a so-called *smoothing* is often applied to the LM in order to prevent zero occurrences of words or high word counts. These two problems may occur because of a limited amount of available training data. Smoothing is a kind of statistical technique which is used to compensate for these two problems and to balance the LM. The simplest way of smoothing would be, for example, to increase all word counts by one and in this way avoid zero word counts. In practice, of course, more sophisticated smoothing algorithms are used (e.g., the so-called Kneser-Ney smoothing method).

Expert Sect. 5.2 Trained N-gram Language Model: The ARPA Format

This LM is trained by VoiceBridge for the LibriSpeech example. Not all data is shown, only a few per N-gram. There are 8140 1-grams, 35,595 2-grams, 49,258 3-grams, etc. The format of the LM is the so-called ARPA format (explained below) which is the standard way of representing an LM.

(continued)

Expert Sect. 5.2 (continued)

```
\data\
ngram 1=8140
ngram 2=35595
ngram 3=49258
ngram 4=49483

\1-grams:
-1.350165        </s>
-99              <s>             -0.667025
-1.991043        a               -0.196317
-4.294041        abandoned       -0.077825
-4.424494        abbe            -0.077825
-4.424494        abduction       -0.077825
-4.424494        ability                 -0.077825
-4.424494        abjectly        -0.077825
-3.821641        able            -0.395152
...
-3.054913        again           -0.141922
...
-3.821641        cold            -0.105992
...
\2-grams:
-1.744562        <s> a           -0.050644
-3.406211        <s> above       -0.024093
-2.629062        <s> after       -0.024093
-3.135568        <s> again       -0.024093
-3.703582        <s> against     -0.024093
-3.471051        <s> ah          -0.024093
-4.133381        <s> ain't       -0.024093
-4.113880        <s> alas        -0.024093
...
\3-grams:
-2.813527        <s> a bed       -0.005870
-2.851221        <s> a brisk     -0.005870
-2.831900        <s> a broken    -0.005870
-2.831900        <s> a circle    -0.005870
-1.835743        <s> a cold      -0.005870
-2.695172        <s> a feeling   -0.005870
-2.009304        <s> a few       -0.005870
...
\4-grams:
-1.099819        <s> a bed quilt
-0.657806        <s> a brisk wind
-1.052603        <s> a broken tip
-0.710015        <s> a circle of
-1.428882        <s> a cold bright
-1.431050        <s> a cold lucid
...
```

The ARPA format first includes a header with the item counts per N-gram. Next, if the LM is trained for 4-gram, then all N-gram levels up to and including the fourth level are shown (i.e., not only the 4-gram) because of the so-called back-off estimate (see Sect. 5.5.2), which is used to estimate the word sequence with a lower order N-gram. In each N-gram, each line first contains the $\log_{10}$ probability ($\log_{10}(P)$) of the next word sequence and then a $\log_{10}$ 'back-off weight' ($\log_{10}(b_w)$), except for the last N-gram (the 4-gram in

(continued)

Expert Sect. 5.2 (continued)

this case) which has no back-off weight (this will be explained in Sect. 5.5.2). It is common practice to represent probabilities with their base-10 logarithmic values in order to have nicer numbers instead of very small ones (e.g., 0.000000235). Note that the $\log_{10}$ (logarithm to base 10) of a number between 0 and 1 is negative; hence, the negative probability values in the LM model.

NOTE: The same type of N-gram LM model can be applied to phones and sub-phones instead of words (each spoken word is decomposed into a sequence of basic sounds called base phones—see the next section).

5.5.1 The Back-off (Lower Order) N-Gram Estimate

It is possible that the training data does not include, for example, "reading this book" but it does include "reading this" and "book" separately. If we are interested in the conditional probability of "reading this book", then we have a problem because it is not included in the LM. In order to solve this problem the so-called back-off estimate was introduced based on statistical theory. The back-off method estimates the conditional probability of an unseen N-gram by using an (N-1)-gram and a so-called back-off weight. If the (N-1)-gram does not exist, then the method goes one level deeper to the (N-2)-gram until the unigram level is reached (single word level).

In probability notation, the back-off calculation for a trigram (3-gram) w_1, w_2, w_3 can be expressed as follows:

$$P(w_3|w_1, w_2) = P(w_3|w_2) \cdot b(w_1, w_2)$$

In the case of "reading this book", $w_1 =$ "reading", $w_2 =$ "this" and $w_3 =$ "book". $P(w_3|w_2)$ is the bigram probability of "this book" and $b(w_1|w_2)$ is the back-off weight of "reading this".

For example, if we want to calculate the conditional probability of "a cold again" using the LM model above (see Expert Sect. 5.2), then we do the following:

1. The 3-gram "a cold again" does not exist in the LM. We look up the $\log_{10}$ back-off weight of "a cold" ($b(w_1|w_2)$ in the formula above) which is -0.005870 .
2. We then look up the bigram "cold again" to obtain $P(w_3|w_2)$. Let us assume that it does not exist (i.e., it is not visible in the example). We look up the $\log_{10}$ back-off weight of "cold" which is -0.105992.

3. We then look up the unigram "again". This can be found in the example and the $\log_{10}$ value is -0.141922.
4. Finally we can calculate P("a cold again") $= 10^{(-0.005870 \; - \; 0.105992 \; - \; 0.141922)} = 0.557$

Note that multiplication is replaced by addition because of the $\log_{10}$ numbers and then the inverse $\log_{10}$, which is the power of 10, is used.

5.5.2   Unknown Words (UNK or OOV)

Unknown words are words which are not in the input LM (input vocabulary) but are found in the training and/or test dataset. We often refer to these unknown words as UNK or OOV (out of vocabulary) words. In order to be able to handle these words elegantly, we replace them during language modeling with the symbol <UNK> or < OOV> and 'pretend' that <UNK> is a known word and model it as any other regular word. In this way we have a conditional probability estimate for all unknown words. This is, of course, an approximation because many different words could be replaced with only one symbol, but it is better than doing nothing with them. Modeling unknown words increases the accuracy of the LM.

5.6   Pronunciation

Each spoken word is decomposed into a sequence of basic sounds called base phones. This sequence of phones is called the pronunciation which is very language and speaker (person) dependent. There are only a limited number of such base phones in each language. For example, in American English with a limited amount of base phones any word can be composed (see Expert Sect. 5.3).

In speech recognition a so-called pronunciation dictionary is used for each natural language, where each word is mapped to a series of phones. Furthermore, a pronunciation model can be trained from a base pronunciation dictionary, which can be used to derive the pronunciation (the sequence of phones) of any word (even unknown words). The pronunciation model learns how words are decomposed into base phones in a specific language. We call this pronunciation modeling. This is similar to the probabilistic language model discussed previously but, instead of words, phones and sub-phones are used in the training of the model. If there is a sequence of letters as input, we search for the most probable phone or sub-phone sequence. Such algorithms are often called *grapheme-to-phoneme* machine learning algorithms.

There are several phonetic alphabets which define how to represent phones in text, for example the International Phonetic Alphabet (IPA) or the ARPAbet for American English (see Expert Sect. 5.3) (Table 5.1).

Table 5.1 The international phonetic alphabet and ARPAbet

| Vowels | | | | Consonants | | | |
| ARPABET | | | | ARPABET | | | |
1-letter	2-letter	IPA	Example(s)	1-letter	2-letter	IPA	Example
a	AA	ɑ	balm, bot	b	B	b	buy
@	AE	æ	bat	C	CH	tʃ	China
A	AH	ʌ	but	d	D	d	die
c	AO	ɔ	story	D	DH	ð	thy
W	AW	aʊ	bout	F	DX	ɾ	butter
x	AX	ə	comma	L	EL	l̩	bottle
N/A	AXR	ɚ	letter	M	EM	m̩	rhythm
Y	AY	aɪ	bite	N	EN	n̩	button
E	EH	ɛ	bet	ɾ	F	f	fight
R	ER	ɝ	bird	g	G	ɡ	guy
e	EY	eɪ	bait	h	HH or H	h	high
I	IH	ɪ	bit	J	JH	dʒ	jive
X	IX	i	roses, rabbit	k	K	k	kite
i	IY	i	beat	l	L	l	lie
o	OW	oʊ	boat	m	M	m	my
O	OY	ɔɪ	boy	n	N	n	nigh
U	UH	ʊ	book	G	NX or NG	ŋ	sing
u	UW	u	boot	N/A	NX	˜ɾ	winner
N/A	UX	ʉ	dude	p	P	p	pie
				Q	Q	ʔ	uh-oh
				r	R	ɹ	rye
				s	S	s	sigh
				S	SH	ʃ	shy
				t	T	t	tie
				T	TH	θ	thigh
				v	V	v	vie
				w	W	w	wise
				H	WH	ʍ	why
				y	Y	j	yacht
				z	Z	z	zoo
				Z	ZH	ʒ	pleasure

Source: https://en.wikipedia.org/wiki/ARPABET

Unfortunately some words may be pronounced in different ways by different people which makes pronunciation modeling and speech recognition a bit more complex. We have to take into account all of these possible variations.

Expert Sect. 5.3 The Pronunciation Lexicon Used by VoiceBridge (AI-TOOLKIT)

The base pronunciation dictionary for English used by VoiceBridge can be seen partially hereunder (such dictionaries exist for all languages). It is used to train a pronunciation model for deriving the pronunciation of any word in English. This is an automatic feature in VoiceBridge. Notice the extended pronunciation alphabet with several ARPAbet symbols divided further into different symbols (for example AE → AE1, AE2 ...). This makes even more accurate modeling possible.

```
a               AH0
abandoned       AH0 B AE1 N D AH0 N D
abbe            AE1 B IY0
abduction       AE0 B D AH1 K SH AH0 N
ability         AH0 B IH1 L AH0 T IY2
abjectly        AE1 B JH EH0 K T L IY2
able            EY1 B AH0 L
abner           AE1 B N ER0
aboard          AH0 B AO1 R D
abolitionism    AE2 B AH0 L IH1 SH AH0 N IH2 Z AH0 M
abolitionists   AE2 B AH0 L IH1 SH AH0 N AH0 S T S
about           AH0 B AW1 T
above           AH0 B AH1 V
abraham         EY1 B R AH0 HH AE2 M
abroad          AH0 B R AO1 D
abruptly        AH0 B R AH1 P T L IY0
absence         AE1 B S AH0 N S
absent          AE1 B S AH0 N T
absolute        AE1 B S AH0 L UW2 T
absolutely      AE2 B S AH0 L UW1 T L IY0
...
```

5.7   Decoding

The third stage in automatic speech recognition (ASR) is the so-called decoding stage in which the acoustic model (see Sect. 5.4) from the previous step is combined with a language model (N-gram; see Sect. 5.5) in order to produce the most probable sequence of words (the transcription) which correspond to the acoustic signal represented by the extracted feature vectors. We call this decoding because we are looking for a sequence of variables which are generated by some kind of observation. In the case of the training of the ASR model, the accuracy of the model is tested with the output of the decoding (output transcription). In the case of inference, the output of the decoding may be fed into another application.

First of all, the trained language model (LM) (trained on millions of sentences in the given language) helps us by expressing how likely it is that a sequence of words is a good sentence in the given language. This will be important when we are generating many sentences and want to decide which sentence has the highest probability of corresponding to the acoustic signal. We, of course, assume that the sentence is spoken according to the grammatical rules of the given language. The LM captures these rules automatically.

From the phone based acoustic model we generate a word model by concatenating the phone based HMMs according to the pronunciation lexicon. The pronunciation lexicon contains the sequence of phones for each word. A pronunciation model may also be trained, which can provide the phone sequence for any word and also for unknown words (which do not exist in the pronunciation lexicon).

The aim of the decoding step is to find a sequence of word models that best describe the input acoustic signal represented by the extracted features (see Sect. 5.3). This is a complex search problem for which we need a specialized search algorithm. We will search for possible word combinations and the LM will help us to decide if the word combination accords with the given language rules. The sentence (word combination) with the highest likelihood provided by the LM will be chosen. The search algorithm is based on a large HMM network which is formed by concatenating all word models which encode all genuine sentences in the given language. How do we decide if a sentence is genuine? With the language model! We always use the same probabilistic search methodology based on finite state models.

5.8   The Accuracy of the Trained ASR Model (WER, SER)

There are two common metrics used for the evaluation of the accuracy of ASR. The first one is the so-called *Word Error Rate* (WER) and the second one is the so-called *Sentence Error Rate* (SER).

The first thing we might think of is to count the number of incorrect words in the decoded transcription and divide this number by the total number of words. This would give the percentage of incorrect words. Unfortunately in the case of ASR this evaluation metric is not good enough because we need to search for three types of errors in the output transcription: *Substitution, Deletion* and *Insertion* errors.

The first type of error is caused by the misidentification of a word. We call this type of error *Substitution* because the correct word is substituted with an incorrect one. For example, the original sentence could be "I am reading this book" and the ASR could wrongly decode from the input acoustic signal "I am eating this book". The word "reading" is substituted with "eating".

The second type of error is caused by not identifying a word (i.e., missing a word). We call this type of error *Deletion*. Using the previous example, the output sentence could be the following: "I reading this book", where the word "am" is missing.

The third type of error is caused by the presence of words in the output transcription which are not in the original spoken text. For example, the output could be "I am also reading this book", where "also" is added or inserted into the sentence and not included in the original spoken text. We call this type of error *Insertion*.

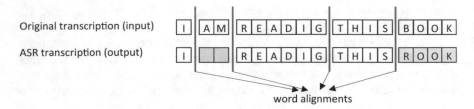

Fig. 5.10 Word alignment for calculating the WER of ASR

How do these errors occur? The reason can be very simple, for example, because of environmental effects (a loud sound, noise near the microphone) the acoustic signal is not optimal and is misinterpreted by the ASR system.

There is a so-called *Minimum Edit Distance* algorithm which first aligns the two word sequences (the input transcription and the output transcription generated by the ASR) and then calculates the number of substitution (S), deletion (D) and insertion (I) errors. The word error rate (WER) can then be calculated using the following simple formula:

$$WER = 100 \cdot \left(\frac{S + D + I}{N} \right) \ [\%]$$

The lower the WER, the better the ASR model!

The correct alignment of the words in the comparison of the input and output sentences is very important, in order to be able to calculate the correct substitution, deletion and insertion errors! Alignment means that we place each correct word in the output transcription 'under' the same word in the input transcription (see Fig. 5.10).

There are two errors in the ASR output on Fig. 5.10, a deletion error because "am" is not recognized and a substitution error because instead of "book" the word "rook" is decoded.

The sentence error rate (SER) can be calculated by counting the number of output sentences with at least one incorrect word (missing, deleted or inserted) and dividing this number by the total number of sentences.

It is interesting to note that the minimum edit distance methodology can be used to compare two words and evaluate how much they differ from each other. In the case of words, the deletion, insertion and substitution 'errors' relate to the letters instead of whole words.

5.9   Practical Training of the ASR Model

We have seen all of the steps needed to build and train an ASR model in the previous sections. Let us now look at the practical steps used in state-of-the-art ASR software systems (like VoiceBridge) for training an ASR model:

1. Train the mono-phone model (see Sect. 5.4) with MFCC features.
2. Align audio with the acoustic model.
3. Train the tri-phone model (see Sect. 5.4) with MFCC features.
4. Re-align audio with the acoustic model and re-train the tri-phone model with added delta+delta-delta features (see Sect. 5.3).
5. Re-align audio with the acoustic model and re-train the tri-phone model with LDA-MLLT or SAT (see Sect. 5.3.1).
6. Re-align audio with the acoustic model and re-train the tri-phone model, etc. Any further extensions of the model may come hereafter.

As you can see the extension of the system at each step is performed on top of the existing system and in between a re-alignment. The repeated alignment is a very important part of the training process. Remember that each feature vector is produced by a specific HMM state (see Sect. 5.4)! The choice of which state produces which feature vector is what we call alignment! We want to make sure that the right part of the acoustic signal (represented by the feature vector) is assigned to the right phones or sub-phones in order to increase ASR accuracy. By training the ASR model with transformed features or with different linguistic units (mono-phone, tri-phone) we change the parameters of the HMM and a re-alignment must be performed before we extend the system!

Part IV
Biometrics Recognition

Chapter 6
Face Recognition

Abstract Recognizing peoples' faces is an important machine learning field. In this chapter you will learn step-by-step how automatic face recognition works. We will focus on the 'how' and the 'why' without going into too much detail about the mathematical formulation (except where it is really necessary). Face recognition can be used in many types of real world applications; for example, as an aid for law enforcement, for security and access control, in smart offices and homes, etc.

6.1 Introduction

The Merriam-Webster dictionary describes biometrics as follows: "the measurement and analysis of unique physical or behavioral characteristics (such as fingerprint or voice patterns) especially as a means of verifying personal identity."

There are many types of biometrics used today, for example, DNA matching, the shape of the ear, eye matching (iris, retina), facial features, fingerprinting, hand geometry, voice, signature, etc. Verifying personal identity may be very important in many applications for law enforcement, security and access control, and even in smart offices and homes where person dependent services may improve processes and everyday life for people.

There have been many types of algorithms and methods used in the past decade for face recognition. It is not the aim of this chapter to review them all, only the current state-of-the-art method which is also available in the AI-TOOLKIT (see Sect. 9.2.7).

Most biometrics identification systems work in a very similar manner and involve two main steps, feature extraction and feature (or pattern) matching. Feature extraction means that we analyze the chosen biometrics (a human face in this case) and extract a collection of features which are necessary to distinguish between different people. The aim is, of course, to limit the extracted information to the minimum amount necessary in order to optimize the machine learning training and prediction phases. Too much information would not only make everything much slower but it would also confuse the machine learning model, which should focus on the features that are really important for distinguishing different people. Feature matching is the

Z. Somogyi, *The Application of Artificial Intelligence*,
https://doi.org/10.1007/978-3-030-60032-7_6

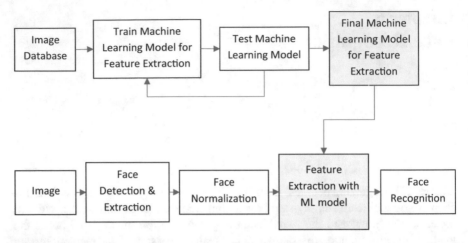

Fig. 6.1 Process flow for building and using a face recognition system

process in which we use the extracted features in order to determine the identity of a person. We usually compare extracted features in a reference database to the input features for recognition.

The main steps of building and using a face recognition machine learning system can be seen on Fig. 6.1 and can be divided into two major tasks:

- Training a machine learning model for feature extraction, and
- Performing face recognition with the help of the trained machine learning model.

Experts often distinguish between face identification and face verification. Face identification is finding a person in a reference database based on her/his face, and face verification is when we assume that a face belongs to a specific person and we want to verify this assumption. In an ideal world face identification and verification are the same, since we can identify the input face and then compare the resultant face ID (the identified person) to verify the assumption. One of the reasons that researchers distinguish between these two types of face recognition is to optimize the verification process. We will not distinguish between face identification and verification in this book and we will simply use the term 'face recognition'.

The two major tasks explained above are further divided into several sub-tasks, as shown on Fig. 6.1. First we need to train a machine learning model based on a huge number of input images (an image database) for feature extraction. The training of such a model may take several days or even weeks and may involve millions of images. The aim is that the ML model (a large scale convolutional neural network (CNN)) learns how to distinguish between the faces of different people. Deep inside of the system the CNN learns which face patterns are important in order to distinguish between different people. We will see in the next section exactly how this happens.

As usual ML model training and testing are both important in order to arrive to a good final ML model.

The face recognition branch of the whole process (the lower part of Fig. 6.1) involves the detection of face(s) in the input image, normalization of the extracted face image (we will see later how and why), feature extraction using the previously trained ML model and, finally, effective face recognition based on the extracted features.

All of these steps will be explained in detail in the following sections.

6.2   Training a Machine Learning Model for Feature Extraction

After many years of face recognition research, recently researchers have been successful in finding a methodology and machine learning model which improved the accuracy of face recognition on several widely used reference face recognition datasets by 30% [73, 74]. This methodology is considered to be the state of the art in face recognition today and will be explained here under.

Machine learning (ML) based automatic face recognition is a complex task because of the possible variations in people's faces and because of the variations in the appearance of the same face (pose, facial expression, cosmetics, accessories, hair, aging, etc.) along with external conditions such as illumination.

We have often talked about generalization error in earlier chapters of the book because it is important that ML models generalize well (e.g., see Sect. 1.2.1). In the case of face recognition, generalization means that our ML model should be able to recognize faces which were not included in the ML training dataset. The aim is, of course, that we can add a picture of any person to a database and that we can identify that person on an image (or video frame = image).

The state-of-the-art ML model used today for face recognition (in the feature extraction phase) is a large scale so-called residual convolutional neural network (RCNN), which "directly learns a mapping from face images to a compact Euclidean space where distances directly correspond to a measure of face similarity" [74]. Very simply put, this means that the inputs of the neural network are face images and the outputs are vectors in an n-dimensional space, where their distance can be simply calculated. Vectors of face images of the same person are very close to each other and vectors of face images of different people are far away from each other in this n-dimensional Euclidean space. As this space is Euclidean, it means that we can use simple clustering algorithms based on the Euclidean distance for grouping the face images (face vectors) of the same person together and grouping faces of different people in different groups. If we know that a specific image (a reference image) is from a specific person, then all the other images in the same group (cluster) are also from that person. In this way, we can easily identify people's faces on images (and video).

A possible representation of the machine learning model can be seen on Fig. 6.2. There are many variations of this RCNN ML model depending on the number of

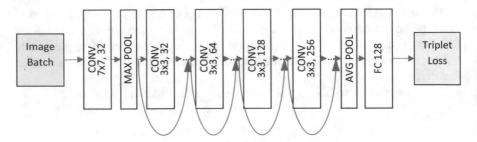

Fig. 6.2 Residual convolutional neural network for feature extraction in face recognition

layers, the number of nodes per layer and where the normalization is applied. The input of the RCNN is a batch of images containing positive and negative samples (we will see later why this is important). A positive sample is an image of a specific person and a negative sample is an image of a different person. The convolutional neural network contains several convolutional layers (CONV) and pooling layers in a similar manner as we have seen in Sect. 2.2.3. On Fig. 6.2, not all of the layers are shown; the dots mean repetition of the same type of layer. There are different configurations possible depending on the number of layers, for example, 18-layers with $4 \times$ [CONV 3×3, 32], $4 \times$ [CONV 3×3, 64], $4 \times$ [CONV 3×3, 128], $4 \times$ [CONV 3×3, 256]; or 34-layers with $6 \times$ [CONV 3×3, 32], $8 \times$ [CONV 3×3, 64], $12 \times$ [CONV 3×3, 128], $6 \times$ [CONV 3×3, 256] etc. Right after each convolutional layer, but before activation, there is a batch normalization (not indicated on the figure) in order to accelerate the learning process and improve accuracy [73]. The output of the neural network is a 128 dimensional vector (indicated with FC 128 on Fig. 6.2) which is the feature vector for each face image embedded into an Euclidean space. The neural network learns to generate such a vector for each face image in the manner required for our face recognition task!

There are two important modifications/differences compared to a standard convolutional neural network:

1. The use of four identity shortcut (residual) connections in the network over each similar group of layers [73]. This is the reason for the name 'residual' convolutional neural network. Researchers have found that such shortcuts in the neural network result in a much faster optimization (learning), a significantly higher accuracy in combination with deeper convolutional neural networks (when using more layers) and a much better generalization accuracy [73]. A better generalization accuracy in the case of face recognition means that the trained model is also good at recognizing face images which were not used during the training of the neural network. The way in which this shortcut connection modifies the neural network can be seen in Fig. 6.3 (blue curved arrows).

2. The *Triplet Loss* calculation at the end of the network during the training of the model. This triplet loss is sometimes also called metric loss. The aim of this extension is to force the neural network to learn directly the Euclidean space for

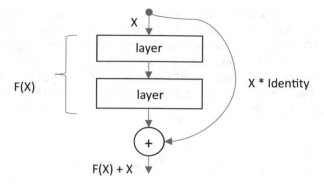

Fig. 6.3 Shortcut connection in a neural network

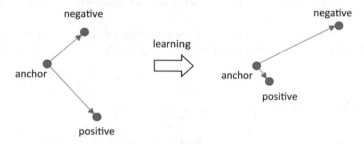

Fig. 6.4 Triplet loss learning

face recognition "such that the squared distance between all faces, independent of imaging conditions, of the same identity is small, whereas the squared distance between a pair of face images from different identities is large" [74]. Instead of using the triplet loss one may also use a pair of positives and negatives in a similar manner, in such way that the neural network learns that images (the extracted feature vectors) from the same person (positives) are close to each other and that images from different people (negatives) are far from each other. What we are doing is very simply guiding the neural network in the direction we need for face recognition in order to use simple clustering of the resultant 128D face vectors (feature vectors).

Figure 6.4 shows how the neural network learns that the anchor face image is close to a positive face image (from the same person) and far from a negative face image (from a different person). The name triplet loss is relates to the three points used. In standard ML training we usually minimize some kind of error but here we minimize the distances between similar feature vectors (anchor - positive) and maximize the distances between different feature vectors (anchor - negative).

In order to be able to calculate the triplet loss (or pairwise loss) we need to carefully select hard positive and hard negative images, where hard positive means very similar images from the same person and hard negative means an image from a very different person.

In order to train a high accuracy face recognition ML model, we need to feed millions of face images to the model in the training phase. The training of such a model may take several days or even weeks depending on the computer used. The aim is that the ML model (our large scale residual convolutional neural network (RCNN)) learns how to distinguish between faces of different people. Deep inside of the system the RCNN learns which face patterns are important in order to distinguish between different people.

6.3   Face Recognition with the Trained Model

After we have trained our RCNN model, which can produce 128D vectors for each face image (feature extraction), we are ready to assemble a professional face recognition system.

There are several important steps in a face recognition process (see the lower part of Fig. 6.1) which will be explained here under in detail.

6.3.1   Face Detection and Extraction

As a first step we need to find automatically all of the faces in the input image and their exact location in order to extract the face images. Face detection is a complex problem because of the many possible face poses, rotations, scales, facial expressions, occlusions, etc.

Many different types of face detection methods have been developed in the past decade, for example, knowledge based methods (a set of rules are defined with what a human face looks like), template matching methods (standard templates are composed which describe a human face), statistical methods (based on the appearance of the shape defined by a set of face landmarks such as eyes, nose, mouth, ears) and face appearance based methods. Appearance based methods, which learn the characteristics of human faces from a large number of input images by using machine learning techniques, are the state of the art at the time of writing this book.

One of the most accurate and fastest face detection methods available today is the so-called histogram of oriented gradients (HOG) method. Some neural network based methods achieve slightly better accuracy but they need much more computer resources and are also much slower than the HOG method.

The HOG algorithm first calculates the local pixel intensity gradients on a dense grid in a sliding window. Then the histogram of gradient directions for each grid cell is calculated. The combined histogram entries for each cell in the grid form the final HOG representation (feature vector) of the sliding window (detection window). The resultant feature vectors from many images (with faces and without faces) are classified with a standard classification machine learning model (e.g., SVM) as being a human face or not. The machine learning model learns which collection of

HOG's represent a human face and which do not (as usual we use hard positive and hard negative images; see Sect. 6.2). The local pixel intensity gradients in an image are very representative of the underlying human face and this is what the ML model learns. It is interesting to note that the same technique can be used for any rigid object detection. The HOG algorithm is explained in detail in Expert Sect. 6.1.

Expert Sect. 6.1 The HOG Algorithm for Detecting Faces in Images
An example implementation of the HOG algorithm will be explained in this section in detail.

Let us first define cells as being 8x8 pixels and create a histogram for each cell with a bin size of 20 (9 bins). For each cell the intensity gradient is calculated and added to the histogram of the cell. Figure 6.5 shows how this can be done for one cell with the following calculation:

$$gradient\ magnitude = \sqrt{(150 - 90)^2 + (150 - 80)^2} = 92$$

$$gradient\ direction = atan\left(\frac{150 - 80}{150 - 90}\right) = 49.$$

Next we distribute the direction of the gradient to both the 40 and the 60 bins (because 49 is between 40 and 60) with the weighted magnitude value on the vertical axis of the histogram:

$$\frac{60 - 49}{40}92 = 25, and\ \frac{49 - 40}{40}92 = 20$$

If we repeat the above calculation for each pixel in the cell (8×8 pixels) then we get the final histogram for the given cell. This results in a 9×1 vector (there are 9 bins and for each bin we have a magnitude value).

Finally we normalize the histograms by taking 2×2 cells (16×16 pixels) together, we call this a *block*. 2×2 cells have $4 \times 9 \times 1$ bin values which is 36×1 in total (we put each 9×1 histogram one after the other). The normalization is simply the division of each value in the 36×1 vector by the square root of the sum of squares for all of the values. By normalizing the histograms we reduce the influence of illumination. If the input image has 100 blocks ($100 \times 16 \times 16$) then the final HOG feature vector will contain $100 \times 36 \times 1 = 3600$ features. This HOG feature vector may represent a specific object, in our case a face.

After the feature vectors have been created we can train a machine learning model for automatically recognizing faces in images. We feed positive and negative feature vectors (remember that a positive image is a face image and a negative image does not contain a face) to an ML classification model (e.g.,

(continued)

SVM) and the ML model will learn which feature vector corresponds to a face and which does not.

Such a HOG face detector may take major pose changes into account in order to increase detection accuracy. For this reason a separate detector is usually used for a frontal face and for a rotated face (side view); these detectors may be combined together in order to find faces in an image with different rotation angles effectively.

After the HOG detector is trained (as explained above) it can be used for detecting a face in images. The face detector performs the same calculations as described above and compares the HOG features in the input image regions with the learned face HOG features at different scales. If the comparison indicates that the region contains a face, then a boundary rectangle can be easily calculated in order to get the exact location of the face in the input image. A bounding rectangle is then used to extract the detected face.

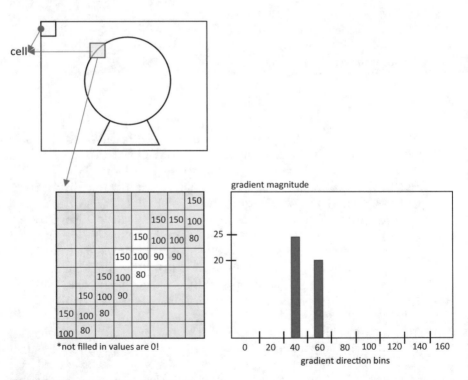

Fig. 6.5 Intensity gradient calculation in a cell

6.3.2   Face Normalization

The next step in face recognition is the normalization of the extracted face image before feature extraction. Normalization means that we transform the detected face image into an approximately frontal face image similar in size to those used during the training of the ML model for feature extraction in order to increase face recognition accuracy. A face may appear in many different poses in an image (side view, downwards looking, etc.). The normalization may involve several steps of face alignment (rotation, transformation, etc.) and scaling.

Face alignment can be performed by using a so-called face landmarking machine learning (ML) model. Face landmarks are the position and shape of the eyes, nose, mouth, ears, eyebrows, etc. (see Fig. 6.6). After such a model is trained it can be used for estimating the face's landmark positions (the pose of the face) from pixel intensities in the input image. And after the face landmarks are detected the required image transformation can be easily calculated by comparing the detected face landmark shape to a centered base frontal face landmark shape. In this way the detected face can be transformed into an approximately frontal face (see Fig. 6.7).

The training and prediction of a face landmarking ML model is somewhat similar to what we have seen in the case of the face detector ML model (a HOG based

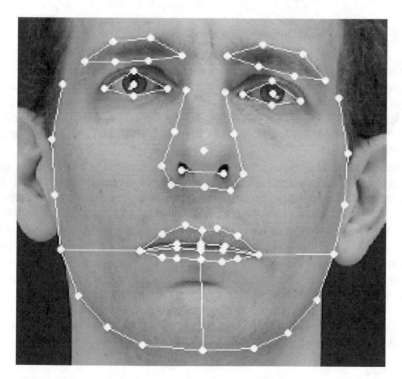

Fig. 6.6 Face landmarks; from [77]

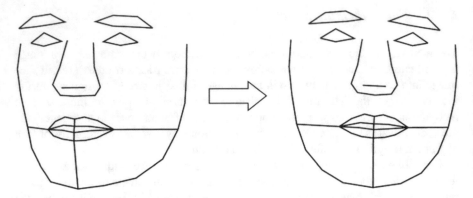

Fig. 6.7   Face alignment; based on [77]

method in which the overall landmark shape is approximated with a regression function); therefore, it will not be explained here in detail. For more information please refer to [75, 76]. It is sufficient to train a landmarking ML model with only a few points which can be used for aligning a face automatically in a few milliseconds.

6.3.3   Feature Extraction and Face Recognition

Before we can perform face recognition we need to build a reference face recognition database with high quality frontal face images of people we want to recognize. The size of the face images should be similar to the size of the images we used during the training of the ML model for feature extraction. Face recognition systems usually extract and scale the face images automatically (AI-TOOLKIT) from selected input images.

The trained residual convolutional neural network (RCNN) can now be used to extract the 128 dimensional (128 numerical values in a vector) feature vector from each detected and normalized face in an input image for recognition. Next we need to extract the 128D feature vector from each image in the reference database. When we have all of the above feature vectors we can simply use a clustering algorithm (see Sects. 1.4 and 2.3) in order to group (cluster) all feature vectors. If the detected image corresponds to one of the reference images, then both images will be grouped into the same cluster because the feature vectors are close to each other in the Euclidean space learned by the ML model (see Sect. 6.2) if they are both face images of the same person. If the detected face (represented by its feature vector) is assigned to a cluster without any other face, then the face is an unknown face (it does not exist in the reference database).

Chapter 7
Speaker Recognition

Abstract As with face recognition, speaker recognition is also an important application in the field of machine learning. Using speaker recognition together with face recognition adds an extra level of certainty (and security) to many applications. There are several tasks in which humans are still much better than machine learning (ML) models, but speaker recognition is an exception because ML models can recognize a speaker better than humans can! In this chapter you will learn how automatic speaker recognition works in detail.

7.1 Introduction

Speaker recognition (recognizing the voice of the person who speaks) may be very useful in many types of real world applications, for example, as an aid for law enforcement, for security and access control, in smart offices and homes, etc.

Do not confuse speaker recognition with speech recognition (the subject of Chap. 5)! In speaker recognition we identify the person who speaks and it is usually independent of the spoken text. In speech recognition we automatically transcribe (comprehend) the contents of the spoken text. Automatic speaker recognition is not easy, but it is easier than speech recognition.

Most biometric identification systems work in a very similar manner and involve two main steps, *feature extraction* and *feature* (or pattern) *matching*. Feature extraction means that we analyze the chosen biometrics (an audio segment containing the human voice in this case) and extract a collection of features which are necessary to distinguish between different people. The aim is to limit the extracted information to the minimum amount necessary in order to optimize the machine learning training and prediction phase. Too much information would not only make everything much slower but it would also confuse the machine learning model, which should focus on the features that are really important for distinguishing between different people. Feature matching is the process in which we use the extracted features in order to determine the identity of a person. We usually compare extracted features in a reference database to the input features for recognition.

© Springer Nature Switzerland AG 2021
Z. Somogyi, *The Application of Artificial Intelligence*,
https://doi.org/10.1007/978-3-030-60032-7_7

Fig. 7.1 The vocal tract;
modified from [81]

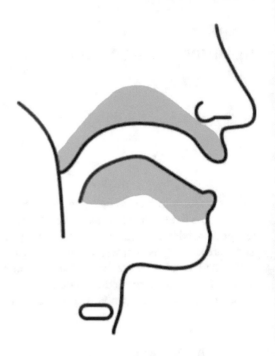

Speaker recognition is based on some acoustic patterns (features) in human speech which are unique to each individual. The uniqueness of these acoustic patterns is due to the unique anatomy of humans (i.e., the shape and size of organs in the mouth—called the vocal tract) and due to learned speech patterns and style (see Fig. 7.1).

Experts often distinguish between speaker identification and speaker verification. Speaker identification is finding a person in a reference database based on her/his voice and speaker verification is when we assume that a voice is from a specific person and we want to verify this assumption. In an ideal world speaker identification and verification are the same, since we can identify the input voice and then compare the resultant voice ID (the identified person) to verify the assumption. One of the reasons that researchers distinguish between these two types of speaker recognition is to optimize the verification process. We will not distinguish between speaker identification and verification in this book and we will simply use the term 'speaker recognition'.

Figure 7.2 shows the major steps in the speaker recognition process.

The first pre-processing step can be used to clean the input audio signal (removing noise, echo, etc.). It is easier to remove unwanted features from an audio signal than try to model them later! This is also true for the audio recording conditions—try to produce as clean a voice recording as possible in all circumstances.

The voice activity detector (VAD) is an extra pre-processing step in order to remove audio without human voice for improving the speed and accuracy of speaker recognition. "Based on experience, close to 30% of the audio frames in a normal

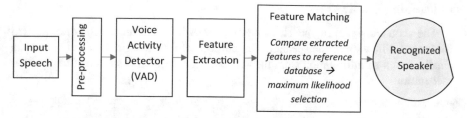

Fig. 7.2 Speaker recognition process

audio recording are silence frames. This means that through silence removal, the recognition process may become faster by the same rate. ... The most efficient silence detection is energy thresholding due to its simplicity and effectiveness. The actual threshold may be variable. Although some even use phone models developed for speech recognition to detect silence. Sophisticated algorithms may be used to estimate it along the time line. However, in principal, it remains quite simple. Once a threshold is known, the signal power is computed and if it falls below a certain threshold, it is considered to be silence. The signal power may be simply computed using Parseval's theorem which means that the total power in a window is the sum of the squares of its sampled values." [80].

In the feature extraction phase we extract a so-called voiceprint from the voice recording (audio signal). A voiceprint is a kind of audio "fingerprint" of the person who is speaking and contains numerical features of the audio signal specific to the speaker. A voiceprint in this book may mean different things, from a simple feature vector to a complex acoustic model.

In the feature matching phase we compare the extracted voiceprint to a reference database of voiceprints which contain the voiceprints of all the people we want to recognize. The voiceprint is defined in such a way that we can easily compare different voiceprints in order to distinguish between different people. The reference voiceprint with the maximum likelihood is selected from the database and the person is identified or the person may be unknown.

Both feature extraction and feature matching will be explained in detail in the following sections.

7.2   Feature/Voiceprint Extraction

Because of the highly variant nature of a voice, acoustic signal feature extraction and thus speaker recognition is a difficult task. The variation in the voice acoustic signal may have many possible causes, including the following:

- Change in voice over time due to aging, health conditions, stress, mood change, etc.
- Background noise

- Recording quality

The acoustic signal of speech contains a lot of information which is not needed for the identification of the speaker. The purpose of feature extraction is to extract a small amount of information from the acoustic signal which is sufficient for speaker recognition.

> TIP: A refresher on the information contained in Sects. 5.2 and 5.3 of this book about acoustic signal processing and feature extraction is recommended.

There are many types of feature extraction schemes and algorithms such as linear prediction coding (LPC) and linear prediction cepstral coefficients (LPCCs) [83, 84], the mel-frequency cepstrum coefficients (MFCCs), etc. The calculation of these features is very similar, and MFCC feature extraction has been explained extensively in Sect. 5.3 of this book.

MFCCs are very useful in speaker recognition because they are based on how the human ear perceives sound varying with frequency; they form the so-called acoustic vector or feature vector per utterance. MFCCs are also less sensitive to noise in the acoustic audio signal than LPCCs. An utterance is a short speech audio segment, for example a spoken word or sentence. Some experts suggest adding the so-called delta and delta-delta features (see Sect. 5.3) to the feature vector.

Let us look at the difference in frequency spectrum for two speakers for the same utterance in Fig. 7.3 in order to understand how MFCC's differentiate between two speakers. The relative location of the peaks in the spectrum is similar for both speakers because they are speaking the same words, but the magnitude and details of the spectrum are speaker dependent. And this uniqueness is what we embed into the feature vectors based on the frequency spectrum.

Next let us look at the graphical representation of feature vectors extracted from different utterances of two speakers. These feature vectors are usually multi-dimensional vectors, but for simplicity and for didactical reasons let us map them into 2D as shown on Fig. 7.4. When the utterances are spoken in a similar manner, the feature vectors are close to each other and they form a group (cluster). When the utterances are spoken differently (e.g., because of stress or mood change) the feature vectors form another group. We call the collection of these feature vector groups per speaker the code book of the speaker. The fact that the feature vectors may form separate groups per speaker makes speaker recognition a bit more difficult. More about this later!

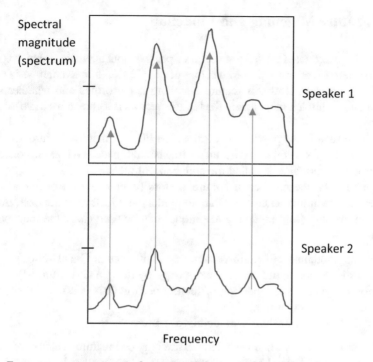

Fig. 7.3 Frequency spectra segment of an utterance for two speakers

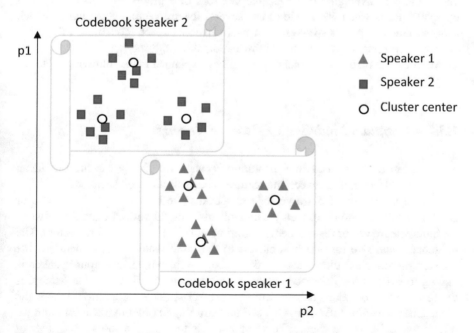

Fig. 7.4 Feature vectors of different utterances for two speakers mapped to 2D

7.3   Feature Matching and Modeling

After we have extracted the feature vectors from the input acoustic signal, we need to compare them to reference feature vectors of people stored in a database in order to identify the speaker. How this feature matching is performed and whether we first create a more complex model from the feature vectors is a choice we have to make at this stage.

We have seen in the previous section and in Fig. 7.4 that the feature vectors of a speaker may form several groups in an n-dimensional space and that the collection of these groups per speaker is called the codebook of the speaker.

Depending on the number of feature vectors (utterances) we decide to use per speaker and depending on how we want to model these feature vectors, several types of feature matching and modeling techniques can be used (only the most common are listed):

- Template modeling of feature vectors (e.g., with Vector Quantization)
- Stochastic modeling of feature vectors (e.g., by using a GMM model)
- Factor analysis + ML modeling of feature vectors (this is often called i-vectors based speaker recognition)
- Neural network embeddings modeling of feature vectors

The simplest case is when we decide to use only one feature vector per speaker or when we combine several feature vectors per speaker together into a super-feature vector by, e.g., averaging or taking the centroid of a group (Fig. 7.4). The most complex case is when we decide to use several feature vectors per speaker, which form several groups per speaker in an n-dimensional space (codebook). By using more feature vectors we have more information about the speaker.

All three feature modeling techniques will be explained in the following sections.

7.3.1   Template Modeling of Feature Vectors

The simplest template modeling technique is when we simply use the Euclidean distance to differentiate between the feature vectors of different speakers.

If there is only one feature vector per speaker (we have extracted only one or combined several into one), then we can calculate the Euclidean distance between the input feature vector for speaker recognition and all reference feature vectors. The reference feature vector which is closest to the input feature vector identifies the speaker. In so-called closed-set speaker recognition, where the input speaker is known to exist in the reference database, this technique will work fine (except if the recording is of poor quality). But in open-set speaker recognition, where the input speaker may be unknown, we need an extra step in order to make sure that we do not make the wrong choice. This extra step is the use of a so-called *Universal Speaker Model* (USM) in order to calculate the likelihood of speaker recognition (such a universal speaker model is sometimes also called a universal background model (UBM) in the case of more complex models). In this simple case the USM is the combination (average) of all reference speaker feature vectors in the database and forms a super-feature vector. After this super-feature vector (USM) is calculated, we can check its distance to the input feature vector for speaker recognition. If this distance is smaller than the minimum distance found above (between the input and the reference speakers) then the speaker is unknown. The universal speaker model is a kind of 'anti'-speaker model since it combines the features of all (or a lot of) speakers, and thus such a speaker does not exist.

In a more complex case, when there are several reference feature vectors per speaker which form several groups (codebooks), unsupervised learning may be used in order to define the different groups (clusters). This technique is often called vector quantization. After the reference clusters are known, it is easy to calculate the Euclidean distance between the centers of the clusters and the input feature vector (s) (or their cluster centers) in order to check to which cluster the input feature vectors are closest to. A universal speaker model may also be used here, in a similar manner as explained in the above simple case, in order to calculate the likelihood of recognition.

The numerical value of this likelihood of recognition may be calculated as follows:

$$\frac{P(F \text{ belongs to speaker } k)}{P(F \text{ does not belong to speaker } k)} = \frac{P(F|\lambda_k)}{P(F|\lambda_U)}$$

where F is a set of feature vectors $F = \{f_n; n = 1,2,\ldots N\}$, λ_k is the speaker model of speaker k (one or more feature vectors or a more complex speaker model explained later), and λ_U is the universal speaker model. P() denotes the probability of the term between the brackets. It can be proved statistically that the probability F does not belong to speaker k can be approximated with the probability that it belongs to the universal speaker model, and we can simply use $P(F|\lambda_U)$ in the denominator. Remember that the universal speaker model is a kind of 'anti'-speaker model since it combines the features of all (or a lot of) speakers, and thus such a speaker does not exist.

In practice we use the so-called log-likelihood ratio by computing the logarithm of the above equation:

$$LLR(F, k) = \log\left(P(F|\lambda_k)\right) - \log\left(P(F|\lambda_U)\right)$$

We usually use a decision threshold (a numerical limit) for accepting or rejecting the hypothesis that F is produced by speaker k. The probabilities are simply calculated from the distances, calculated as explained above, where closer means higher probability.

7.3.2   Stochastic Modeling of Feature Vectors

A stochastic process or model has a random probability distribution or pattern that can be estimated by using statistical techniques but cannot be determined precisely. In a stochastic or statistical speaker model, the speaker is treated as a random source producing the observed feature vectors.

"Within the random speaker source, there are hidden states corresponding to characteristic vocal-tract configurations. When the random source is in a particular state, it produces spectral feature vectors from that particular vocal-tract configuration. The states are called hidden because we can observe only the spectral feature vectors produced, not the underlying states that produced them. Because speech production is not deterministic (a sound produced twice is never exactly the same) and spectra produced from a particular vocal-tract shape can vary widely due to coarticulation effects, each state generates spectral feature vectors according to a multidimensional Gaussian probability density function" [81].

Remember that the feature vectors from a particular speaker may form several separate groups in an n-dimensional space. This is the reason why we need several Gaussian distributions to model them and not only one. We call the model based on a collection of Gaussian distributions a Gaussian mixture model (GMM). More information about GMMs and HMMs can be found in Appendix. A GMM is a simplified HMM; the simplification comes from the fact that "we are primarily concerned with text-independent speech, we simplify the statistical speaker model by fixing the transition probabilities to be the same, so that all state transitions are equally likely." [81].

As usual the GMM model parameters are estimated in an unsupervised manner by using the expectation-maximization (EM) algorithm (see section "The Hidden Markov Model (HMM)" in Appendix). The input to the EM algorithm is the extracted feature vectors. We train, in this way, a separate GMM model for each speaker.

Speaker recognition can then be performed by calculating the likelihood of each GMM speaker model when compared to the input feature vectors. The GMM speaker model with the maximum likelihood is chosen and the speaker is identified [81]:

$$\widehat{s} = argmax \sum_{t=1}^{T} \log \left(P(x_t|\lambda_s) \right)$$

where T is the number of feature vectors and $P(x_t|\lambda_s)$ is the probability that feature vector x_t is from speaker s, represented by the λ_s speaker model. Note that we sum the log probabilities for each input feature vector!

In straightforward closed-set speaker recognition the above methodology will work fine. Closed-set means that we are sure that the input speaker exists in the collection of reference speakers. In the case of open-set speaker recognition (or verification) where an unknown speaker may be the input, we need to add an extra step to the speaker recognition process. This extra step was explained in the previous section about template modeling and concerns the matching of the input feature vectors for speaker recognition with a so-called universal background model (UBM). Note that we use here the term UBM instead of USM (see Sect. 7.3.1) but we're talking about something very similar. If the likelihood of the UBM is higher than the likelihoods of the reference models in the database, then the input feature vectors for speaker recognition come from an unknown speaker.

The UBM is a GMM model trained in the same way as the individual GMM speaker models but by using the feature vectors of all (or many) of the speakers in the reference database. The choice of the feature vectors (type of speaker, number of speakers, etc.) is usually not important, but one should train the UBM on a large number of different speakers (since we want to create a kind of anti-speaker model, see Sect. 7.3.1). Some researchers suggest the training of UBMs per gender, but in practice this is not very useful since it is usually not permissible to ask the gender of the speaker or it is not practical. It is, however, useful to train separate UBMs per language.

The log-likelihood ratio explained in the previous section can then be used to test speaker identity with the following extension because of the number of feature vectors [81]:

$$\log\left(P(F|\lambda_k)\right) = \frac{1}{T} \sum_{t=1}^{T} \log\left(P(x_t|\lambda_k)\right)$$

$$\log\left(P(F|\lambda_U)\right) = \log\left(\frac{1}{B} \sum_{b=1}^{B} P(F|\lambda_b).\right)$$

where $P(F|\lambda_b)$ is calculated in a similar manner to the first eq. ($P(F|\lambda_k)$), F is the set of feature vectors, T is the number of feature vectors, $P()$ denotes probability, λ_k is a GMM speaker model for speaker k, λ_U is the universal background GMM model and B is the number of feature vectors used to train the UBM.

The log-likelihood ratio is then calculated with the same equation as in the previous section:

$$LLR(F,k) = \log\left(P(F|\lambda_k)\right) - \log\left(P(F|\lambda_U)\right)$$

Note: It is interesting to mention that universal background models (UBMs) are sometimes also used as a starting point for individual speaker models in order to decrease the time and data needed for training individual speaker models. In this scenario the trained UBM parameters (GMM!) are adapted to each speaker and in this way all individual speaker models are created. The

(continued)

resultant individual speaker models are called single speaker adapted Gaussian
mixture models. It is an art to adapt the UBM to each individual speaker well
and not all researchers agree that this provides a good individual speaker
model. The problem is deciding which parts of the UBM should be kept—
will we lose too much from the UBM or the opposite, use too much, and get a
more general model instead of a model for the individual speaker?

7.3.3   Factor Analysis + ML Modeling of Feature Vectors (I-Vectors)

In the last several years so-called *Factor Analysis* has been extensively researched as
a replacement and extension of feature extraction in speaker recognition. According
to what we know today, factor analysis based speaker recognition is considered to be
the state of the art and provides the best accuracy. It is, however, important to
mention that this improvement in accuracy is not very significant or in many cases
negligible for clean acoustic signals (containing little or no noise and background
sound), and some researchers have even reported slightly better accuracy with the
MFCC based GMM models explained in the previous sections. If you can obtain
clean speech recordings or you can clean the recordings (remove noise and background sound), simple MFCC based GMM or template models may be sufficient. It
is always better to use clean acoustic signals instead of trying to model noise and
background sound together with human speech! However, sometimes it is not
possible to obtain clean speech recordings and then more complex i-vector based
speaker recognition systems may be very useful.

In some complex languages, for example Asian languages, the combination of
MFCC's + delta + delta-delta features with pitch may provide better accuracy in the
case of GMM and template models. In these languages the i-vector based speaker
recognition model may also increase accuracy.

Figure 7.5 shows the simplified i-vectors based speaker recognition process.

The i-vectors based speaker recognition process starts with training the UBM and
an i-vectors extractor model by using many input speech recordings. The i-vectors
extractor is trained with the so-called Baum-Welch algorithm based on the UBM and
by using the same speech data as was used for training the UBM. The mathematical
equation used for the i-vector extractor is the following [88]:

$$M = m + Tw$$

where M is a speaker dependent super-vector, m is a speaker and channel
independent super-vector (from the UBM super-vector), T is a rectangular matrix

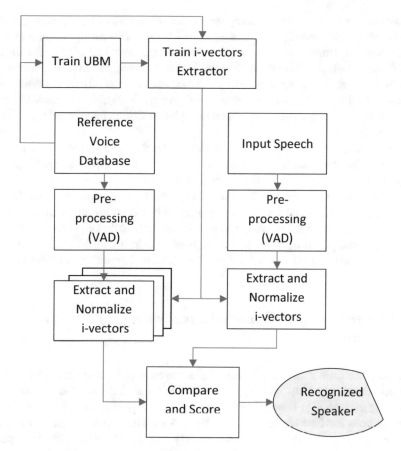

Fig. 7.5 i-vectors based speaker recognition process

and w is a random vector (the i-vector) with a standard normal distribution (w contains the so-called total factors).

The pre-processing step is the same as explained in the previous sections (VAD is the voice activity detector in order to remove the non-speech part of the acoustic signal).

In the extract and normalize i-vectors step we first use the trained i-vectors extractor model to extract the i-vectors from the input utterance and then we normalize them. The normalization of the i-vectors usually happens with linear discriminant analysis (LDA) and within class covariance normalization (WCCN). The normalization is important because it minimizes the false acceptance and false rejection rates of the model (WCCN) and it maximizes the between-class variance and minimizes the intra-class variance (LDA) [88] and thus increases accuracy.

The last step in the i-vector speaker recognition process is the comparison of the reference i-vectors with the input i-vectors and the scoring of each pair. The reference i-vector with the best score infers the recognized speaker. Different techniques can be used for the comparison and scoring step. For example, with an SVM classification model, or with simple cosine distance scoring, or with probabilistic linear discriminant analysis (PLDA) scoring, etc. The SVM classification model learns which i-vector belongs to which speaker, the cosine distance can directly be used to measure (score) the i-vectors (smaller distance means better score), and PLDA provides a speaker variability model which separates the information characterizing a speaker from all other information and thus helps to identify a speaker [88].

Many variations of this i-vectors based speaker recognition process exist depending on the features, algorithms, etc. used. For more information about i-vectors based speaker recognition you may read [88–91].

It is interesting to note that the researchers who reported this technique [85, 86] stated that i-vectors based speaker recognition is based on the so-called joint factor analysis (JFA), but other researchers correctly pointed out that it is closer to principal component analysis [80]. Joint factor analysis was introduced by [87] but it was found to be too complex and not optimal for speaker recognition in recent research.

7.3.4   Neural Network Embeddings Modeling of Feature Vectors

There has also been a lot of research in the last several years towards finding a neural network based machine learning model which can be used for automatically extracting embeddings from speech recordings, in a similar manner to that which we have seen for face recognition (see Sect. 6.2). The extracted embeddings then replace the i-vectors (and MFCCs) completely in the previously explained speaker recognition processes.

The advantages of such a system are that there would be much less noise and background sound sensitivity, the deep neural network could learn to differentiate human voice from everything else, and it could differentiate better between the speech of different people. Despite the fact that some researchers have reported some small success (0.5%–2% accuracy increase [91]), there has not yet been a breakthrough and major improvement in this field. The reason is that it is much more difficult to extract the appropriate speaker dependent embeddings from a speech signal than to extract person dependent embeddings from a face image. Some researchers have also reported that the above mentioned accuracy increase is only in the case of using short utterances (<10–15 s), but the accuracy of the model decreases when the utterance length increases (to even below that of other types of models) [91]. This may signal a fundamental problem because normally one would expect better accuracy when using more data.

The major disadvantages of this model are that it needs much more computational resources and it is much slower than the other models we have seen in the previous sections. Also much more speech data is necessary for training the neural network model compared to other models.

Part V
Machine Learning by Example

Chapter 8
Machine Learning by Example

Abstract In the earlier chapters of this book we have seen how machine learning works and what the different machine learning techniques are. This chapter will explain how to apply these machine learning techniques to real-world problems: automatic classification (clustering) of an unknown dataset, dimensionality reduction of large datasets, recommendation models, anomaly detection, root cause analysis, engineering applications of supervised learning regression, predictive maintenance, image recognition, detection of different diseases, business process improvement, etc. Not only the specific applications will be explained but also tips and tricks and important sector-dependent machine learning recommendations. After reading this chapter you will be able to apply the machine learning techniques explained in the previous chapters to your own machine learning problems, even if they are different from the examples presented here.

8.1 Introduction

The application of machine learning to real-world problems is often a challenge. Not only because of the large number of different types of machine learning models available but also because of the many real-world problems machine learning can be applied to. This chapter will introduce the application of machine learning to a commonly occurring selection of real-world problems. Although we of course can not elaborate on all possible real-world problems in this book, the selected problems will also be helpful in applying machine learning to many other problems which are not explained here.

We will use the AI-TOOLKIT in all of the examples, which can be downloaded free of charge from the website mentioned at the beginning of the book. You may of course also use another software which supports the machine learning models used in the examples.

© Springer Nature Switzerland AG 2021
Z. Somogyi, *The Application of Artificial Intelligence*,
https://doi.org/10.1007/978-3-030-60032-7_8

8.2 How to Automatically Classify (Cluster) a Dataset

We saw in Sect. 3.3 how to calculate the performance of clustering (unsupervised learning) models and how to determine the optimal number of clusters by using internal criterion based performance measures and the elbow method. This section will present several simple examples with different types of datasets and with different types of clustering models with the aim of explaining the advantages and disadvantages of the different models. Clustering a dataset is often difficult because there is no single unsupervised learning model which is suitable for all types of datasets. We often need to try several clustering models and choose the one which best fits the problem at hand. The examples in this section will highlight several possible difficulties with datasets and will show which clustering model can handle which problem.

We will use the Fundamental Clustering Problems Suite [51] which is a collection of low dimensional and simple datasets with known classification. Each dataset is special in some way and contains some kind of difficulty for the clustering model.

8.2.1 The Datasets

We will use six datasets. The dataset called 'Hepta' is a normal dataset with seven clearly separable clusters and thus without difficulties, and with 212 data points (records) and three dimensions (columns). 'Hepta' is shown on Fig. 8.1.

'ChainLink' has 1000 data points, three dimensions and two clusters, and it has the difficulty that *the clusters overlap* (linearly not separable). 'ChainLink' is shown on Fig. 8.2.

'Lsun' has 400 data points, two dimensions and three clusters, and it has the difficulty of having *different variances and inter cluster distances*. 'Lsun' is shown on Fig. 8.3.

'EngyTime' has 4096 data points, two dimensions and two clusters, and it has the difficulty of having an *overlapping Gaussian mixture data*. 'EngyTime' is shown on Fig. 8.4.

'TwoDiamonds' has 800 data points, two dimensions and two clusters, and it has the difficulty of that the *cluster border is defined by density*. 'TwoDiamonds' is shown on Fig. 8.5.

'Target' has 770 data points, two dimensions and six clusters, and it has the difficulty of *outliers* (noise). 'Target' is shown on Fig. 8.6.

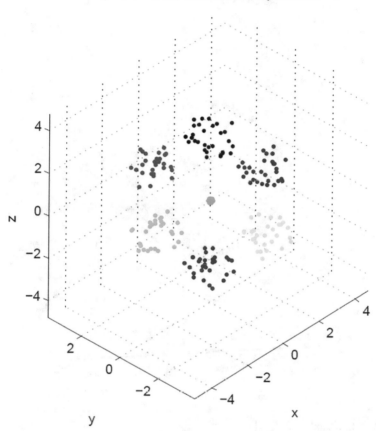

Fig. 8.1 'Hepta' dataset [51]

8.2.2   *Clustering*

The first step is to create an AI-TOOLKIT database with the "Create New AI-TOOLKIT Database" command on the Database tab on the left taskbar. Save the database in a directory of your choice. The second step is to import all data in the database created in the previous step with the "Import Data into Database" command. You can import all data into one database in different tables. *Do not forget to indicate the number of header rows (if any) and the correct zero based index of the decision column*!

Next we must create one by one the AI-TOOLKIT project files for each of the clustering models. Use the "Open AI-TOOLKIT Editor" command and then insert the model template with the "Insert ML Template" button. Each model has different

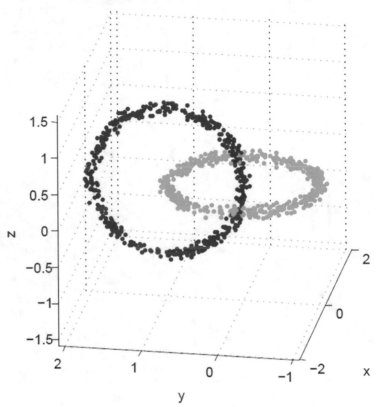

Fig. 8.2 'ChainLink' dataset [51]

parameters. As an example, the k-Means clustering project file for the 'Hepta' dataset can be seen in Table 8.1.

For a detailed explanation of the parameters of each clustering model see Sect. 2.3 in Chap. 2.

The optimized parameters for each model can be seen in Table 8.2. You can use these parameters to replicate the results shown later (Sect. 8.2.3).

8.2.3 *Analysis of the Results*

All of the results are presented in Tables 8.3 and 8.4. As expected all models can handle the 'Hepta' dataset well and they all have 100% accuracy. 100% accuracy means that all of the models can reproduce the labeling (clustering) of the dataset

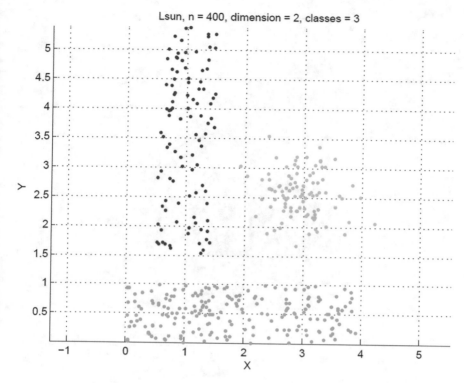

Fig. 8.3 'Lsun' dataset [51]

without knowing the original labels! For an explanation of all performance measures used in the results see Sect. 3.3 in Chap. 3.

8.2.3.1   ChainLink Dataset

'ChainLink' has 1000 data points, three dimensions and two clusters, and it has the difficulty that the clusters overlap (linearly not separable). If we look at the 2D plot of the dataset on Fig. 8.2 it is understandable that this is a very difficult dataset for most of the models. There is, however, one model which can handle this type of dataset well. DBScan achieves 100% accuracy. MeanShift and k-Means cannot handle the dataset at all and the hierarchical model achieves a very low accuracy (68%).

8.2.3.2   EngyTime Dataset

'EngyTime' has 4096 data points, two dimensions and two clusters, and it has the difficulty of slightly overlapping Gaussian mixture data. MeanShift and k-Means can

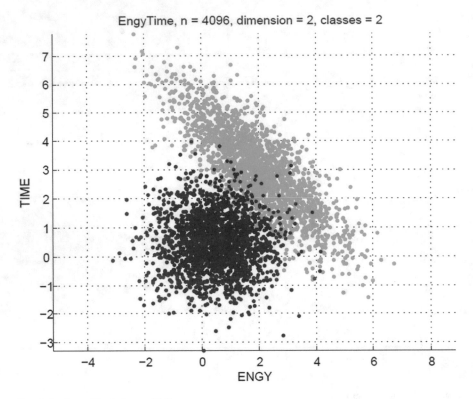

Fig. 8.4 'EngyTime' dataset [51]

handle this type of data well achieving 95.6% and 95% accuracy respectively. DBScan cannot handle this type of data at all. The hierarchical model has a low 74% accuracy.

8.2.3.3 Lsun Dataset

'Lsun' has 400 data points, two dimensions and three clusters, and it has the difficulty of different variances and inter cluster distances. DBScan is the best model to handle this kind of data with 100% accuracy. All of the other models achieve only 73–75% accuracy.

8.2.3.4 Target Dataset

'Target' has 770 data points, two dimensions and six clusters, and it has the difficulty of outliers (noise). This is a special dataset because of the outliers. One of the models, the DBScan model, is specialized in detecting outliers and DBScan achieves

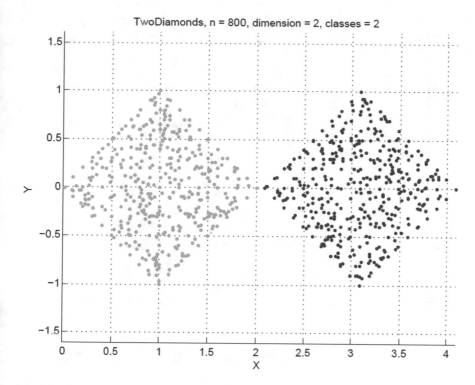

Fig. 8.5 'TwoDiamonds' dataset [51]

nearly 99% accuracy despite the very difficult and noisy dataset. The other models achieve only around 70% accuracy.

The DBScan model in the AI-TOOLKIT groups all of the outliers into the last cluster and this is why there are three predicted class counts and two identified classes (see the dashed line in the results table—'Target' dataset). The reference class count is six because the outliers are in three separate groups (see the plot of the data!).

8.2.3.5 TwoDiamonds Dataset

'TwoDiamonds' has 800 data points, two dimensions and two clusters, and it has the difficulty of having the cluster border defined by density. All models can handle this type of data very well (~100% accuracy) except DBScan. DBScan does not work well with clusters which have varying density.

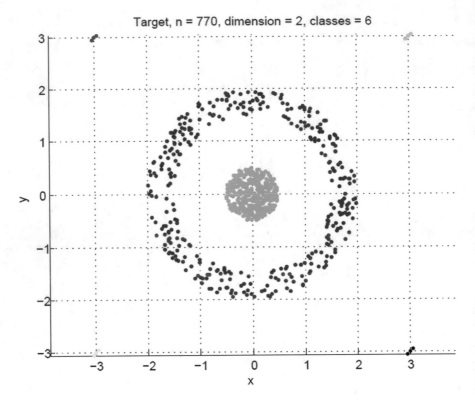

Fig. 8.6 'Target' dataset [51]

Table 8.1 k-Means clustering project file

```
model:
  id: 'ID-_lvdypA-Jt'
  type: CKMEANS
  path: 'cluster.sl3'
  params:
    - clusters: 7
    - iterations: 1000
    - projections: 0
  training:
    - data_id: 'Hepta'
    - dec_id: 'decision'
  test:
    - data_id: 'Hepta'
    - dec_id: 'decision'
  input:
    - data_id: 'Hepta_input_data'
    - dec_id: 'decision'
  output:
    - data_id: 'Hepta_output_data'
    - col_id: 'decision'
```

Table 8.2 Optimized parameters

Dataset	k-Means	Hierarchical	DBScan	MeanShift
Hepta	clusters: 7 iterations: 1000 projections: 0	clusters: 7 ModelType: PairwiseMaximum Linkage	Epsilon: 0.1 MinPoints: 20	UseKernel: true Radius: 0.0 MaxIterations: 1000 KernelBandwidth: 0.1 KernelType: GaussianKernel
ChainLink	Can not handle dataset.	clusters: 2 ModelType: PairwiseAverage Linkage	Epsilon: 0.1 MinPoints: 20	Can not handle dataset.
EngyTime	clusters: 2 iterations: 1000 projections: 0	clusters: 2 ModelType: PairwiseMaximum Linkage	Can not handle dataset.	UseKernel: true Radius: 0.0 MaxIterations: 1000 KernelBandwidth: 0.5 KernelType: GaussianKernel
Lsun	clusters: 3 iterations: 1000 projections: 0	clusters: 3 ModelType: PairwiseCentroid Linkage	Epsilon: 0.102 MinPoints: 20	UseKernel: true Radius: 0.0 MaxIterations: 1000 KernelBandwidth: 0.36 KernelType: GaussianKernel
Target	clusters: 3 iterations: 1000 projections: 0	clusters: 3 ModelType: PairwiseMaximum Linkage	Epsilon: 0.1 MinPoints: 20	UseKernel: true Radius: 0.4 MaxIterations: 1000 KernelBandwidth: 0.51 KernelType: GaussianKernel
TwoDiamonds	clusters: 2 iterations: 1000 projections: 0	clusters: 2 ModelType: PairwiseAverage Linkage	Can not handle dataset.	UseKernel: true Radius: 0.0 MaxIterations: 1000 KernelBandwidth: 1.0 KernelType: GaussianKernel

8.2.4 General Conclusions

This section has proven what was explained in the other chapters of the book about unsupervised learning (clustering). We need several types of models, based on different types of similarity metrics, in order to be able to handle all types of data. This is also the reason why the AI-TOOLKIT contains all of these different types of models. When you start with clustering an unknown dataset then you should be prepared to carry out a few trial and error runs with each model. The AI-TOOLKIT helps you in this process by minimizing the work necessary to define and run a model.

Table 8.3 Clustering results, part 1

Dataset	Performance	DBScan	MeanShift	k-Means	Hierarchical
	class count	2	N/A	N/A	2
	ref. class count	2	N/A	N/A	2
	pred. class count	2	N/A	N/A	2
	accuracy [%]	100	N/A	N/A	67.90
ChainLink	Silhouette	0.167	N/A	N/A	0.209
	Calinski-Harabasz Index	271.459	N/A	N/A	248.892
Difficulty: linearly not separable	Xu Index	-11.323	N/A	N/A	-11.271
	Global Purity [%]	100	N/A	N/A	67.90
	Global Precision [%]	100	N/A	N/A	54.47
	Global Rand Index [%]	100	N/A	N/A	56.36
	Global Recall [%]	100	N/A	N/A	76.97
	Global F1 [%]	100	N/A	N/A	63.80
	class count	N/A	2	2	2
	ref. class count	N/A	2	2	2
	pred. class count	N/A	2	2	2
	accuracy [%]	N/A	95.61	95.04	73.97
EngyTime	Silhouette	N/A	0.425	0.424	0.354
	Calinski-Harabasz Index	N/A	2713.7	774.106	3026.122
Difficulty: Gaussian mixture	Xu Index	N/A	-7.254	-5.845	-7.120
	Global Purity [%]	N/A	95.61	95.04	73.97
	Global Precision [%]	N/A	91.48	90.34	59.09
	Global Rand Index [%]	N/A	91.60	90.58	61.49
	Global Recall [%]	N/A	91.73	90.86	74.63
	Global F1 [%]	N/A	91.60	90.60	65.95
	class count	7	7	7	7
	ref. class count	7	7	7	7
Hepta	pred. class count	7	7	7	7
Difficulty: No (clearly defined clusters)	accuracy [%]	100	100	100	100
	Silhouette	0.702	0.702	0.702	0.702
	Calinski-Harabasz Index	520.724	391.983	16.865	520.724
	Xu Index	-8.776	-7.995	-3.615	-8.776
	Global Purity [%]	100	100	100	100

Table 8.4 Clustering results, part 2

Dataset	Performance	DBScan	MeanShift	k-Means	Hierarchical
Lsun Difficulty: different variances and inter cluster distances	class count	3	4	3	3
	ref. class count	3	3	3	3
	pred. class count	3	4	3	3
	accuracy [%]	100	75.00	74.25	73.25
	Silhouette	0.477	0.565	0.493	0.476
	Calinski-Harabasz Index	963.125	3047.509	109.919	1078.789
	Xu Index	-5.469	-5.894	-3.649	-5.546
	Global Purity [%]	100	75.00	74.25	73.25
	Global Precision [%]	100	68.47	63.44	60.69
	Global Rand Index [%]	100	65.59	72.35	70.80
	Global Recall [%]	100	62.25	61.30	61.90
	Global F1 [%]	100	65.21	62.35	61.29
Target Difficulty: outliers (noise)	class count	2	6	3	3
	ref. class count	6	6	6	6
	pred. class count	3	6	3	3
	accuracy [%]	98.83	69.35	71.68	74.41
	Silhouette	0.276	0.405	0.421	0.397
	Calinski-Harabasz Index	0.503	128.051	187.005	349.120
	Xu Index	-6.213	-5.369	-5.363	-6.136
	Global Purity [%]	98.83	69.35	71.69	74.42
	Global Precision [%]	99.96	54.92	56.62	61.38
	Global Rand Index [%]	99.98	58.39	60.32	62.34
	Global Recall [%]	N/A	79.12	N/A	N/A
	Global F1 [%]	99.98	64.83	65.46	65.82
TwoDiamonds Difficulty: cluster border defined by density	class count	N/A	2	2	2
	ref. class count	N/A	2	2	2
	pred. class count	N/A	2	2	2
	accuracy [%]	N/A	100	100	99.75
	Silhouette	N/A	0.631	0.631	0.630
	Calinski-Harabasz Index	N/A	2377.116	2387.268	2379.856
	Xu Index	N/A	-7.697	-7.702	-7.700

CASE STUDY 2

8.3   Dimensionality Reduction with Principal Component Analysis (PCA)

We saw in Chap. 4 (Sect. 4.3.2.2) that principal component analysis (PCA) can be used to reduce the dimensionality of the dataset and through this reduce the training time needed for our machine learning models. This may also improve the accuracy of the machine learning model.

NOTE: This example is included in the AI-TOOLKIT. You can open the example project in the default location ('\PCA\Iris.yaml').

This example will explain the application of PCA to the so-called Iris dataset [68] using the AI-TOOLKIT.

The Iris dataset contains four measured attributes (cm) of three types of iris plants. For each type of plant 50 measurements are included (measured on collected plants) which results in 150 data records. The dataset contains four columns (features; dimensionality = 4) of data and one more column containing the *class* labels (iris plant types). 'Iris-setosa' with class label 0, 'Iris-versicolor' with class label 1 and 'Iris-virginica' with class label 2. The four features are the four attributes of the plants (see Table 8.5). The Iris dataset is a very simple and small dataset, which is of course not the usual target of PCA but it is very useful to explain how it works. The dataset contains a difficulty for the PCA algorithm because two of the classes are not linearly separable from each other (*virginica* and *versicolor*).

NOTE: The textual class labels are converted to integer numbers in this example. This is called integer encoding. For more information about this subject read Sect. 4.3.2.2.

We will reduce the dimensionality of the dataset from four to two (two columns or features) and we will also check how much variance is retained in the resultant dataset. Variance retained is a quality measure of PCA because it indicates how much information from the original dataset is still included in the resultant reduced dataset (see Sect. 4.3.2.2). We want as much information retained as possible in order to have enough information in the resultant dataset for our main machine

Table 8.5 The Iris dataset

Sepal length	Sepal width	Petal length	Petal width	Class
5.1	3.5	1.4	0.2	0
4.9	3	1.4	0.2	0
4.7	3.2	1.3	0.2	0
4.6	3.1	1.5	0.2	0
5	3.6	1.4	0.2	0
5.4	3.9	1.7	0.4	0
4.6	3.4	1.4	0.3	0
5	3.4	1.5	0.2	0
4.4	2.9	1.4	0.2	0
4.9	3.1	1.5	0.1	0
5.4	3.7	1.5	0.2	0
4.8	3.4	1.6	0.2	0
4.8	3	1.4	0.1	0
4.3	3	1.1	0.1	0
…	…	…	…	…

learning model (e.g., SVM, Neural Network, etc.). PCA is only a pre-processing step.

As we will see in the next section the dimensionality reduction from four to two dimensions works well in this example. In the case of other datasets you may have to carry out several trial and error PCA runs in order to determine the best degree of reduction, also taking the variance retained in the resultant dataset into account!

8.3.1 How to Apply PCA with the AI-TOOLKIT

8.3.1.1 Start a New Project

Open the AI-TOOLKIT editor with the 'Open AI-TOOLKIT Editor' button. Insert a new PCA template ('Insert ML Template' button). You will find the template in the Applications section with the name 'Dimensionality Reduction (PCA)'. In the inserted template change the line 'D:\mypath\MyDB.sl3' to 'Iris.sl3' because we will put the database in the next step in the same folder as the project file (no absolute path is needed). All of the other default parameters are fine (see Table 8.6)! *Save the project file!*

Table 8.6 Iris PCA example AI-TOOLKIT project file

```
model:
 id: 'ID-RyuNJyUSGx'
 type: PCA
 path: 'Iris.sl3'
 params:
  - NewDimension: 2
  - VarianceToRetain: -1
  - Scale: 0
  - PCAType: PCA_EXACT
 training:
  - data_id: 'mytable'
  - dec_id: 'decision'
 output:
  - data_id: 'mytable_out'
  - dec_id: 'decision'
```

All comments (green text) are removed from the project file for clarity. The comments in all of the templates explain the most important rules of the machine learning models. More information about this model can be found in Sect. 9.2.3.11.

8.3.1.2 Create a New Database and Import the Data

On the Database tab use the 'Create New AI-TOOLKIT Database' button with the plus sign on it. Select the folder where the project was saved in the previous step and name the database to 'Iris.sl3'.

Next import the Iris dataset (included with the AI-TOOLKIT in CSV format in the PCA example folder: 'iris.csv'). Use the 'Import Data Into Database' button. In the import module which appears (see Fig. 8.7) enter a **comma** (,) into the 'Delimiters' field (note that several delimiters may be present, here a 'tab' and a comma but only the comma is used). Indicate that there is **one header row** and that the **fourth column is the decision column** (class). Select the above created database with the SEL button. And finally push the 'Import' button.

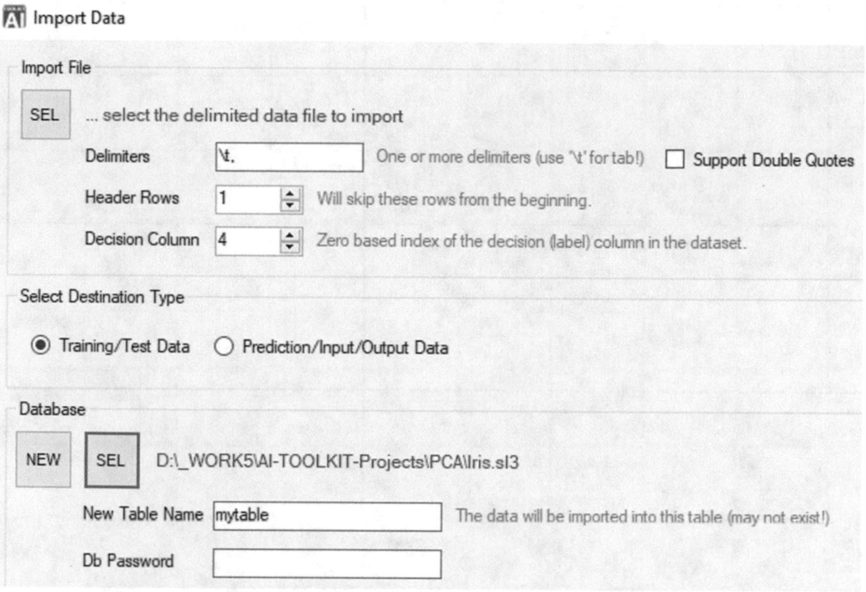

A̅ Import Data

Import File

| SEL | ... select the delimited data file to import |

Delimiters | \t. | One or more delimiters (use '\t' for tab!)   ☐ Support Double Quotes

Header Rows | 1 | ⬍   Will skip these rows from the beginning.

Decision Column | 4 | ⬍   Zero based index of the decision (label) column in the dataset.

Select Destination Type

⦿ Training/Test Data   ○ Prediction/Input/Output Data

Database

| NEW | SEL | D:_WORK5\AI-TOOLKIT-Projects\PCA\Iris.sl3

New Table Name | mytable | The data will be imported into this table (may not exist!)

Db Password | |

Fig. 8.7 Import Data module

8.3.1.3   Train the Model (Apply PCA)

Next you can train the model with the 'Train AI Model' button. A question will appear on whether you want the output data table to be created automatically. Answer **Yes**! When the model training process is finished the retained variance will be indicated in the Output Log at the bottom (97.76%). You can look at the results and also export the reduced dataset in CSV format from the Database Editor (Open Database Editor in the Database tab). Open the database Iris.sl3 in the editor and go to the Browse Data tab. Select 'mytable_out'. The third small button on the toolbar next to the table name is the 'Export to CSV' button.

8.3.2   Analysis of the Results

The plot of the two principal components (new columns or features) can be seen on Fig. 8.8. The retained variance in the new reduced dataset is 97.76% (reported by the AI-TOOLKIT). The three classes are very clearly separable in 2D with a few outliers. The choice for the reduction to two dimensions was for this reason.

The resultant reduced dataset in Table 8.7 (only a subset is shown) can be used to train a machine learning model instead of the original data.

This very simple example shows how PCA can be used to reduce the dimensionality of large datasets. There is of course a much greater number of dimensions in the datasets where PCA should be applied, but the principles and the application steps are the same!

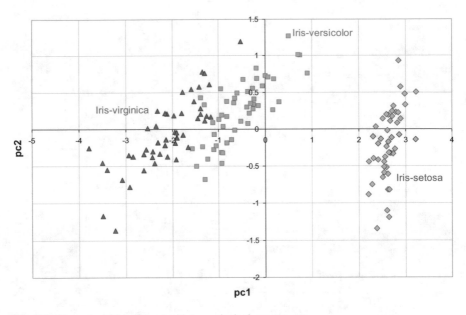

Fig. 8.8   The reduced Iris dataset with two principal components

Table 8.7 The reduced Iris dataset	Class	pc1	pc2
	0	2.68421	−0.32661
	0	2.71539	0.169557
	0	2.88982	0.137346
	0	2.74644	0.311124
	0	2.72859	−0.33393
	0	2.2799	−0.74778
	0	2.82089	0.082105
	0	2.62648	−0.17041
	0	2.88796	0.570798
	0	2.67384	0.106692
	0	2.50653	−0.65194
	0	2.61314	−0.02152
	0	2.78743	0.22774
	...	...	...

It is not always possible to reduce the dimensionality of the dataset without losing a significant amount of variance (knowledge) in the data and therefore you may decide to not apply PCA.

> TIP: PCA should only be applied when it is really necessary or useful because usually more knowledge in the data means better machine learning model! PCA reduces the knowledge (variance) in the data.

CASE STUDY 3

8.4 Recommendation AI

The aim of this case study is to explain how machine learning models work, which can be used for making automatic recommendations. An example recommendation system will also be explained step by step and in detail.

Making a recommendation means that we recommend one or more items to potential users. The recommended items may be many different things, for example, physical products which are being sold (e.g., cars, smart phones, etc.) or articles, web pages, documents, etc. There are many recommender systems in use today, for example, for recommending books, movies, clothing, holiday destinations, etc., and usually they help to increase revenue and/or help users to find the most relevant, interesting and/or important information or product.

As usual in this book we will focus on the understanding of the methodologies and algorithms used without going into too much detail. Complex mathematical algorithms will not be explained in detail but after reading this section you will understand how they work and why they are applied. This can be very valuable to beginners on this subject but even experts may find helpful information.

Furthermore, the easy application of state-of-the-art recommendation models using the AI-TOOLKIT will be explained. These models may be very useful to experts or to potential users of professional recommendation systems.

Recommendation machine learning models work with explicit and/or implicit feedback data collected from users while they are interacting with items (products, documents, etc.).

Explicit feedback is when the user provides some kind of rating or like/dislike of the items he/she is interacting with. There are many types of rating scales, for example, the five-star rating in which one star means low appreciation and five stars mean a very highly appreciated product. All of these ratings can be expressed on a numerical scale (e.g., 1,2,3,4 and 5; or 0 and 1 for dislike and like).

Implicit feedback is when the user does not directly provide some kind of rating but we collect information about user actions, for example, buying a product, viewing a document or web page, etc.

The basic principle behind how recommender machine learning models work is that *correlation exists between how different users appreciate similar items, how different items are appreciated by similar users and the combination of the two* (joint correlation). The user's appreciation is expressed with explicit or implicit feedback.

These correlations or behaviors can be learned by a machine learning model based on the collected explicit and/or implicit feedback data from the users (user + item + feedback).

There are two main types of state-of-the-art machine learning recommendation models:

- Collaborative filtering (CF)
- Content-based (CB)

There are other types of models and methods for creating recommendations, but these two are the best performing methods available today. There are also some recommendation methods which do not use machine learning but the user's choices to make a selection, we call these knowledge based recommendations.

As we will see in the following sections both models (CF and CB) have their advantages and disadvantages but in general we may say that CF models are much more powerful because they use data from all users, and eventually they may even be combined with some content information. CB models are usually not used alone but in combination with other methods and models for several reasons. For example, because they always tend to recommend items with known descriptions (properties) and thus new items with different descriptions are never recommended (the reason will be obvious after reading Sect. 8.4.2). CB models usually use *only* data from one user to make recommendations for that user, which is also a disadvantage as the data from other users is neglected.

We call the combination of several types of recommendation models *hybrid* models, which try to make use of the advantages of the different models. Such a combination may be very simple, for example, generating recommendations with several standalone recommendation models and then combining the results into a final recommendation list.

8.4.1   Collaborative Filtering (CF) Model

Collaborative filtering models use the explicit and/or implicit user feedback from all users to predict the rating of not yet rated items for all users. This model uses the collected feedback data from all users (collaborative), but it usually does not use content information (description of the items). When the predicted explicit (ratings) or implicit (user actions) feedback is known, a top-k number of recommendations can be made to any user.

8.4.1.1   The User's Feedback

Collaborative filtering models may use explicit and/or implicit feedback data in the form of a triplet consisting of a user ID, an item ID and the feedback value. These triplets form a three dimensional space which can be represented in a matrix or table.

Each row in the table represents a user, each column represents an item and the values in the table represent the feedback value. Table 8.8 shows the two types of feedback. The explicit feedback is the rating provided by the users. The implicit feedback is the indication of users' interaction with the items (for example viewing a product page); the value 1 indicates an interaction and 0 indicates no interaction. Implicit feedback may of course also count the number of user interactions (e.g., viewing, downloading, buying, etc.) but in this case it is just an indication of any interaction. The data (triplets) collected can be seen under the two tables—the first number is the user ID, the second number is the item ID and the last number is the rating.

The feedback matrix in Table 8.8 (sometimes called a utility matrix) contains a lot of empty cells. We call this a sparse matrix. These empty or unknown cells are the ratings which we want to predict. For example, we know that the user with ID 2 (user 2) likes the item with ID 1 (item 1) very much (rating = 5, blue cell) but we do not know what user 2 thinks about item 2 (green cell). If we can predict this then we can make a clever recommendation to the user about some new items for him/her.

Table 8.8 Explicit and implicit feedback matrix

		Items						Items		
ID	**1**	**2**	**3**	**4**		**ID**	**1**	**2**	**3**	**4**
1		2		3		1		1		1
2	5	?	1	2		2	1		1	1
3		1	2			3		1	1	
4	2			4		4	1			1
5		3	5			5		1	1	

(Users label on the left side of both tables)

Explicit Feedback	Implicit Feedback
The collected input data:	The collected input data:
1, 2, 2	1, 2, 1
1, 4, 3	1, 4, 1
2, 1, 5	2, 1, 1
2, 1, 3	2, 1, 1
...	...

There is an important difference between these two types of user feedback. Explicit feedback allows both negative and positive feedback (rating $= 1$ is, for example, strongly negative feedback and rating $= 5$ is strongly positive feedback or preference), but implicit feedback allows only positive feedback (we only count positive interactions with the items). In both types of feedback empty/unknown elements do not mean negative feedback!

8.4.1.2 How Does the Collaborative Filtering (CF) Model Work?

After the sparse user-item matrix is created, several techniques can be used to train a model which can be used to predict missing implicit or explicit feedback (the empty values in the matrix). The state-of-the-art method at the time of writing this book is the so-called *Latent Factor* method.

The latent factor method uses a combination of matrix factorization, dimensionality reduction and objective function optimization to approximate the feedback matrix (all values) with the product of two smaller matrices containing the so-called latent factors. The basic principle behind the method is based on the joint correlations between the different users and between the different items.

How does it work? Let us denote the input sparse feedback matrix with F. F can be decomposed into two smaller matrices U (which refers to users) and P (which refers to products/items). U and P are not sparse, they are completely filled in!

$$F \approx U \cdot P^T$$

Matrix U has k number of latent vectors (columns). The rows of U are called user latent factors. U can be thought of as the extension of F to the right (see Fig. 8.9). The T in the above equation is the transpose of the matrix (if we didn't transpose P then it would look like U and we would not get F by multiplying U and P!). This is a very basic matrix multiplication rule (if we have a matrix with size 5 x 3 and we multiply this matrix with another matrix with size 3 x 4 then we get a matrix of size 5 x 4; in this example there are five users and four items and $k = 3$).

Matrix P^T is very similar to U but it contains the item latent factors (and vectors).

The latent factor method estimates U and P, which then can be used to calculate the full (completely filled in) matrix F as it is shown on Fig. 8.9 ($F_{ui} = 7 = (1 \times 2) + (1 \times 1) + (2 \times 2)$). Please note that only the calculated values are shown on Fig. 8.9, all other values are not shown!

How are U and P estimated? The estimation of these two matrices can be performed in different ways. One of the methods is by minimizing the sum of squared errors on the estimates of the known values of F (known users' feedback).

Fig. 8.9 Feedback matrix factorization

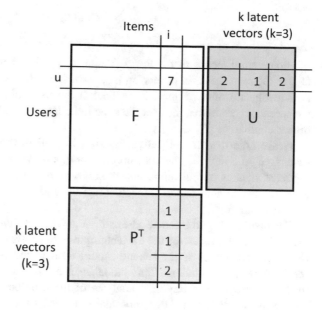

$$min\,O = \frac{1}{2} \sum_{n,m \in G} \left(F_{nm} - U_n P_m^T\right)^2$$

Where n is the user-wise index in F and U, and m is the item-wise index in F and P. '$n,m \in G$' in the above matrix equation means that we only take the values of n and m which are pointing to a given (known) rating (or feedback) in the original sparse input F matrix. We can of course only make calculations with the known values! A known value means that there is a user-item interaction recorded. The size of U depends on the chosen value of k (the number of latent vectors) and on the number of users. The size of P depends on the chosen value of k and on the number of items. Higher k values mean higher precision but slower processing. After a specific k value the optimization does not improve significantly anymore. The value of k is application dependent.

'O' in the equation is often called the *Objective Function* which must be minimized in this case. This minimization (optimization problem) can be performed, for example, with the so-called gradient descent family of algorithms (more information can be found in Sect. 2.2.2.3). Please note that for simplicity some terms (i.e., regularization and optional bias term) are not added to the above objective function, which is meant to prevent over-fitting (improve generalization).

In the case of implicit feedback this objective function is modified by adding confidence levels (C) and by binarizing (B) the implicit feedback values:

$$min\ O = \frac{1}{2} \sum_{n,m \in G} C_{nm} \left(B_{nm} - U_n P_m^T\right)^2$$

Binarization means that if $F_{nm} > 0$ then $B_{nm} = 1$, otherwise $B_{nm} = 0$. The term C_{nm} is the confidence in observing B_{nm}, which can be used to boost confidence in high feedback values (if implicit feedback is the count of positive events) since then we are more certain that the user likes the item. Higher feedback values will result in higher confidence (C_{nm}).

Please note that this objective function is used in the AI-TOOLKIT's WALS (weighted ALS) implicit model (more about this later). Where ALS means alternating least squares optimization, and the algorithm alternates between re-computing user-factors and item-factors (the latent factors) and each step lowers the value of the objective function.

The matrices U and P can also be estimated with a different objective function and method. The other state-of-the-art technique is the *Bayesian Personalized Ranking* (BPR) model, which is the second model used in the AI-TOOLKIT for implicit feedback recommendations. This model uses a so-called maximum posterior estimator derived from a Bayesian analysis of the problem (probabilistic model; see Expert Section 8.1) and it does not take into account the magnitude of the implicit feedback but just the binarized form (1 means feedback or interaction with the item = the user likes the item, and 0 means no feedback). If there is no feedback it does not mean dislike! The objective function is the area under the ROC curve (see Sect. 8.4.1.3 for information about ROC) which is maximized (gradient descent method; see Sect. 2.2.2.3).

Expert Section 8.1: Bayesian Analysis of the Recommendation Problem
The aim of this expert section is to help you understand how state-of-the-art recommendation models like Bayesian personalized ranking (BPR) work. Since Bayesian analysis is based on probability theory it is assumed that you know what probability and conditional probability mean (only very basic knowledge is needed).

Let us assume that we have collected explicit ratings with the value of 1, which means that the user likes an item, and with the value of 0, which means that the user dislikes an item. From the collected data we assemble a feedback matrix F with m rows representing users and n columns representing items. This is the explicit feedback matrix explained before. Let us index the users with the letter 'j' and the items with the letter 'i'.

Bayesian analysis is often the basis for classification models. In our case the user rows in F are the records in a classification dataset and the items (columns) are the features. This is exactly the same as we have seen in supervised classification (discussed in the first two chapters of this book) but in the case of a recommendation dataset there can be several missing entries

(continued)

Expert Section 8.1 (continued)

(unrated items) in any row which need to be classified. In normal supervised classification there is only one decision variable which needs to be classified instead of several.

The first step in the Bayesian analysis of the recommendation problem is the definition of the problem in probabilistic terms. Since the main aim is the classification of the missing entries, or in other words the estimation of the ratings, this is what we first need to define. Let us assume that the list of recommendations will be made based on the ranking of the ratings.

The probability that rating r_{ji} equals 1 (user j likes item i) with the condition that we know the rating values F_j of user j (row j) can be written as follows:

$$P(r_{ji} = 1 | \text{known ratings } F_j)$$

Please note that the vertical line in the expression means that the left side is evaluated with the condition defined on the right side.

We can say this in another way: we can estimate the probability of whether user j likes item i by analyzing other known ratings of user j.

The probability that user j dislikes item i, if other ratings of user j are known, can be expressed in a similar manner:

$$P(r_{ji} = 0 | \text{known ratings } F_j)$$

According to basic probability rules the sum of these two probabilities is 1.

Let us generalize these two equations by replacing the classification (rating) values 1 and 0 with c_L ($c_1 = 0$, $c_2 = 1$):

$$P(r_{ji} = c_L | \text{known ratings } F_j)$$

According to Bayes' rule of conditional probabilities we can simplify this expression which will take the following form:

$$P(A/B) = \frac{P(A) \cdot P(B/A)}{P(B)}$$

Our probability equation can then be written as follows:

(continued)

Expert Section 8.1 (continued)

$$P(r_{ji} = c_L/knownratingsF_j) = \frac{P(r_{ji} = c_L) \cdot P(knownratingsF_j/r_{ji} = c_L)}{P(knownratingsF_j)}$$

Since we are looking for the rating c_L (like or dislike) which maximizes the above probability (we want to determine the value of c_L which has the highest probability) and because the denominator does not depend on c_L we can eliminate the denominator! We can then estimate the rating of item i by user j ($\hat{r}_{ji}$) as follows:

$$\hat{r}_{ji} = argmax_{c_L} P(r_{ji} = c_L/knownratingsF_j)$$
$$= argmax_{c_L} \left[P(r_{ji} = c_L) \cdot P(knownratingsF_j/r_{ji} = c_L) \right]$$

Where $argmax_{cL}$ is an operation (function) that is looking for the value of c_L which maximizes the expression.

$P(r_{ji} = c_L)$ in the above equation is often called the prior-probability—the probability that user j rated item i with the rating c_L, which is the fraction of the users who specified the rating c_L for item i. And this leads us to how probabilities are calculated:

$$P(r_{ji} = c_L) = \frac{count^i_{c_L}}{\sum_{l=1}^{L} count^i_{c_l}}$$

where $count^i_{cL}$ is the number of users who specified the rating of c_L for item i, and the denominator is the number of users who specified any rating. For example, if we have nine users and we want to calculate the probability that the tenth user will rate an item with 1, then we count the number of users who rated the same item with 1 and divide by the number of users who specified any rating for the same item. Let us say that (from the nine users) eight users rated the same item and three of these users rated the item with 1. The probability that the tenth user will rate the same item (i) with 1 is then P $(r_{10i} = 1) = 3/8 = 0.375$. There is a 37.5% chance that the tenth user will rate item i with 1!

In practice the above equation is often modified with a correction term (called Laplacian smoothing) in order to prevent over-fitting:

$$P(r_{ji} = c_L) = \frac{count^i_{c_L} + \alpha}{\sum_{l=1}^{L} count^i_{c_l} + L\alpha}$$

where alpha is a constant, often chosen to be 1.

(continued)

Expert Section 8.1 (continued)

Let us go back to our final probability equation:

$$\widehat{r}_{ji} = argmax_{c_L}\left[P(r_{ji} = c_L) \cdot P(knownratingsF_j/r_{ji} = c_L)\right]$$

Remember that $argmax_{c_L}$ is an operation (function) that is looking for the value of c_L which maximizes the expression—we are looking for the rating of item i with the highest probability. We know now how to calculate the first term by using the collected input rating data. The second term can also be easily calculated by using the input data as follows:

$$P(knownratingsF_j/r_{ji} = c_L) = \prod_{k \in F_j}P(r_{jk}/r_{ji} = c_L)$$

The expression on the right means that we are multiplying several probabilities together which results in the fraction of users who rated the kth item (any rating value) with the condition that the same user is rated the ith item with c_L. The multiplication operation comes from basic probability theory and will become clear when we go through the example presented below. Basically we are looking for the rating behavior of the users; if one user rated an item (any item) in the same way as the user for whom we want to determine the rating, then there is a high probability that this user will rate the selected item in a similar manner.

The final probability equation which can be calculated with the collected input rating data is as follows:

$$\widehat{r}_{ji} = argmax_{c_L}\left[P(r_{ji} = c_L) \cdot \prod_{k \in F_j}P(r_{jk}/r_{ji} = c_L)\right]$$

Next we will look at a simple example and will go through all of the calculation steps. We will follow a simplified calculation procedure in order to make each step as clear as possible. The optimization of the above equation (extended with some other terms) in real problems is a task for an optimization algorithm but the principle is the same.

Bayesian Analysis Example

The input data with the collected ratings is shown in Table 8.9. There are seven users and five items. We want to decide if we should recommend item 4 (I4) to user 7 (U7). The ratings have the same meaning as in the explanation above, 1 means that the user likes the item and 0 means that the user dislikes the item.

(continued)

Expert Section 8.1 (continued)

Let us first calculate the probability that user 7 (U7) likes item 4 (I4) as shown in Table 8.10.

All values are calculated in a similar manner. Let us go through the first column. The value 3 is the count of 0's where the user dislikes I1 but likes (=1) I4. We count these 0's because user 7 rated I1 with 0! The value 4 comes from counting all existing ratings equal to 0 because user 7 rated I1 with 0. By dividing the value 3 by the value 4 we get 0.75. What we are doing here is looking at the rating behavior of other users and comparing it to user 7's rating behavior!

The column with the ratings for I4 (fourth column) is special because this is the item for which we want to give a recommendation. This is the prior probability in the final probability equation $P(r_{ji} = c_L)$. We calculate this by counting the number of likes (=1) and dividing this number by the total number of existing ratings. There are four 1's and in total 6 ratings, $4/6 = 0.6667$.

We do the same for all columns and then we multiply all probabilities to get the probability of user 7 liking item 4.

The probability that user 7 dislikes item 4 can be calculated in a similar manner as shown in Table 8.11.

If we normalize the calculated probabilities in order to satisfy the basic rule that the sum of these probabilities (for like and dislike of item 4 (I4) by user 7 (U7)) must be equal to 1, then we get a 98% probability that user 7 likes item 4 and a 2% probability that he/she dislikes it. This is calculated as follows: like I4 = 98% = 100 * 0.15 * (1/(0.15 + 0.003125)) and dislike I4 = 2% = 100 * 0.003125 * (1/(0.15 + 0.003125)).

This is the simplest form of Bayesian analysis of the recommendation problem. There are many different more complex versions of the algorithm with different objective functions. For example, the Bayesian personalized ranking (BPR) model in the AI-TOOLKIT is based on a similar principle, but it first creates user specific (personalized) pair-wise preferences between a pair of items (in the form of a table). Each value in the pair-wise preferences table indicates whether a user prefers item i over item j. The Bayesian analysis is then applied to these tables per user. The objective function is also different because the area under the ROC curve is maximized. It is not the aim of this book to explain this algorithm in detail, please refer to the introductory article about BPR in [30].

8.4.1.3 Evaluating the Accuracy of the Recommendation Model

The main aim of making recommendations is to present a list of recommended items to the user. We call this list the top-k recommendation list where k is chosen according to the requirements of the application. We often calculate the missing

Table 8.9 Bayesian Analysis Example Part I

Items / Users	I1	I2	I3	I4	I5	Like I4?
U1	1	0	1	0	0	Dislike
U2	1	0	0	1	0	Like
U3	0	1	0	0	0	Dislike
U4	0	0	0	1	1	Like
U5	0	1	0	1	0	Like
U6	0	0	0	1	1	Like
Recommend U7	0	0	0	?	0	?

Table 8.10 Bayesian Analysis Example Part II

LIKE	Like I4 if I1 = 0?	Like I4 if I2 = 0?	Like I4 if I3 = 0?	Like I4?	Like I4 if I5 = 0?	
Count this rating where I4 = 1	3	3	4	4	2	
Count all known ratings equal to this one	4	4	5	6	4	
P	P1(I1=Dislike|Like)	P2(I2=Dislike|Like)	P3(I3=Dislike|Like)	P4(I4=Like)	P5(I5=Dislike|Like)	P1*P2* P3*P4* P5
	0.7500	0.7500	0.8000	0.6667	0.5000	0.1500
Note	The probability that the user dislikes I1 but likes I4. U7 dislikes I1 (=0)!	The probability that the user dislikes I2 but likes I4. U7 dislikes I2 (=0)!	The probability that the user dislikes I3 but likes I4. U7 dislikes I3 (=0)!	The probability that the user likes I4	The probability that the user dislikes I5 but likes I4. U7 dislikes I5 (=0)!	

ratings in the feedback matrix first and then deduce the top-k list by ranking the items according to their rating. We can evaluate the accuracy of both the estimated ratings and also the estimated top-k list.

There are also some other logical requirements for a top-k recommendation list; for example, the recommended items should not be known by the user (no recorded rating or interaction) and they should be diverse and different from what the user already knows, etc. These requirements may also be evaluated but we will concentrate on the accuracy of the ratings and the top-k recommendation list in this section.

The evaluation of the accuracy of a recommendation model is very similar to the evaluation of the accuracy of supervised regression models. The accuracy of

Table 8.11 Bayesian Analysis Example Part III

DISLIKE	Dislike I4 if I1 = 0?	Dislike I4 if I2 = 0?	Dislike I4 if I3 = 0?	Dislike I4?	Dislike I4 if I5 = 0?	
Count this rating **where** **I4 = 0**	1	1	1	2	3	
Count all known ratings equal to this one	4	4	5	6	4	
P	P1(I1=Dis-like|Dislike)	P2(I2=Dis-like|Dislike)	P3(I3=Dis-like|Dislike)	P4(I4=Dis-like)	P5(I5=Dis-like|Dislike)	P1*P2*P3*P4*P5
	0.2500	0.2500	0.2000	0.3333	0.7500	0.003125
Note	The probability that the user dislikes I1 and dislikes I4.	The probability that the user dislikes I2 and dislikes I4.	The probability that the user dislikes I3 and dislikes I4.	The probability that the user dislikes I4.	The probability that the user dislikes I5 and dislikes I4.	Dislike

predicted ratings can be estimated with the root mean squared error (RMSE) which can be calculated with the following well known equation:

$$RMSE = \sqrt{\frac{\sum e^2}{n}}$$

The term 'e' (error) is the deviation of the predicted ratings from the original known ratings in the input dataset ($\hat{r}_{ji}-r_{ji}$), and n is the number of deviations we calculate. This can, of course, only be calculated for the ratings that are given in the input dataset. Lower RMSE values mean better accuracy!

The accuracy of the top-k recommendation list can be estimated by comparing the order of the items in the predicted top-k list with the order of the same items in the original input dataset (ordered according to the known ratings). This comparison can be performed, for example, by calculating the correlation between the predicted and original recommendation lists. It is also possible to calculate the top-k ranking correlation for all users and then average the results to get the global correlation coefficient. For example, if the order of the elements is identical in the predictions and in the original dataset, ranked according to the known ratings, then the correlation coefficient will be equal to 1. In practice, a more sophisticated approach is used, but the principle is the same. For example, some methods may combine the rating and the ranking measures and weight the results according to the magnitude of the rating (utility based methods) because the order of the highly rated elements is more important than the order of the poorly rated elements.

Another method for estimating the user specific accuracy of the top-k recommendation list is by calculating the precision-recall-FPR metrics and the area under the ROC curve (see also Chap. 3 for more information).

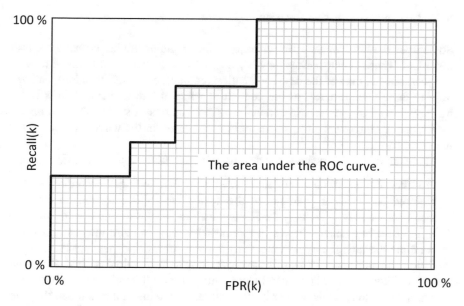

Fig. 8.10 Example ROC curve

The precision is the number of well identified items in the top-k list divided by the number of recommendations (k). Higher values are better and the precision ranges from 0 to 1.

The recall is the number of well identified items in the top-k list divided by the total number of recommendations given by the user. Higher values are better.

The false positive rate (FPR) can be calculated by dividing the number of wrongly identified items in the top-k list by the total number of items the user had no interaction with (no rating is given in the input). FPR can also be described as the fraction of the unrated items (input dataset) which were recommended (i.e., the erroneous recommendation of irrelevant items according to the input dataset). Remember that we base our accuracy calculation on the original known ratings! Lower values are better.

The so-called ROC curve is the plot of the FPR(k) values (horizontal axis) against the Recall(k) values (vertical axis) for varying values of k (see Fig. 8.10). The area under the ROC curve, often called the AUC, may be used to evaluate the accuracy of the recommendation model, and it is also often used as an objective function since larger AUC values mean better accuracy (maximizing optimization). The AUC takes values between 0 and 1 (with 1 indicating the best accuracy). A ROC curve can be created per user and also for all users (all metrics are calculated per user and then combined).

8.4.1.4 Context-sensitive Recommendations

Sometimes the circumstances in which recommendations are being made are also important. These circumstances may be, for example, a specific time frame (day of the year, season, time of the day), or location (country, city, GPS location), etc. We refer to these circumstances as the context in which the recommendation is being made. For example, in the case of a clothing merchant the season of the recommendation may be important since some clothes are meant for the winter and others for the summer, etc.

There are two main methods for dealing with necessary contextual information:

1. Pre-filter the input data (collected feedback, e.g., ratings) according to the required context. For example, in the case of the above mentioned clothing merchant we may select only the items which are meant for the specific season.
2. In the case of smaller datasets (per context) it may be more interesting to first combine the ratings for all contexts to make a context independent recommendation and then make a final selection according to the required context. For some types of items there will be stand-alone ratings for each item in each context (for example, clothing in the winter is different from clothing in the summer, they are different items), but in many cases the items will be the same in each context (for example, in some location dependent recommendations). If some items are the same in each context, then we must combine their ratings into one rating. There is a difference between combining explicit feedback values and implicit feedback values. Explicit feedback (ratings) may be best averaged through all contexts for the same item, whereas implicit feedback (e.g., the count of events) may be best summed through all contexts for the same item for obvious reasons.

8.4.1.5 Incorporating Item Properties and User Properties

In collaborative filtering recommendation models we normally do not use the properties (description) of the items and the users but only the explicit or implicit feedback provided by the users. In content-based recommendation models (see Sect. 8.4.2), however, we use the description of the items (item property) in combination with the feedback from the user to generate recommendations for that one user, but only the feedback from one user is used.

It could be interesting to incorporate item and/or user properties, for example, in the form of keywords from the description of the items into our collaborative filtering model which uses the feedback from all users. Why would this be interesting? Because then we could provide more information which could improve the accuracy of the model, or we could reduce the cold start problem (not yet having enough ratings).

Item properties could be, for example, the color of an item, the style of a book, music, clothes, etc. User properties could be, for example, the gender of the user, age range, profession, etc.

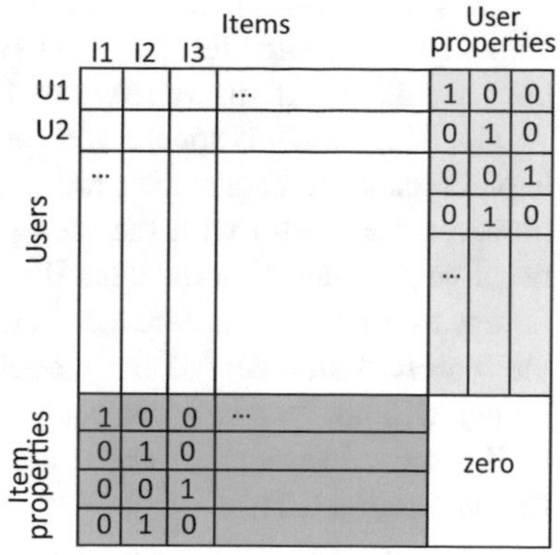

Fig. 8.11 Item properties
and user properties added to
the feedback matrix

There is a simple way of doing this by extending the feedback matrix with the chosen properties. Item properties may be added as new rows to the feedback matrix and user properties as new columns (see Fig. 8.11). If user properties are also added to the feedback matrix then we must make sure that these items are not selected as recommended items! You could, for example, request more recommendations and remove the user properties if they exist.

Properties may be added as binary data, where 1 means that the property belongs to the item and 0 means that it does not belong to the item.

8.4.1.6 AI-TOOLKIT Collaborative Filtering (CF) Models

The AI-TOOLKIT supports several types of CF models which are optimized for explicit or implicit feedback.

The explicit feedback models are alternating least squares (ALS) models (see Sect. 8.4.1.2) with two types of optimization algorithms:

- Batch singular value decomposition (SVD) learning
- Regularized SVD with stochastic gradient descent optimizer

The algorithms predict user ratings for unknown user-item pairs and provide a list of recommendations which does not include items already rated by the user.

The implicit feedback models are the following:

- Weighted ALS (WALS)
- Bayesian personalized ranking (BPR) with stochastic gradient descent optimizer

These algorithms predict the preference of the user with regard to the item. Higher values mean higher preference. They also provide a list of recommendations which does not include items already rated by the user.

The BPR model is specialized for implicit feedback, but it may also be used with explicit feedback data since a rating indicates an interaction with an item. One must, however, be careful with the rating values because low rating values may mean a negative preference for the item (dislike) but the implicit model would take these low ratings as a positive preference (like)! In order to prevent this problem you can use the 'cutoff' parameter of the model and set it to a rating value below which the ratings will not be taken into account!

For more information about these algorithms please refer to Sect. 8.4.1.2 and Expert Section 8.1.

8.4.2   Content-Based (CB) Recommendation Model

In content-based models the description of the items (called content) together with explicit or implicit user feedback are used to generate recommendations. The basic principle here is that we search for similar items (based on the description) to those already highly appreciated (high rating) by the user. This model uses data from only one user for making the recommendations!

The following steps summarize how content-based recommendations work:

- We select keywords from the description of each item. This is often called feature selection or extraction. Some keywords may be the same for some items (overlapping). For example, if we recommend articles, some articles may contain the same keywords which make them a bit similar.
- We create a training dataset by first converting each keyword to a numerical value (e.g., by using the keyword's frequency of occurrence; more about this later!). Then each item can be represented by putting the numerical value of each keyword into a separate column (these are the features in the dataset). We know exactly which keyword belongs to which column. We call this a vector representation of each description, which is often used in information retrieval tasks. Finally we add one more column with the numerical value of the user's feedback (e.g., rating) of the specific item.
- We use a supervised classification or regression model to learn the user behavior (model training). The rating (or implicit feedback) column is the decision variable (label or class).
- The trained model can then be used to predict the user's rating (decision) of any item with similar keywords.

The basic principle in content-based recommendation models is that items with similar keyword combinations (attributes) will have a high correlation per user which is expressed by a similar rating.

The trained content-based recommendation model is often called the 'user profile' because it represents the rating behavior of the user concerning the different attributes of the items (keywords).

We can use the built-in supervised machine learning models in the AI-TOOLKIT for learning content-based recommendation models. Because these models are extensively explained in the other chapters of this book, we will not discuss these here in more detail.

The content-based recommendation model has several disadvantages compared to the collaborative filtering model:

- New items with unknown keywords in their descriptions will not be recommended because the model first needs to learn how the user behaves—how he/she rates items with similar keywords.
- The model works with one user at a time and therefore it must first learn the user's behavior. New users will not be able to get recommendations. This is the so-called cold-start problem.

For the above reasons content-based models are mostly used in combination with collaborative filtering models. We call such systems hybrid recommendation models, which try to make use of the advantages of several different models.

8.4.2.1 Feature Extraction and Numerical Representation

The most important parts of content-based recommendation models are feature extraction and how these features (keywords) are converted to a numerical representation, because if we do not select the right features or do not adequately convert them into a numerical representation, then the training of the model will fail or the model will provide poor recommendations.

Feature extraction is application specific and must be performed with domain specific knowledge. The importance of keywords and item attributes is application dependent. For example, in the case of a clothing merchant 'red shirt' could be very important but in other types of application it may be unimportant. It may also be possible to use a natural language processing (NLP) system to learn the relative importance of words in an item description automatically.

How the numerical representation of a keyword is calculated is also application dependent. If the keyword refers to attributes then these may be binary encoded with 1 for the existence of the attribute and 0 for non-existence. For example, if the attribute is color, then if the item has that color we set the value to 1, otherwise to 0. In other cases when longer item descriptions (e.g., documents) are used, the so-called TF-IDF method may be used.

TF-IDF Representation of a Keyword

TF refers to the term frequency which is *the occurrence count of the keyword in the description of an item divided by the total number of words in the description. IDF* refers to the inverse document frequency which is *the total number of descriptions divided by the number of descriptions in which the keyword appears*. Instead of the word 'description' we may also use the word 'document' in all of these definitions.

The numerical representation of a keyword can be calculated by multiplying the TF and IDF values. The reason for multiplying with IDF is that in this way we can down-weight frequently occurring words which appear in all descriptions! For example, if there are 10 documents and the word 'long' appears in all descriptions then its IDF value is $10/10 = 1$, but if another word, for example 'shirt', only appears in one document then its IDF value is $10/1 = 10$. This means that the numerical representation of 'shirt' will be much higher than that of 'long'. IDF normalizes the numerical representation according to importance.

In order to dampen the effect of very high occurrence counts we usually transform the TF and IDF values with a function which has a damping effect, for example, the square root or logarithm function. Damping means that high values are reduced; for example, if we have two occurrence counts, 10 and 100, and we take the logarithm of both of them, then we get a much smaller difference between these two counts because $\log(10) = 1$ and $\log(100) = 2$.

The final numerical representation of a keyword w_i, which we could call normalized document frequency (OFN), with all of the above transformations and normalizations, can be written as follows:

$$OFN(w_i) = \log_{10}(TF_i) \cdot \log_{10}(IDF_i)$$

In some literature you may also find $[1 + \log_{10}(TF_i)]$ instead of just $\log_{10}(TF_i)$. You may experiment with any type of damping function you think may be useful in your application.

8.4.3   Example: Movie Recommendation

In this section we will look at a complete movie recommendation example using the AI-TOOLKIT. We will use the MovieLens "ml-20 m" dataset available at http://movielens.org. This dataset describes 5-star rating activity (explicit user feedback) from MovieLens, a movie recommendation service. It contains 20,000,263 ratings across 27,278 movies (titles + several genres per movie). The data were created by 138,493 users between January 09, 1995, and March 31, 2015. The data in the files are written as comma-separated values with a single header row. The users were selected at random for inclusion. Their IDs have been anonymized. Only movies with at least one rating are included in the dataset.

All ratings are contained in the file 'ratings.csv' with user ID, movie ID, rating and timestamp. Ratings are made on a 5-star scale, with half-star increments (0.5

Table 8.12 Ratings example

userID	movieID	rating	timestamp
1	2	3.5	1112486027
1	29	3.5	1112484676
1	32	3.5	1112484819
1	47	3.5	1112484727
1	50	3.5	1112484580
1	112	3.5	1094785740
1	151	4	1094785734
1	223	4	1112485573
1	253	4	1112484940
...	...	...	...

stars–5.0 stars). We will only use the user ID, movie ID and rating, and not the timestamp which will need to be removed. Just as a side note about the timestamp, how could we use this extra information? It gives us the *context which* we discussed in Sect. 8.4.1.4; however, to keep this example simple, we will not include it here.

The first ten lines from the ratings file look like as it is shown in Table 8.12.

After reading this section you will be able to make movie recommendations to yourself or to others. The only thing which is needed for this is to add your ratings of several films to the database (with a new user ID) and then you will be able to recommend films from this database (> 27,000 films). You can of course also further extend the database. You will also be able to use the AI-TOOLKIT to make professional recommendations to clients.

8.4.3.1 Step 1: Create the Project File

The project file structure is very similar to the other project files used in the AI-TOOLKIT (see Table 8.13). Please note that all comments are removed from the project file for simplicity. If you are new to the AI-TOOLKIT then please read Sect. 9.2.2 first!

This project file was created almost completely automatically by the AI-TOOLKIT. The steps to reproduce this project file are as follows:

1. Start a New Project.
2. Use the Insert ML Template command on the AI-TOOLKIT tab.
3. Go to the Applications section in the dialog window which appears and select the "Recommendation with Explicit Feedback (Collaborative Filtering (CFE))" template and then the Insert command.
4. You will have to change some of the parameters such as the path and database table names. This will be explained in the following sections!
5. Save the project file.

Table 8.13 Movie Recommendation AI-TOOLKIT Project

```
model:
  id: 'ID-LACerqbang'
  type: CFE
  path: 'ml20m.sl3'
  params:
    - Neighborhood: 5
    - MaxIterations: 100
    - MinResidue: 1e-5
    - Rank: 0
    - Recommendations: 10
  training:
    - data_id: 'ratings'
    - dec_id: 'decision'
  test:
    - data_id: 'ratings'
    - dec_id: 'decision'
  input:
    - data_id: 'input_data'
    - dec_id: 'decision'
  output:
    - data_id: 'output_data'
    - col_id: 'userid'
    - col_id: 'itemid'
    - col_id: 'rating'
    - col_id: 'recom1'
    - col_id: 'recom2'
    - col_id: 'recom3'
    - col_id: 'recom4'
    - col_id: 'recom5'
    - col_id: 'recom6'
    - col_id: 'recom7'
    - col_id: 'recom8'
    - col_id: 'recom9'
    - col_id: 'recom10'
```

8.4.3.2   Step 2: Create the AI-TOOLKIT Database and Import the Data

As a pre-processing step you may create a new CSV file from 'ratings.csv' (included with the movielens ml-20 m data) by removing the timestamp column. This can only be done with software which allows the editing of so many records of data. If you are not able to do this then import all data into the database and use the built-in database editor to remove the timestamp column.

Go to the Database tab and use the 'Create New AI-TOOLKIT Database' command. Indicate the path and name of the database file. In this example we use 'ml20m.sl3' as the database name. Next use the 'Import Data into Database' command and follow the steps below in order to import the training data:

- In the Import Data dialog window which appears select the delimited data file to import with the SEL button. This is the original 'ratings.csv' or your modified version.
- In the Delimiters field add a comma after the '\t' sign. It will look like this: '\t,'. The ratings.csv file is comma separated. You may also remove the tab separation sign ('\t') if you wish but this is not necessary.
- Indicate that there is one header row in the Header Rows field.
- Indicate the zero based index of the decision (rating) column. In ratings.csv the rating column is the third column and thus the zero based index is 2!
- Select the newly created database 'ml20m.sl3' with the SEL button.
- Change the New Table Name field to 'ratings'.
- Use the 'Import' command to import the data.

The 'Import Data Into Database' module is also explained in Chap. 9!

As a post processing step we must remove the timestamp column from the database and at the same time we can also check that everything is correct.

Next, on the Database tab use the 'Open Database Editor' command and the built-in database editor will open. Please check Chap. 9 for more information about the database editor!

Use the 'Open Database' command in the editor and navigate to the above created 'ml20m.sl3' database and open it. *Click on the ratings table* on the Database Structure tab to expand all fields. You will see four columns: decision, col1, col2 and col3. Click on the 'Modify Table' command (this is only visible if you select the table name), select the field named 'col3' (this is the last column with the timestamp data) and use the 'Remove Field' command. Finally click OK. The result will appear as is shown on Fig. 8.12 except that the 'input_data' and 'output_data' tables will not yet exist.

Next go to the Browse Data tab and select the 'ratings' table (drop down control with a small arrow on it) and check the data. It may take a few seconds to show the data because there are more than 20 million records! The editor is very fast.

You may also check the other automatically created table which will be used to store the trained machine learning model. Make sure that you do not modify the table!

IMPORTANT: We do not create the 'input_data' and 'output_data' tables because they will be created automatically by the AI-TOOLKIT. You must make sure that when you train the recommendation model you click *yes* when asked whether to create these tables automatically!

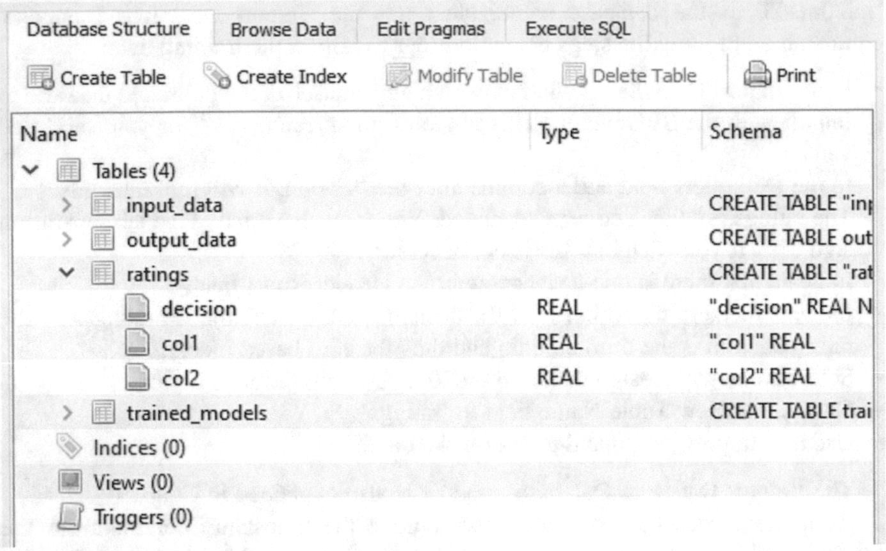

Database Structure	Browse Data	Edit Pragmas	Execute SQL	
🗐 Create Table	🗞 Create Index	🖺 Modify Table	🗐 Delete Table	🖨 Print

Name	Type	Schema
∨ 🏢 Tables (4)		
> 🏢 input_data		CREATE TABLE "in
> 🏢 output_data		CREATE TABLE out
∨ 🏢 ratings		CREATE TABLE "rat
🗋 decision	REAL	"decision" REAL N
🗋 col1	REAL	"col1" REAL
🗋 col2	REAL	"col2" REAL
> 🏢 trained_models		CREATE TABLE trai
🗞 Indices (0)		
🖵 Views (0)		
🗐 Triggers (0)		

Fig. 8.12 Database editor

8.4.3.3   Step 3: Finalize the Project File

First of all, change the path parameter in the project file to 'ml20m.sl3'. There is no need for an absolute path if the project file and the database are in the same directory.

Then change the maximum number of iterations (MaxIterations) to 100 and the number of recommendations (Recommendations) to 10.

In the **'training:' section** change the 'data_id' to the name of the database table containing the training data—in this example this is called 'ratings'.

In the **'test:' section** change the 'data_id' to the name of the database table containing the training data ('ratings') as above. We will use the same data for testing for simplicity.

In the **'input' section** change the 'data_id' to 'input_data'.

Lastly, change the 'data_id' to 'output_data' in the **'output' section**. Next change the third 'col_id' from 'decision' to 'rating'. And finally add nine more recommendation columns with col_id's: recom2, recom3... see Table 8.13. All fields must have a unique name! Please note that the order of the output columns is very important for this model. The first col_id must be the user ID, then the item ID, then the rating and finally all of the recommendations in separate columns (if recommendations are requested)! The template guides you through this requirement.

IMPORTANT: There are some **important requirements** concerning the data (training, test, input and output) for collaborative filtering models! This model requires that the data is in a specific order in the database tables (training, test and input): **The first column must be the rating (decision), the second column must be the user ID and the third column must be the item ID** (i.e., movie ID in this example)! The decision (rating) column will be automatically the first column if you import the data; therefore, when you import your data make sure that you select the right decision column in the import module! After importing the data the second and third columns must be the user ID and item ID in this order! Therefore, make sure that you always have the user ID first and then the item ID in the original CSV file, and that there are no other columns in between or before!

The effective user ID's and item ID's (the values) in the test and input datasets must all exist in the training dataset! Only users and items that were present in the original training dataset can be tested or predicted/recommended!

The **output data table** may be the same as the input data table but you must make sure that the recommendation columns are defined if recommendations are requested! The contents of the recommendation columns are not important because they will be overwritten. The user ID and item ID columns in the output definition will not be used if the input data table is chosen as output (because there is already a user ID and item ID column in the input) but they must be defined in the output section!

The order of the output columns is very important for this model. The first col_id must be the user ID, then the item ID, then the rating (decision) and finally all of the recommendations in separate columns! The only exception to this rule is when the input data table is the same as the output data table because then the order of the columns is: decision column (rating), user ID, item ID and finally the recommendation columns (if recommendations are requested).

In all tables, except for the output data table, the 'col_id' definitions (column selection) may be omitted and then all columns are used in the data table.

The data in the **input data table**'s decision (rating) column must be empty or NULL. This signals the program to evaluate that record! The recommendation columns (if they exist in the input because it is also used as output) may contain data but will be replaced.

8.4.3.4   Step 4: Train the Recommendation Model

IMPORTANT: We did not create the 'input_data' and 'output_data' tables because they will be created automatically by the AI-TOOLKIT. You must make sure that when you train the recommendation model you click *yes* when asked whether to create these tables automatically! You may of course also create these tables manually before training the model, but it is much easier to let the software do it for you.

When you click the Train AI Model button, the software will check the project file and database for inconsistencies and check that all database tables exist. The training data table is required and the model training cannot proceed without it. If the input and output data tables do not exist then the software will offer to create them. The database column descriptions must be well defined in the project file!

Each step in the model training process will be logged in the output log. When the training of the recommendation model is ready, the model will be saved into the database together with the root mean squared error (RMSE). The model is now ready for making recommendations!

You can check the saved model and model parameters in the 'trained_models' table in the database as shown on Fig. 8.13.

The accuracy of the model (RMSE = 0.772) is very good if we consider that the main aim of the model is not to predict the real rating values but the order of the ratings and that there are a huge number of data points (calculated on the whole training dataset of more than 20 million ratings). The training process takes about 5–6 min on an Intel Core i5–6400 @2.7 GHz PC, which is very fast.

8.4.3.5   Step 5: Make a Recommendation

In order to make recommendations we must first add at least one input record to the 'input_data' table, which was created automatically when the model training process

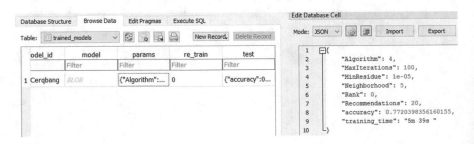

Fig. 8.13 Database Editor: 'trained_models' table

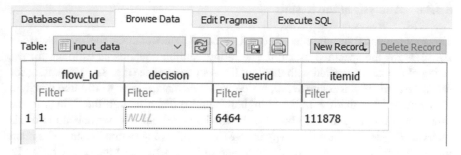

Fig. 8.14 Database Editor: 'input_data' table

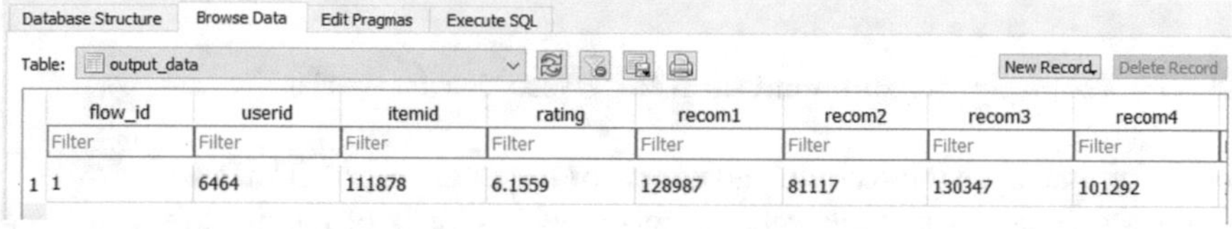

Fig. 8.15 Database Editor: 'output_data' table

started. Open the database editor on the Database tab, then open the database 'ml20m.sl3' and navigate to the 'input_data' table as shown on Fig. 8.14.

Add a new record by using the 'New Record' command, and then fill in the User ID and Item ID of your choice as above. Leave the decision value as NULL.

The software will only evaluate the record if the decision column (which corresponds to the rating column in an explicit collaborative model) is empty or NULL! You can set the decision cell to NULL at any time by first selecting it and then using the 'Set as NULL' command (top-right button) followed by the 'Apply' command right-middle button.

The 'flow_id' column is the identifier of each input record and therefore may not be removed (it is set automatically to a unique value)!

It is also possible to add new records to the 'input_data' table with another type of software or script automatically and let the ML Flow module make the recommendations automatically/continuously. Please read Sect. 9.2.5 on ML Flow for more information.

After you have completed the above steps use the 'Write Changes' command in order to save the database.

Next go back to the AI-TOOLKIT tab and click the 'Open AI-TOOLKIT Editor' command to reopen the project file. If you click the 'Predict with AI model' button the software will load the trained model and make all open recommendations (where the decision value is empty or NULL). The results can be seen in the database in the automatically created 'output_data' table as shown on Fig. 8.15.

You may of course also read out the results with another type of software or script from the AI-TOOLKIT database automatically (SQLite database).

8.4.3.6   Analysis of the Results

The recommendation columns (recom1, recom2, etc.) contain the recommended item IDs for the specified user ID! The rating column contains an estimate for the rating of the item specified with the item ID. As you can see (Fig. 8.15) the rating for this user-item ID pair is above 5.0, which indicates that the user would like this item very much. Note that the rating is higher than 5.0 (even though the ratings in the input are in the range of 1–5) because the model results are not normalized into this interval. This is also the reason why the RMSE of 0.7 is a very good result! The main aim of the recommendation model is to determine which items the user would like more than others and not what the exact rating would be in the range of 1–5. The ratings could, of course, be normalized into this interval.

8.4.3.7   Example: Recommendation for User with ID 46470

Let us fully analyze the recommendations for a specific user with ID 46470.

But first let us look at the statistics to see which kinds of movies this user rated with 5 stars. This user rated 4094 movies in total, from which there are 213 movies with 5 stars with the characteristics shown in Table 8.14 (some movies may span several genres).

Table 8.14 User ID46470 5 stars Movie Statistics

Genre	Count
Comedy	93
Drama	70
Action	68
Romance	55
Thriller	46
Sci-Fi	41
Animation	37
Crime	36
Fantasy	26
Children	22
Documentary	15
Mystery	13
Musical	10
Western	3

Table 8.15 Recommendations

Movie Title	Genres	How another user rated this movie? (*user ID→rating*)
Otakus in Love (2004)	Comedy|Drama|Romance	122868 → 4.5
Kevin Smith: Too Fat For 40 (2010)	Comedy	74142 → 5.0
Sierra, La (2005)	Documentary	95614 → 5.0
Summer Wishes, Winter Dreams (1973)	Drama	30317 → 5.0
Between the Devil and the Deep Blue Sea (1995)	Drama	114009 → 5.0
Smashing Pumpkins: Vieuphoria (1994)	Documentary|Musical	31122 → 5.0
Kevin Smith: Sold Out—A Threevening with Kevin Smith (2008)	Comedy|Documentary	74142 → 5.0
Abendland (2011)	Documentary	3127 → 5.0
Tales That Witness Madness (1973)	Comedy|Horror|Mystery|Sci-Fi	46396 → 5.0
Rewind This! (2013)	Documentary	74142 → 5.0

The recommendations for this user generated by using the AI-TOOLKIT (as explained above) can be seen in Table 8.15.

Nearly all of the recommendations are in the top five genres that the user liked most and all of the films were rated highly by other users. This seems to be a very good set of recommendations, and it is based only on ratings data collected from users. We have not used any movie genre, description or user properties and we seem to have found the right sort of films to be recommended. Here you can see the power of collaborative filtering and the value of good data!

CASE STUDY 4

8.5 Anomaly Detection and Root Cause Analysis

Detecting anomalies, and finding the root cause of the anomaly, is an important application in the field of machine learning in nearly all sectors and disciplines. An anomaly may mean different things in different applications, for example, fraudulent use of credit cards or suspicious transactions in the financial sector, a specific disease or the outbreak of a disease in healthcare, the signs of intrusion in a computer network, a fault in a production system or product in the manufacturing industry, an error in a business process, etc.

Machine learning discovers patterns in the data and therefore it is well suited for discovering unusual patterns which are the signatures of anomalies.

Anomaly detection can be combined with root cause analysis (RCA) in a machine learning (ML) model, or ML anomaly detection may assist traditional RCA techniques in finding the root cause. Machine learning automated RCA may also be useful for the following reasons:

- To perform complex RCA several domain experts are often needed, who may not always be available or may be expensive or difficult to deploy. In this case ML automated RCA can also be considered as a knowledge management tool.
- ML automated RCA may save a lot of time because traditional RCA projects may often take even several days.

There are three main types of machine learning models (methods) which can be used for anomaly detection and/or RCA:

- Supervised,
- Unsupervised and
- Semi-supervised learning based methods.

8.5.1   Supervised Learning Based Anomaly Detection and Root Cause Analysis

Remember from earlier chapters that for supervised learning we need labeled (classified) data. In the case of anomaly detection, the data must contain data records labeled as 'normal' and data records labeled as 'anomaly'. We can combine anomaly detection with root cause analysis by defining several types of anomaly labels which all identify a specific root cause. For example, in the case of the root cause analysis of a defective product in a production process, we could define root causes as "wrong material", "handling error", "machine error", etc. instead of just using "wrong product".

In anomaly detection datasets there are usually many more normal data records than anomalous data records (we call this class imbalance in the data) which must be taken into account, for example, in the following two ways:

- By using performance metrics which are not sensitive to imbalance in the data. More information can be found about this subject in Chap. 3.
- By over sampling of the anomalous data records with one of the following methods:

  - Randomly selecting and using the same anomalous data record several times.
  - Generating extra artificial anomalous data records in some way by respecting the class boundaries defined by the machine learning model.

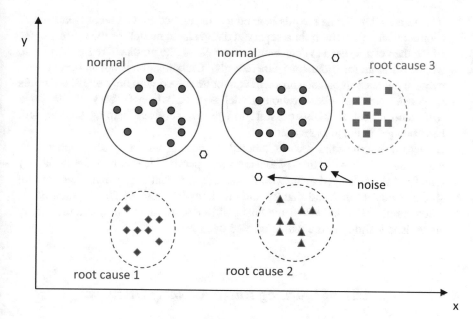

Fig. 8.16 Anomaly detection and root cause analysis (2D)

Supervised machine learning based anomaly detection and root cause analysis is a very powerful technique, but collecting and labeling the data is a lot of work and must be done by domain experts. You may also consider using a semi-supervised method (see later) in order to decrease the amount of work needed for labeling the data.

More information about supervised learning and about the available supervised learning models in the AI-TOOLKIT can be found in Chaps. 1, 2 and 9.

A 2D representation of an anomaly detection and root cause identification classification example can be seen on Fig. 8.16. Machine learning models usually work in higher dimensions, but this 2D example shows the principle of operation.

8.5.2 Semi-supervised Learning Based Anomaly Detection

If there is a limited amount of labeled data then we cannot use a supervised learning model, but we still have a couple of other options:

- If there are a lot of data records labeled as normal but no data records labeled as anomalous then we can train a supervised regression model for the normal cases which then can be used to detect anomalous cases (the predicted regression value will be different for anomalous data records). Classification is not an option here, because the class imbalance problem cannot be solved with only one class unless a special one-class classification model is used which is very similar to a regression model. It is also possible to use unsupervised learning but this will be explained in the next section.
- If there are some data records labeled as normal and also some labeled as anomalous, then we can first try to use an unsupervised learning model to classify (label) as yet unlabeled data (with the assumption that the normal items are all similar to each other and that anomalous items are also similar to each other, where similarity is defined by the unsupervised learning method) and then apply supervised learning to the whole labeled dataset.

8.5.3   Unsupervised Learning Based Anomaly Detection

In the case where we have no labeled data at all, we can use unsupervised learning models to cluster (group) the data records. If we assume that there are many more normal data records than anomalous data records, and that these are clearly separated from each other, then we can identify clusters which contain anomalous records. Each anomaly cluster may be the result of a specific root cause. Root causes may not be derived directly but the results may help to identify them more easily.

Depending on the type of unsupervised learning model different similarity metrics may be used for grouping data records together. It is recommended to try several unsupervised learning models and choose the model which best fits the specific problem and data!

TIP: Very often dimensionality reduction with PCA (see Sect. 8.2) may help to identify anomalous records more easily. In lower dimensions, and after removing the correlation between the features (this is what PCA does), the differences between normal and anomalous clusters may be more distinct.

More information about unsupervised learning and about the available unsupervised learning models in the AI-TOOLKIT can be found in Chaps. 1, 2 and 9.

8.5.4 Special Considerations in Anomaly Detection

Anomalies may also be context or sequence dependent. Context dependency here means the same as in Sect. 8.4.1.4 (e.g., time or location) and context dependent anomalies may also be handled similarly (e.g., by pre-filtering the data).

Sequence dependency (sometimes called collective anomaly) means that not just one specific record is anomalous but a whole sequence of records. We need a different model than those explained in the previous sections or we can use a data pre-processing step to generate one anomalous record from a sequence of records. For example, if a sequence of actions (with corresponding data records) a1, a2 and a3 is detected which signals an anomaly, then we may combine these records into one record by moving a window with a record length of three through the dataset. Some of the features may just be averaged or summed up (in the window) and the action value may be named, e.g., a123. The same technique may also be used for time series data where, for example, a specific value during a given time period may signal an anomaly (e.g., heart beat failure), but the same value is not an anomaly if it only appears once.

Noise in the data may cause problems in anomaly detection if, for example, there are not enough records with all types of anomalies. A noise record is a normal record with higher variation on some parameters and therefore may be identified as an anomaly. Some machine learning models are not sensitive to noise and thus better suited for data with a lot of noise (see, e.g., unsupervised learning). Another solution is the pre-processing of the data and the removal of the noise before anomaly detection is applied.

8.5.5 Data Collection and Feature Selection

One of the most important parts of anomaly detection and root cause analysis is the data collection and feature selection phase. It is very important to select the right features which can be used to distinguish between normal and anomalous phenomena. In most cases (especially if root cause analysis is integrated) domain experts must design the data collection and feature selection process, and they also must take care of the labeling of the data records if supervised learning models are used.

Usually a traditional structured and systematic approach is used to investigate anomalies and their root causes (when appropriate) and in order to determine the best set of features used in the input data for the machine learning model.

> TIP: It is often possible to simulate a real situation or environment and introduce all forms of potential anomalies and record the attributes (responses) of the system. It is much easier and faster to study the signatures of anomalies

(continued)

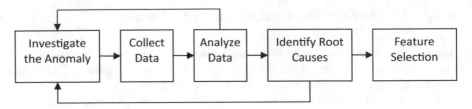

Fig. 8.17 Data collection and feature selection

> in this way than to monitor real systems. Try to think about a solution for modeling your real situation in some way and try to simulate potential anomalies.

The structured feature selection process can be seen on Fig. 8.17 and contains the following steps:

1. The first step is to investigate the anomalies. Usually a project team is assembled for the investigation containing domain experts.
2. The second step is to collect all available data (problem, incident, failure, etc.) based on the first investigation step. You can drop data which do not seem to be useful but report its existence because it may need to be added back in later.
3. The third step aims to organize the collected information and data, and identify the features which are informative for the anomaly (often called causal factors). Traditional analysis techniques such as cause and effect trees, timelines and causal factor charts, etc. may be used to identify all informative features and their root causes. It is not the aim of this book to explain these techniques in detail.

    Note that not all data is informative when identifying an anomaly and too much data may confuse (deteriorate) the machine learning model. Categorical values may need to be converted to numerical values (see Sect. 4.3.2.2); dimensionality reduction may be necessary if there are a lot of features (see Chap. 2), etc. We often need to go back to the investigation and data collection steps in order to collect more data if we do not have enough information yet.
4. The fourth step, if root cause analysis is integrated into the final solution, is to use traditional root cause identification techniques to identify the final root causes of the anomalies (based on the analysis step).
5. A final feature selection step selects features which are associated with most of the root cause or causal factors. You can drop features which do not seem to be useful but make sure that they can be added back in if needed.

The above structured process may need to be repeated if the performance of the machine learning model is not satisfactory. More data may need to be added or data may need to be removed.

8.5.6 Determining an Unknown Root Cause

The first question that many people ask themselves when they hear about AI-aided root cause analysis is, can you determine an unknown root cause using machine learning? If the root cause of a problem isn't known, then we cannot teach the machine learning model because we also do not know the root cause yet! So the answer to this question is, in most cases, no, because we do not yet have a general artificial intelligence which can deduce the unknown root cause of a problem. But it is possible to use machine learning to detect an anomaly and to deduce the near-root cause by analyzing the correlation between how the unsupervised machine learning model groups data points and the features in the dataset. Often there is a correlation between one or more features and the separated anomalous records which may point us in the direction of the unknown root cause. Traditional root cause analysis techniques are, however, still needed in most cases, but we can automate the root cause analysis, after the possible root causes are known, with the techniques explained in the previous sections.

It is also possible, but it is a complex exercise, to build a chain of machine learning models which try to imitate a kind of human deduction process in order to find an unknown root cause. This is an active field of research which partially tries to close the gap between a general artificial intelligence and what we have today. The AI-TOOLKIT has built-in support for this kind of chain model with the ML-FLOW functionality, which allows the connection of several machine learning models.

You may of course use the information provided in this book and try to develop a clever way to find the unknown root cause of a problem based on the data you collect!

> TIP: In Sect. 8.7 there is another example of root cause analysis in the case of predictive maintenance with machine learning! The supervised learning model is able to identify the root cause of machine failure and it is also able to suggest an appropriate time for predictive maintenance!

8.5.7 Example: Intrusion Detection in a Military Environment (U.S. Air Force LAN): Anomaly Detection and Root Cause Analysis

In this example we will train a supervised machine learning model which will be able to identify an attack (anomaly) to an ICT network and it will also be able to indicate the root cause (source) of the attack.

The DARPA intrusion detection evaluation program [34] provided several gigabytes of tcp-dump data of seven weeks of network traffic, which was processed into

Table 8.16 List of attack root causes

Attack category	Attack root cause	Number of data records
Normal (no attack)	–	812814
dos	back	967
	land	18
	neptune	242148
	smurf	3006
	tear drop	917
	pod	205
r2l	ftp write	7
	imap	11
	multi hop	7
	phf	7
	spy	7
	warez client	893
	guess password	52
	warez master	19
probe	nmap	1553
	port sweep	3563
	satan	5018
	ip sweep	3723
u2r	load module	8
	perl	7
	root kit	9
	buffer overflow	29

about five million connection records [34]. The data contains normal network traffic and four main categories of attacks:

- Denial-of-service (DOS); e.g., "smurf".
- Unauthorized access from a remote machine (R2L); e.g., guessing password.
- Unauthorized access to local super-user (root) privileges (U2L); e.g., buffer overflow attack.
- Surveillance and probing (PROBE); e.g., port-scan.

The list of the root causes of the attacks can be seen in Table 8.16.

If you are not an ICT expert the root causes may not be familiar to you. The principle here is to identify as precisely as possible the source of the attack, and the root causes shown in Table 8.16 are known sources of attacks in different categories.

The main difficulty in this problem is the extraction of the right features from the tcp-dump raw data (a huge amount of all kinds of connection and communication information logged per time step). This is the data collection and feature extraction step introduced in the previous section.

In order to extract the right features we must compare (analyze) the raw data containing normal connections with the raw data containing attack connections and identify patterns of events which occur frequently in attack connections but not in normal connections. These patterns of events define the necessary features which must be extracted from the raw data and which may still need to be extended with so-called content features suggested by domain experts (logical attributes which indicate suspicious behavior). This structured way of thinking and analysis of the problem is the same from all kinds of problems!

By examining and comparing the raw data for normal and attack connections it was found that time based traffic features can be extracted from every two second periods of the connection records. Connections that have the same host or the same

Table 8.17 Basic features of individual TCP connections [34]

Short feature name	Description	Type
Duration length	(number of seconds) of the connection	Number
Protocol type	Type of protocol, e.g. tcp, udp, etc.	Categorical
Service	Network service on the destination, e.g., http, telnet, etc.	Categorical
Source bytes	Number of data bytes from source to destination	Number
Destination bytes	Number of data bytes from destination to source	Number
Flag	Normal or error status of the connection	Categorical
Land	1 if connection is from/to the same host/port, 0 otherwise	Number
Wrong fragment	Number of "wrong" 'fragments'	Number
Urgent	Number of urgent packets	Number

Table 8.18 Content features per connection (domain knowledge) [34]

Short feature name	Description	Type
Hot	Number of "hot" indicators	Number
Number of failed logins	Number of failed login attempts	Number
Logged in	1 if successfully logged in, 0 otherwise	Categorical
Number of compromised	Number of "compromised" conditions	Number
Root shell	1 if root shell is obtained, 0 otherwise	Categorical
Su attempted	1 if "su root" command attempted, 0 otherwise. This is a UNIX command which switches to the root user (may be the super user), accesses root privileges.	Categorical
Number of root	Number of "root" accesses	Number
Number of file creations	Number of file creation operations	Number
Number of shells	Number of shell prompts	Number
Number of access files	Number of operations on access control files	Number
Number of outbound commands	Number of outbound commands in an ftp session	Number
Is hot login	1 if the login belongs to the "hot" list, 0 otherwise	Categorical
Is guest login	1 if the login is a "guest" login, 0 otherwise	Categorical

Table 8.19 Traffic features computed using a two second time window [34]

Short feature name	Description	Type
Count	Number of connections to the same host as the current connection in the past two seconds	Number
The following features refer to the same host connections (see description above):		
syn error rate	% of connections that have "SYN" errors. The 'SYN' is the first packet sent from a client to a server; it literally asks a server to open a connection with it.	Number
rej error rate	% of connections that have "REJ" errors. Connection attempt rejected.	Number
Same service rate	% of connections to the same service.	Number
Different service rate	% of connections to different services.	Number
Service count	Number of connections to the same service as the current connection in the past two seconds	Number
The following features refer to the same service connections (see description above):		
Service syn error rate	% of connections that have "SYN" errors.	Number
Service rej error rate	% of connections that have "REJ" errors. Connection attempt rejected.	Number
Service different host rate	% of connections to different hosts.	Number

Table 8.20 LAN intrusion example data records

```
record 1:
0,tcp,http,SF,236,1721,0,0,0,0,0,1,0,0,0,0,0,0,0,0,0,0,13,13,0.00,0.00,0.00,0.0
0,1.00,0.00,0.00,255,255,1.00,0.00,0.00,0.00,0.00,0.00,0.00,0.00,normal.

record 2:
0,tcp,private,REJ,0,0,0,0,0,0,0,0,0,0,0,0,0,0,0,0,0,0,265,11,0.00,0.00,1.00,1.0
0,0.04,0.06,0.00,255,11,0.04,0.07,0.00,0.00,0.00,0.00,1.00,1.00,neptune.

record 3:
0,icmp,ecr_i,SF,1032,0,0,0,0,0,0,0,0,0,0,0,0,0,0,0,0,0,511,511,0.00,0.00,0.00,0
.00,1.00,0.00,0.00,255,255,1.00,0.00,1.00,0.00,0.00,0.00,0.00,0.00,smurf.
```

service in the two second window frame were extracted with their statistical properties (protocol behavior, service, etc.). It was also found that probing attacks (the fourth attack type) were much slower and thus took longer than two seconds. For this reason a separate but similar set of features were extracted for a connection window of 100 seconds from probing attacks data. It was also discovered that other types of attacks (e.g., R2L) involved only a single connection.

By using domain knowledge it is still possible to extend these features with features which define suspicious behavior, e.g., the number of failed login attempts, the number of file creations, etc., which we call content features.

Tables 8.17, 8.18, and 8.19 summarize the final features selected and the contents of the input data file to our supervised machine learning model (which will be introduced later).

Three records from the input data file can be seen in Table 8.20. The first record is a normal transaction. The second and the third records are attacks caused by "neptune" and "smurf" (denial-of-service attacks). All records are labeled (last column) with their root cause (in the case of an anomalous (attack) record) or with "normal". The machine learning model will learn the patterns in the data and will learn to identify which transactions are normal and which are attacks, and for which reason (root cause) these attacks occurred. This is the essence of supervised machine learning aided root cause analysis!

After the data collection and feature selection steps have been completed and we have the final input data, we can select a supervised machine learning model and start the training of the model.

Please note that textual categorical values will need to be converted to numbers, this can be done automatically with the AI-TOOLKIT (which will be explained later).

8.5.7.1   Training a Machine Learning Model for the Detection of a LAN Intrusion and its Root Cause (AI-TOOLKIT)

Step 1: Choose a Machine Learning Model

There are several supervised machine learning models available in the AI-TOOLKIT. It is very easy to use one of the available templates and to build a new model. The choice of the model depends on several things such as the structure of the input data, the amount of input data, the existence of noise in the data, etc. You will need some experience with building machine learning models in order to make the best choice, but even then it is often a process of trial and error. Usually the SVM and neural network models are the best initial choices.

Let us choose the SVM model for this example. Designing an SVM model is much easier than designing a neural network because the AI-TOOLKIT includes an SVM parameter optimization module. It is not possible to design or optimize the architecture of a neural network automatically. There is ongoing research in this area but a good solution has not yet been discovered. These algorithms basically try several architectures in a clever way and search for the best performance.

Step 2: Create the Project File

The project file in Table 8.21 is created almost completely automatically by the AI-TOOLKIT. The steps to reproduce this project file are as follows (except for the green comments which were removed):

1. Start a New Project.

Table 8.21 'idlan' Project File

```
model:
  id: 'ID-TXLdUgyRzJ'
  type: SVM
  path: 'idlan.sl3'
  params:
    - svm_type: C_SVC
    - kernel_type: RBF
    - gamma: 0.05
    - C: 80.0
    - cache_size: 1000
    - max_iterations: 100
  training:
    - data_id: 'idlan'
    - dec_id: 'decision'
  test:
    - data_id: 'idlan'
    - dec_id: 'decision'
  input:
    - data_id: 'input_data'
    - dec_id: 'decision'
  output:
    - data_id: 'output_data'
    - col_id: 'decision'
```

2. Use the 'Insert ML Template' command on the AI-TOOLKIT tab.
3. Go to the Supervised Learning section in the dialog window which appears and select the "Support Vector Machine (SVM) Classification/Regression" template and then the 'Insert' command.
4. You will have to change some of the parameters such as the path and database table names. This will be explained in the following sections!
5. Save the project file.

For more information about the AI-TOOLKIT project file please see Chap. 9.

Step 3: Create the AI-TOOLKIT Database and Import the Data
There are nearly five million records in the input data file which first must be imported into an AI-TOOLKIT database created for this project. Each project may use a separate database in order to better structure the work per project.

Go to the Database tab and use the 'Create New AI-TOOLKIT Database' command. Indicate the path and name of the database file. In this example we will use 'idlan.sl3' as the database name. Next use the 'Import Data Into Database' command and the steps as follows to import the training data:

- In the Import Data dialog window which appears select the delimited data file to import (for example, 'idlan.csv') with the SEL button.
- In the Delimiters field add a comma after the '\t' sign. It will look like this: '\t,'. The idlan.csv file is comma separated. You may also remove the tab separation sign ('\t') if you wish but this is not necessary.
- Indicate the zero based index of the decision (rating) column. In idlan.csv the rating column is the 42nd column and thus the zero based index is 41!
- Select the newly created database 'idlan.sl3' with the SEL button.
- Change the New Table Name field to 'idlan'.
- Select the "Automatically Convert Categorical or Text values" option. This is necessary because there are several categorical values in the input data file (for example, protocol type, root cause, etc.). All categorical values will be automatically converted to numbers. More information about this can be found in Sect. 4.3.2.2.
- Use the 'Import' command to import the data.
- A dialog box will appear in which the conversion of the categorical values can be set. Leave all of the default values (integer encoding) and push OK.

NOTE: The 'Import Data Into Database' module is also explained in Chap. 9!

At this point the data is ready to be used with the machine learning model.

Next you can check if the data was imported successfully. On the Database tab use the 'Open Database Editor' command. The built-in database editor will open. Check Chap. 9 for more information about the database editor! Use the 'Open Database' command in the editor and navigate to the 'idlan.sl3' database and open it. *Click on the idlan table* on the Database Structure tab to expand all fields. You will see many columns: decision, col1, col2, col3... You can also look at the data on the "Browse Data" tab.

Step 4: Finalize the Project File

There are five parameters which are important for this type of model:

```
- kernel_type: RBF
- gamma: 0.05
- C: 80.0
- cache_size: 1000
- max_iterations: 100
```

For the "kernel type" usually the radial basis function (RBF) is the best choice. For this kernel, the gamma and the C parameters must be defined. Setting the values of these two parameters is not trivial but the built-in SVM parameter optimization module helps to determine the optimal values. Normally a subset of the data is used for the optimization of the parameters because otherwise the optimization takes too long, but the software will also limit the data for optimization to a maximum of 1e6 records. The values 0.05 and 80.0 are determined with the optimization module.

The cache_size is set to 1GB (1000 Mb) because of the huge amount of data. You will need at least 8GB RAM in order to train this model with so much data! Having 16GB RAM is recommended.

The max_iterations parameter is set to 100. You could set this number to −1 so as not to limit the number of iterations or to a higher value than 100 but the training will take much longer.

Step 5: Train the Model

IMPORTANT: We did not create the 'input_data' and 'output_data' tables because they will be created automatically by the AI-TOOLKIT. You must make sure that when you train the recommendation model you click *yes* when asked whether to create these tables automatically! You may of course also create these tables manually before training the model, but it is much easier to let the software do it for you.

When you click the Train AI Model button, the software will check the project file and database for inconsistencies and check that all database tables exist. The training data table is required and the model training cannot proceed without it. If the input and output data tables do not exist, then the software will offer to create them. The database column descriptions must be well defined in the project file!

Each step in the model training process will be logged in the output log. When the model training is complete the model will be saved into the database together with a performance report (also visible in the output log at the bottom of the screen). The model is now ready for detecting attacks and their root cause!

You can check the saved model and model parameters in the 'trained_models' table in the database.

The training of the model takes about 140 min on an Intel Core i5–6400 PC.

Step 6: Analysis of the Performance of the Trained Machine Learning Model
If the performance of the machine learning model is not satisfactory, then we may need to change some of the parameters, choose another type of model (e.g., a neural network) or change the input data. This is the reason why it is important to pay attention to the performance evaluation. For more information about the contents of the performance evaluation see Chap. 3.

In the case of an SVM model, which we are using in this example, one of the parameters that influences the performance of the model is "max_iteration". The software will warn you if this value is too low while training the model with messages similar to "[WARNING] Reaching max number of iterations = 100". If you set max_iteration to −1 the software will optimize the solution until it still improves significantly, but this may take much longer.

Table 8.22 Performance results after 100 iterations—part 1

Confusion Matrix [predicted x original] (number of classes: 23):   Number of iterations: 100   PART1

ORIGINAL

PREDICTED	normal	bufferoverflow	loadmodule	perl	neptune	smurf	guesspasswd	pod	teardrop	portsweep	ipsweep
normal	969646	33	9	3	1	1346	53	5	0	80	6430
bufferoverflow	0	0	0	0	0	0	0	0	0	0	0
loadmodule	0	0	0	0	0	0	0	0	0	0	0
perl	0	0	0	0	0	0	0	0	0	0	0
neptune	2082	0	0	0	1072011	0	0	0	0	0	0
smurf	0	0	0	0	0	2806539	0	0	0	0	0
guesspasswd	0	0	0	0	0	0	0	0	0	0	0
pod	1	0	0	0	0	0	0	259	0	0	0
teardrop	0	0	0	0	0	0	0	0	979	0	0
portsweep	3	0	0	0	5	0	0	0	0	10331	0
ipsweep	958	0	0	0	0	1	0	0	0	2	6051
land	0	0	0	0	0	0	0	0	0	0	0
ftpwrite	0	0	0	0	0	0	0	0	0	0	0
back	0	0	0	0	0	0	0	0	0	0	0
imap	0	0	0	0	0	0	0	0	0	0	0
satan	91	0	0	0	0	0	0	0	0	0	0
phf	0	0	0	0	0	0	0	0	0	0	0
nmap	0	0	0	0	0	0	0	0	0	0	0
multihop	0	0	0	0	0	0	0	0	0	0	0
warezmaster	0	0	0	0	0	0	0	0	0	0	0
warezclient	0	0	0	0	0	0	0	0	0	0	0
spy	0	0	0	0	0	0	0	0	0	0	0
rootkit	0	0	0	0	0	0	0	0	0	0	0
Precision:	98%	0%	0%	0%	100%	100%	0%	100%	100%	98%	79%
Recall:	100%	0%	0%	0%	100%	100%	0%	98%	100%	99%	48%
FNR:	0%	100%	100%	100%	0%	0%	100%	2%	0%	1%	52%
F1:	99%	0%	0%	0%	100%	100%	0%	99%	100%	99%	52%
TNR:	100%	100%	100%	100%	100%	100%	100%	100%	100%	100%	100%
FPR:	0%	0%	0%	0%	0%	0%	0%	0%	0%	0%	0%

Table 8.23 Performance results after 100 iterations—part 2

Confusion Matrix [predicted x original] (number of classes: 23):

Number of iterations: 100   PART 2

ORIGINAL

PREDICTED	land	ftpwrite	back	imap	satan	phf	nmap	multihop	warezmaster	warezclient	spy	rootkit
normal	21	8	2203	8	1920	4	1682	7	20	1020	2	9
bufferoverflow	0	0	0	0	0	0	0	0	0	0	0	0
loadmodule	0	0	0	0	0	0	0	0	0	0	0	0
perl	0	0	0	0	0	0	0	0	0	0	0	0
neptune	0	0	0	4	0	0	0	0	0	0	0	0
smurf	0	0	0	0	10	0	0	0	0	0	0	0
guesspasswd	0	0	0	0	0	0	0	0	0	0	0	0
pod	0	0	0	0	0	0	0	0	0	0	0	0
teardrop	0	0	0	0	0	0	0	0	0	0	0	0
portsweep	0	0	0	0	148	0	11	0	0	0	0	0
ipsweep	0	0	0	0	6	0	623	0	0	0	0	0
land	0	0	0	0	0	0	0	0	0	0	0	0
ftpwrite	0	0	0	0	0	0	0	0	0	0	0	0
back	0	0	0	0	0	0	0	0	0	0	0	0
imap	0	0	0	0	0	0	0	0	0	0	0	0
satan	0	0	0	0	13808	0	0	0	0	0	0	0
phf	0	0	0	0	0	0	0	0	0	0	0	0
nmap	0	0	0	0	0	0	0	0	0	0	0	0
multihop	0	0	0	0	0	0	0	0	0	0	0	0
warezmaster	0	0	0	0	0	0	0	0	0	0	0	0
warezclient	0	0	0	0	0	0	0	0	0	0	0	0
spy	0	0	0	0	0	0	0	0	0	0	0	0
rootkit	0	0	0	0	0	0	0	0	0	0	0	1
Precision:	0%	0%	0%	0%	99%	0%	0%	0%	0%	0%	0%	100%
Recall:	0%	0%	0%	0%	87%	0%	0%	0%	0%	0%	0%	10%
FNR:	100%	100%	100%	100%	13%	100%	100%	100%	100%	100%	100%	90%
F1:	0%	100%	0%	0%	93%	100%	100%	100%	0%	0%	0%	18%
TNR:	100%	100%	100%	100%	100%	100%	100%	100%	100%	100%	100%	100%
FPR:	0%	0%	0%	0%	0%	0%	0%	0%	0%	0%	0%	0%

Table 8.24 Global performance metrics

Accuracy:	99.62%
Error:	0.38%
Cohen's Kappa:	99.34%

Table 8.25 General performance metrics

Accuracy:	99.91%
Error:	0.09%
Cohen's Kappa:	99.85%

Table 8.26 Per class performance metrics

	portsweep	ipsweep	nmap
Precision:	100%	99%	97%
Recall:	100%	99%	92%
FNR:	0%	1%	8%
F1:	100%	99%	95%
TNR:	100%	100%	100%
FPR:	0%	0%	0%

Table 8.22 and 8.23 how the performance evaluation results of the model if the maximum number of iterations is limited to 100 (the table is split in two because of its size). The diagonal of the confusion matrix is colored with blue; these values are the correct predictions. The global performance metrics are shown in Table 8.24

The global performance metrics show a very good performance but these values are misleading in the case of several of the root causes. We can see this if we look at the per class performance metrics! The two tables show with a green color the root causes which are identified well and with a light red color the root causes which are not identified well. For example, there are 30 cases of "bufferoverflow" in the training dataset but the model identifies all of them as "normal" transactions for buffer overflow. We can see this in the first row of the confusion matrix and also in the 0% precision and recall values! All of these mistakes are linked to the root causes which are not well represented in the dataset, or in other words, there are very few cases of these root causes in the input training dataset! Both the limited number of iterations (100) and the class imbalance in the data are the reasons for the low per class performance measures in this case of several root causes!

So how do we correct this problem? There are several possible solutions. First of all, the most obvious solution is to increase the number of iterations. Secondly, it is also possible to oversample the records with the classes which are not represented well in the dataset and/or down sample the other cases (e.g., normal cases). Resampling (oversampling, down sampling, etc.) is a built-in feature in the AI-TOOLKIT (see the import data module). Another solution to this problem could be the selection of another machine learning model, e.g., neural network.

By increasing the iteration limit from 100 to 1000 there is a significant improvement in the per class performance metrics and the general performance metrics also improve. The new general performance metrics can be seen in Table 8.25

The significantly improved new per class performance metrics can be seen in Table 8.26

By just increasing the iteration limit to 1000, three more attack types can be almost perfectly identified. The iteration limit can still be increased, and it can even be set to unlimited number of iterations (-1) which means that the model training will only stop if there is no significant improvement. There are quite a lot of warning messages, "[WARNING] Reaching max number of iterations = 1000.", with the 1000 iteration limit, which means that the performance of the model can still be improved by allowing more iterations!

Step 7: Detect Attacks and their Root Cause
The trained model can be used for detecting attacks and their root cause by feeding similar records to the model as were used for training of the model. The prediction process is very simple and can be summarized as follows:

- Import the input data into the database. Make sure that you select Prediction as the target and not training! This can be done with the Import module—manually in the database editor or with any compatible external software which can update a SQLite database. The input tables for prediction have a special format besides what is defined in the project file (see "input" in Step 2) because they must contain a "flow_id" column with a unique integer value and the decision column must be empty or NULL (this signals to the software that the records must be evaluated). If you have allowed the software to create an input data table automatically when starting the training of the model, then you already have the correct format in the database! Make sure that you choose the right decision column (the same as for the training data) if you use the import module!
- Use the Predict with AI Model button on the toolbar to start the prediction. All open records (where the decision column is empty or NULL) will be evaluated.
- You will find the results in the decision column of the input data table.

IMPORTANT: If there are textual categorical values in the input data for prediction and you add the input data manually, then make sure that you convert the categorical values with the same conversion method as for the training data! The software shows the conversion map in the output log (it can be saved!) and it can also be found in the database in the cat_map table in the record identified by the data_id, which is the training data table name. *Be careful with one-hot and binary encoding because these add extra columns to the data table!*

Resample the Data in Order to Improve Performance (Removing Class Imbalance)

One of the reasons for the low per-class accuracy is the class imbalance (there are many more data records from some of the classes than from others) in the data. Due to class imbalance the machine learning model 'focuses' more on majority examples than on minority ones, and this causes a low minority class performance. We can correct this imbalance by resampling the data. Resampling the data means that we add new data records to minority classes and/or remove data records from majority classes. More information about resampling can be found in Sect. 4.3.2.6 in Chap. 4.

The AI-TOOLKIT has built in support for semi-automatic resampling of the data with several techniques:

- Undersampling with TOMEK links removal. This technique can also be considered as data cleaning.
- Oversampling with borderline SMOTE (SMOTEB). This is an advanced oversampling technique which may also improve ML performance because of the improved separation of the different classes.
- Random oversampling and undersampling.

The best resampling performance can be achieved by applying several of these techniques to the LAN Intrusion dataset in several steps. The first logical step is the TOMEK links removal, which will remove noise or confusing points on the boundary between two classes. The second step is the random undersampling of the minority classes. The AI-TOOLKIT allows us to choose the maximum number of majority data records which should stay in the dataset. In this example we will choose 100,000 random data records from each majority class. The next logical step is the use of the borderline SMOTE (SMOTEB) over sampler on the minority classes. The AI-TOOLKIT allows us to choose the minimum number of minority data records which should be in the dataset. In this example we will choose 100,000 minority data records from each majority class in order to balance them with the majority classes! As a final step we will use the random over sampler to fill in the gaps where SMOTEB could not create enough data records and make sure that we have 100,000 data records from each class. This is a complex resampling scheme. Let us look at the ML performance results after retraining the SVM model (1000 iterations) with the new resampled dataset. Tables 8.27 and 8.28 show the detailed results.

As you can see in Tables 8.27 and 8.28 the SVM model was able to learn most of the attack signatures except for 'multihop' and 'warezmaster' which are most probably overlapping regions/records. This can be seen in the confusion matrix because a huge number of 'multihop' records are classified as 'warezmaster' (55,103). By removing one of these two attack types the other one will most

Table 8.27 ML performance results—1000 iterations—part 1

Confusion Matrix [predicted x original] (number of classes: 23):

RESAMPLED1 Number of iterations: 1000 PART1

ORIGINAL

PREDICTED	normal	bufferoverflow	loadmodule	perl	neptune	smurf	guesspasswd	pod	teardrop	portsweep	ipsweep
normal	98942	0	0	0	0	2	0	0	0	4	2
bufferoverflow	31	100000	0	0	0	0	0	0	0	0	0
loadmodule	15	0	100000	0	0	0	0	0	0	0	0
perl	0	0	0	100000	0	0	0	0	0	0	0
neptune	106	0	0	0	100000	0	0	0	0	0	1
smurf	0	0	0	0	0	99997	0	0	0	0	0
guesspasswd	2	0	0	0	0	0	100000	0	0	0	0
pod	0	0	0	0	0	0	0	100000	0	0	0
teardrop	0	0	0	0	0	0	0	0	100000	0	0
portsweep	1	0	0	0	0	0	0	0	0	99992	0
ipsweep	8	0	0	0	0	0	0	0	0	0	99926
land	0	0	0	0	0	0	0	0	0	0	0
ftpwrite	3	0	0	0	0	0	0	0	0	0	0
back	14	0	0	0	0	0	0	0	0	0	0
imap	0	0	0	0	0	0	0	0	0	0	0
satan	686	0	0	0	0	1	0	0	0	3	31
phf	0	0	0	0	0	0	0	0	0	0	0
nmap	61	0	0	0	0	0	0	0	0	0	41
multihop	0	0	0	0	0	0	0	0	0	0	0
warezmaster	4	0	0	0	0	0	0	0	0	0	0
warezclient	87	0	0	0	0	0	0	0	0	0	0
spy	1	0	0	0	0	0	0	0	0	0	0
rootkit	39	0	0	0	0	0	0	0	0	0	0
	normal	bufferoverflow	loadmodule	perl	neptune	smurf	guesspasswd	pod	teardrop	portsweep	ipsweep
Precision:	100%	98%	98%	100%	100%	100%	100%	100%	100%	100%	100%
Recall:	99%	100%	100%	100%	100%	100%	100%	100%	100%	100%	100%
FNR:	1%	0%	0%	0%	0%	0%	0%	0%	0%	0%	0%
F1:	99%	99%	99%	100%	100%	100%	100%	100%	100%	100%	100%
TNR:	100%	100%	100%	100%	100%	100%	100%	100%	100%	100%	100%
FPR:	0%	0%	0%	0%	0%	0%	0%	0%	0%	0%	0%

Table 8.28 ML performance results—1000 iterations—part 2

Confusion Matrix [predicted x original] (number of classes: 23):

ORIGINAL · RESAMPLED1   Number of iterations: 1000   PART2

PREDICTED	land	ftpwrite	back	imap	satan	phf	nmap	multihop	warezmaster	warezclient	spy	rootkit
normal	0	0	6	0	145	0	8	0	0	48	0	0
bufferoverflow	0	0	2	0	0	0	0	2368	0	29	0	0
loadmodule	0	0	0	1	0	0	0	1585	0	0	0	0
perl	0	0	0	0	1	0	0	0	0	0	0	0
neptune	0	0	0	0	0	0	0	0	0	0	0	0
smurf	0	0	0	0	1	0	0	0	0	0	0	0
guesspasswd	0	0	0	0	0	0	0	0	0	0	0	0
pod	0	0	0	0	0	0	0	0	0	0	0	0
teardrop	0	0	0	0	0	0	0	0	0	0	0	0
portsweep	0	0	0	0	0	0	0	0	0	0	0	0
ipsweep	0	0	0	0	1	0	19	0	0	0	0	0
land	100000	0	0	0	0	0	0	0	0	0	0	0
ftpwrite	0	100000	0	0	0	0	0	0	0	0	0	0
back	0	0	99992	0	0	0	0	0	0	0	0	0
imap	0	0	0	100000	0	0	0	0	0	0	0	0
satan	0	0	0	0	99825	0	37	0	0	0	0	0
phf	0	0	0	0	0	100000	0	0	0	0	0	0
nmap	0	0	0	0	0	0	99929	0	0	0	0	0
multihop	0	0	0	0	2	0	0	40944	0	0	0	0
warezmaster	0	0	0	0	0	0	0	55103	100000	0	0	0
warezclient	0	0	0	0	1	0	0	0	0	99923	0	0
spy	0	0	0	0	0	0	0	0	0	0	100000	0
rootkit	0	0	0	0	24	0	7	0	0	0	0	100000
	land	ftpwrite	back	imap	satan	phf	nmap	multihop	warezmaster	warezclient	spy	rootkit
Precision:	100%	100%	100%	100%	99%	100%	100%	100%	64%	100%	100%	100%
Recall:	100%	100%	100%	100%	100%	100%	100%	41%	100%	100%	100%	100%
FNR:	0%	0%	0%	0%	0%	0%	0%	59%	0%	0%	0%	0%
F1:	100%	100%	100%	100%	100%	100%	100%	58%	78%	100%	100%	100%
TNR:	100%	100%	100%	100%	100%	100%	100%	100%	98%	100%	100%	100%
FPR:	0%	0%	0%	0%	0%	0%	0%	0%	3%	0%	0%	0%

Table 8.29 Global performance metrics

Accuracy:	97.37%
Error:	2.63%
C.Kappa:	97.25%

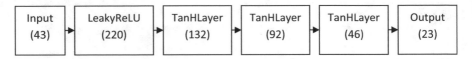

Fig. 8.18 Intrusion detection neural network model architecture

probably be able to be predicted well. Another solution (and probably the most logical one) is to increase the number of data records in these classes because there are only 7 and 19 records from 'multihop' and 'warezmaster' respectively (the ML model doesn't have enough information about their signature and oversampling does not help either).

The global performance metrics of the model can be seen in Table 8.29.

Training a Neural Network Model for the Same Task
In order to demonstrate the use of a neural network model instead of an SVM model we will design and train a supervised neural network classification model for detecting intrusions. We will use the resampled (balanced) dataset from the previous example!

Designing a good neural network model is often more difficult than building an SVM model because of the many choices we have to make about the architecture and optimization parameters of the neural network.

Please read Sect. 2.2.2.4 of Chap. 2 about designing a neural network first, if you have not done so already! The architecture of the neural network explained below is designed according to these guidelines. The placement of the different neural network layer types is also important! This is explained in Sect. 2.2.2.2.

After a few trials the neural network shown on Fig. 8.18 has been chosen and trained. The numbers below each layer show the number of nodes. Please visit the above mentioned sections of the book to read about the different layer types used.

The AI-TOOLKIT project file can be seen in Table 8.30.

The Adam optimizer has been chosen with 2,300,000 iterations (equal to the number of data records) and with a batch size of 64. It is possible that this is not the most optimal neural network architecture for this problem but it is good enough for this example.

The global performance metrics of the trained neural network model can be seen in Table 8.31.

Table 8.30 'idlan_final' Project File

```
model:
  id: 'ID-nKrMQsjesG'
  type: FFNN1_C # feedforward neural network for classification
  path: 'idlan_final.sl3'
  params:
    - layers:
      - Linear: # input connection  => 43 + 1 (bias)
      - LeakyReLU:
          nodes: 220  # chosen as 44 x 5
          alpha: 0.03
      - Linear:
      - TanHLayer:
          nodes: 132  # chosen as 44 x 3
      - Linear:
      - TanHLayer:
          nodes: 92  # chosen as 23 x 4
      - Linear:
      - TanHLayer:
          nodes: 46  # chosen as 23 x 2
      - Linear: #output connection   => 23
    - iterations_per_cycle: 2300000
    - num_cycles: 200
    - step_size: 5e-5
    - batch_size: 64
    - optimizer: SGD_ADAM
    - stop_tolerance: 1e-5
    - sarah_gamma: 0.125
  training: # training data
    - data_id: 'idlan' # training data table name
    - dec_id: 'decision' # the decision column ID.
  test: # test data
    - data_id: 'idlan' # test data table name
    - dec_id: 'decision'
  input: # ML Flow input data
    - data_id: 'input_data' # input data table name
    - dec_id: 'decision'
  output:
    - data_id: 'output_data' # output data table name (for the prediction)
    - col_id: 'decision' # the column where the output will be written
```

Table 8.31 Global performance metrics

Accuracy:	98.14%
Error:	1.86%
C.Kappa:	98.05%

Table 8.32 Per class performance metrics (1)

	normal	bufferoverflow	loadmodule	perl	neptune
Precision	92%	100%	99%	100%	100%
Recall	85%	100%	98%	100%	100%
FNR	15%	0%	2%	0%	0%
F1	88%	100%	99%	100%	100%
TNR	100%	100%	100%	100%	100%
FPR	0%	0%	0%	0%	0%

Table 8.33 Per class performance metrics (2)

	land	ftpwrite	back	imap	satan	phf
Precision	100%	100%	89%	100%	100%	100%
Recall	100%	100%	100%	100%	92%	100%
FNR	0%	0%	0%	0%	8%	0%
F1	100%	100%	94%	100%	96%	100%
TNR	100%	100%	99%	100%	100%	100%
FPR	0%	0%	1%	0%	0%	0%

Table 8.34 Per class performance metrics (3)

	smurf	guesspasswd	pod	teardrop	portsweep	ipsweep
Precision	99%	100%	100%	100%	100%	100%
Recall	100%	100%	100%	100%	100%	100%
FNR	0%	0%	0%	0%	0%	0%
F1	100%	100%	100%	100%	100%	100%
TNR	100%	100%	100%	100%	100%	100%
FPR	0%	0%	0%	0%	0%	0%

Table 8.35 Per class performance metrics (4)

	nmap	multihop	warezmaster	warezclient	spy	rootkit
Precision	100%	100%	86%	99%	100%	98%
Recall	100%	83%	100%	100%	100%	100%
FNR	0%	17%	0%	0%	0%	0%
F1	100%	91%	93%	100%	100%	99%
TNR	100%	100%	99%	100%	100%	100%
FPR	0%	0%	1%	0%	0%	0%

The per-class performance metrics can be seen in Tables 8.32, 8.33, 8.34, and 8.35.

There are some differences compared to the SVM model results. The 'multihop' and 'warezmaster' results are significantly better, but the results for normal (no intrusion) and 'back' records are a bit worse. The global accuracy is also a bit higher in the case of the neural network model. One of the reasons for this difference is that the neural network model was probably trained for too long and the optimizer was focusing on improving the accuracy. In this case it would probably be better to stop the iterations earlier with a lower global accuracy (we call this early stopping) and have a better 'normal' and 'back' per class performance. We always make such decisions based on the given problem. Over-fitting can be prevented by an early stopping strategy!

CASE STUDY 5

8.6 Engineering Application of Supervised Learning Regression

Electric motors with a permanent magnet (permanent magnet synchronous motor) are used in many applications such as electric vehicles, money changing machines, industrial robots, pumps, compressors, servo drives, HVAC systems, laundry machines, amusement park equipment, refrigerators, microwave ovens, vacuum cleaners, etc. In high torque and speed operations there is a significant temperature rise in the magnets which causes a decrease in torque, due to a decrease in the flux density, and it may also cause safety issues. For these reasons it is important to know the magnet surface temperature during operations; safety issues can be prevented and the torque can be optimally controlled. Measuring this temperature in operational conditions is very difficult and very expensive (i.e., it involves rotating parts in a small closed environment), but it was discovered that some other easily and inexpensively measured parameters can be used to train a machine learning model, which then can be used to estimate the surface temperature of the magnet.

In this example we will train a supervised regression model which then can be used to estimate the rotor temperature (which is equal to the magnet temperature) of a synchronous electric motor with a permanent magnet in operational conditions.

The input data to the machine learning model (around one1 million records) is the laboratory measurements data presented in Table 8.36. [52]:

Before using any dataset it is important to review it and clean it! For more information about all tasks concerning the input data see Chap. 4.

Table 8.36 Measured parameters

Measured parameter	Range/Notes
Ambient temperature as measured by a thermal sensor located closely to the stator.	$[-8.57, 2.97]$
Coolant temperature (the motor is water cooled; measurement is taken at outflow)	$[-1.43, 2.65]$
Voltage d-component	$[-1.66, 2.27]$
Voltage q-component	$[-1.86, 1.79]$
Motor speed	$[-1.37, 2.02]$
Torque induced by current.	$[-3.35, 3.02]$
Current d-component	$[-3.25, 1.06]$
Current q-component	$[-3.34, 2.91]$
Permanent magnet surface temperature representing the rotor temperature (this was measured with an infrared thermograph unit)	$[-2.63, 2.92]$ *Decision variable. Regression!*
Stator yoke temperature measured with a thermal sensor.	$[-1.83, 2.45]$
Stator tooth temperature measured with a thermal sensor.	$[-2.07, 2.33]$
Stator winding temperature measured with a thermal sensor.	$[-2.02, 2.65]$

As usual the first step is to create an AI-TOOLKIT database with the "Create New AI-TOOLKIT Database" command on the Database tab on the left taskbar. Save the database in a directory of your choice. The second step is to import all data into the database created in the previous step with the "Import Data into Database" command. Do not forget to indicate the number of header rows (if any) and the correct zero based index of the decision column (eight in this example)! Next we must create the AI-TOOLKIT project file. Use the "Open AI-TOOLKIT Editor" command and then insert the chosen model template with the "Insert ML Template" button. In this example we will use a supervised SVM model.

8.6.1   Parameter Optimization

Each ML model has its own parameters. For SVM models there is a built-in parameter optimization module in the AI-TOOLKIT. For a regression SVM model (the rotor temperature is the decision variable) the svm_type is 'EPSILON_SVR'. This model uses the kernel_type, gamma, C, p, cache_size and max_iterations parameters. The degree and coef0 parameters are not used in this model because the kernel_type is RBF. More information about this model type can be found in Sect. 2.2.1 of Chap. 2! All algorithm and parameter options are explained there in detail.

The ML model for this example with the optimal parameters can be seen in Table 8.37.

Table 8.37 SVM model

```
model:
  id: 'ID-uHXdJRNxyH'
  type: SVM
  path: 'em.sl3'
  params:
    - svm_type: EPSILON_SVR
    - kernel_type: RBF
    - gamma: 15.0
    - C: 6.31
    - p: 0.359
    - cache_size: 1000
    - max_iterations: 5000
  training:
    - data_id: 'em'
    - dec_id: 'decision'
  test:
    - data_id: 'em'
    - dec_id: 'decision'
  input:
    - data_id: 'em_input_data'
    - dec_id: 'decision'
  output:
    - data_id: 'em_output_data'
    - col_id: 'decision'
```

Before starting the optimization set the max_iterations parameter to a lower value (e.g., 100 or 1000) because the model will be trained many times and this will reduce the time needed for the optimization. An unlimited number of iterations can be chosen with max_iterations = −1! The cache_size (in MB's) is important in the case of large datasets because it may speed up the calculations significantly.

The parameter optimization can be started on the left sidebar with 'Svm Parameter Optimizer'. The module will search for the optimal parameters and at the end it will train the model with the optimal parameters and with an unlimited number of iterations. You will have to adjust the project file manually with these optimal parameters (see the log file)! IMPORTANT: Retrain the model with the optimized parameters for full performance evaluation!

TIP: In the case of a very large data file like this, the training of the model after the parameter optimization may take too long (set automatically to an unlimited number of iterations). In this case you may stop the process with the 'Stop operation' button or just exit and restart the software.

You may of course also train a neural network model for the same purpose and check if the accuracy is better.

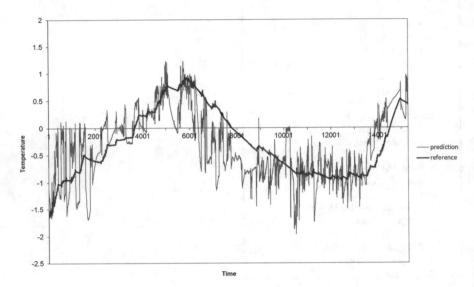

Fig. 8.19 Electric motor rotor temperature prediction with 1000 iterations

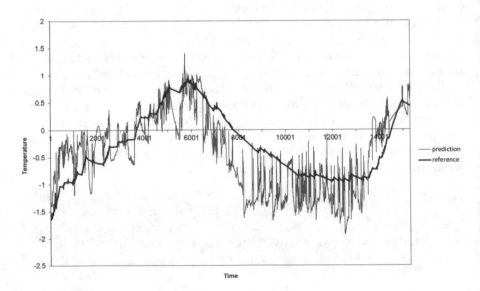

Fig. 8.20 Electric motor rotor temperature prediction with 2000 iterations

8.6.2 Analyzing the Results

The final total mean squared error (MSE) on the whole training dataset (nearly one million records) with a maximum number of 1000 iterations is 0.222, and with a maximum number of 2000 iterations is 0.18. There is a significant improvement in the MSE and also in the prediction results if we increase the maximum number of iterations from 1000 to 2000. The rotor temperature prediction results for one of the electric motors (ID 72) can be seen on Figs. 8.19 and 8.20. We could still increase the number of iterations significantly and obtain an even better prediction result. As mentioned before, a neural network model could also be developed for the same purpose.

Remember that normally we should divide the dataset into training and test datasets and test the results on data which is not seen during the training! This remains as an exercise for the reader.

TIP: Open the project database in the built-in database editor and go to the 'Browse data' tab. Select the output data table (it is 'em_output_data' in this example). There is a small save button near the table name which can be used to save the table in CSV format ('save the table as currently displayed'). This file can then be opened and analyzed in MS Excel.

CASE STUDY 6

8.7 Predictive Maintenance by Using Machine Learning

In many industries the reliability of machines is very important. In aerospace, transportation, manufacturing, utilities, etc. complex machines containing many components undergo periodic inspection and repair (preventive maintenance). The main challenge is to schedule preventive maintenance and component replacement in an optimal way such that the machines can work reliably and the components are not replaced too early. Reliability, high asset utilization and operational cost reduction are, in short, the aims of each company in these industries.

By using machine learning and historical data we can train a model which can predict when the next failure will occur and thus when preventive maintenance should be scheduled. We call these kinds of machine learning models predictive maintenance machine learning (PMML) models. There are two main types of PMML models:

- Regression models predict the remaining useful lifetime (RUL) of the machine or components.
- Classification models predict the failure within a pre-defined time period (time window).

In order to build a useful PMML model we need to go through some important steps which are summarized hereunder:

- Data collection
- Feature engineering
- Data labeling
- Defining the training and test datasets
- Handling imbalance in the data

All of these steps will be explained in detail in the following sections.

8.7.1   Data Collection

The data may come from different sources and usually contains failure history, maintenance history, machine operating conditions and usage, machine properties and operator properties.

8.7.1.1   Failure History: Time Series Data

Failure history data contains the failures of the whole machine and/or of its components. Such data can be found in maintenance records (e.g., parts replacement). Anomalies in the data may also be used as the failure history if no other data is available. Anomalies are special events which are represented by, for example, sudden changes in the data which can be detected (see Sect. 8.5).

It is crucial that there are a sufficient number of normal records and failure records in the data because otherwise the machine learning model will not be able to learn what normal is and what failure is!

Typically, there are two types of failure history data (Table 8.38), either containing a time stamp and machine id, which indicate the failure of the specific machine at the specific time step, or containing a time stamp, machine id and a

Table 8.38   Failure history data formats

Data type format 1			Data type format 2		
Time	Machine id		Time	Machine id	Reason

Table 8.39   Maintenance history data

Data format		
Time	Machine id	Action or error, etc.

failure reason in the form of a categorical variable or id. The failure reason may also be, for example, the ID of the component which caused the failure.

8.7.1.2 Maintenance History: Time Series Data

Maintenance history data contains information about replaced components, repairs, error codes, etc. Typically, a time stamp is recorded with the machine ID and the repaired component or maintenance action in the form of a categorical variable (Table 8.39).

8.7.1.3 Machine Operating Conditions and Usage: Time Series Data

This data, often called telemetry data, is collected by sensors connected to different parts of the machine and to its environment in operating conditions.

Typically, the time stamp, machine ID and several columns of operating conditions are included which may be numerical or categorical data (Table 8.40). Categorical data will need to be converted to numerical data.

8.7.1.4 Machine Properties: Static Data

Machine properties data (Table 8.41) contains all kinds of useful information about the machine. This is especially useful if the dataset contains data from different machines. The data may be numerical or categorical.

8.7.1.5 Operator Properties: Static Data

Operator properties data (Table 8.42) contains all kinds of useful information about the operator such as experience, education, etc. This may, for example, be useful if the failure is caused by not having enough knowledge or the lack of training. The

Table 8.40 Machine operating conditions and usage data

Data format				
Time	Machine id	cond1	cond2	...

Table 8.41 Machine properties static data

Data format			
Machine id	property1	property2	...

Table 8.42 Operator properties data

Data format			
Operator id	property1	property2	...

data may be numerical or categorical. The operators are usually linked to the telemetry data (machine operating conditions).

8.7.2   Feature Engineering

After we have collected all necessary data we must combine them into one synchronized dataset which can be fed into the machine learning model. We call this step feature engineering because we are building a dataset from features fabricated from the collected data. This is often a complex process in the case of a predictive maintenance model and the performance of the model will entirely depend on it.

The method for combining the collected data into the final dataset is usually very similar but of course business case and data dependent. Remember that the aim is to predict when the next failure of the machine will occur by using historical data.

There are two types of data, time series and static data. Static data can usually be simply combined with the other data by grouping them per machine ID. For example, if the maintenance history is defined with "time | machine ID | component", and the machine properties are defined with "machine ID | property 1 | property 2...", then we can simply add the static machine properties per machine ID as follows: "time | machine ID | component | property 1 | property 2..."

In the case of times series data we need to aggregate the data according to some pre-defined rules based on the business case. In the next section the data aggregating technique will be explained in detail and in a later section it will also be demonstrated as part of an example.

8.7.2.1   Aggregating Data in a Time Window

We usually want to predict machine failures in a future time period (time window) based on a historical time period. The data may be collected with a frequency of seconds, minutes, hours, etc. and we need to aggregate it into a pre-defined time period based on the business case. The evolution of the features in the time window is captured by the aggregated values. The machine learning model will learn which aggregated values result in a failure in the next time window. For example, if we want to predict whether a machine will fail in the next 24 h period, then we can use a time window of 24 h and label the aggregated records which fall into a 24 h window just *before* a failure occurs as FAILURE and all other records as NORMAL. It is of course business case dependent as to how long the time window should be. Sometimes 24 h is appropriate but sometimes we need to use a longer period, for example, to allow for a longer period of supply of repair parts. If it takes one week to get repair parts then we need to predict a failure much earlier in time.

Aggregating the data in a time window sometimes means calculating the average of all of the values or the sum, the standard deviation, variance, etc., or using a combination of these. Where combination means that extra columns (features) will be added to the same record with the different aggregates (average, the standard deviation... etc.). Each aggregate captures a specific property of the signal evolution in the time window. Which aggregates are needed depends on the business case and is often determined by trial and error.

By defining the time window we are actually telling the machine learning model how far in the past it has to look back for decision making. We often call this looking back period the 'lag' period. The time series aggregated in such a lag period are then called the lag features. There are two types of data in predictive maintenance, numerical and categorical. Lag features are created with numerical data. Categorical data is usually encoded as an integer and, for example, its most frequent value is used as a feature in the lag window.

Calculating the aggregate is very simple and can be summarized as follows:

- Choose the time window size (W).
- Choose the step size (S).
- Choose the type of aggregate(s) to use (average, count, cumulative sum, standard deviation, variance, maximum, minimum, etc.). Several aggregates may be chosen and added as features (extra columns in the record).
- Calculate the aggregate(s) for the first time window.

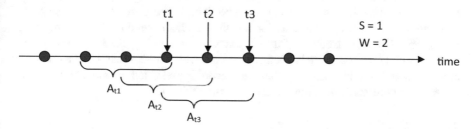

Fig. 8.21 Aggregate time window

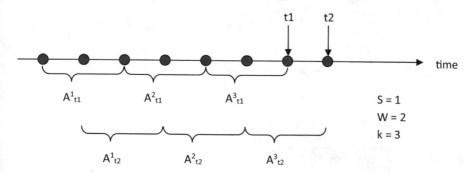

Fig. 8.22 Several aggregate time windows in the past

- Shift the time window by the step size (S) and repeat from the previous step until the end of the data (the final time window) is reached (Fig. 8.21).

It is also possible to capture trend changes further in the past by using k number of time windows in the past instead of just one. These are usually not overlapping windows because otherwise we would use the same information several times, but there may be an overlap with the next windows in the future. Figure 8.22 shows a possible configuration with $W = 2$, $S = 1$ and $k = 3$. Where t1 and t2 are two time steps in the final aggregated data file and A^1_{t1}, A^2_{t1} and A^3_{t1} are extra features (columns) in the data file for time step t1. This method captures sudden level changes in the signal instead of just smoothing them out with one aggregate.

8.7.3   Labeling the Data

Remember that there are two types of predictive maintenance machine learning models, classification and regression. In the case of a classification model we usually choose a time window (W) and label all aggregate records in the time window just before a failure occurs with, for example, FAILURE and other records with NOR-MAL (encoded as the integers 1 and 0 respectively). We usually choose the same time window for the aggregate as for the failure check window. For example, if $W = 3$ and $S = 1$ the labeling of the data can be seen on Fig. 8.23. The records at time steps t2, t3, t4 and t5 are labeled with failure which means that when a machine learning model is trained and a similar data record is presented the model will report a possible future machine failure in the next time window.

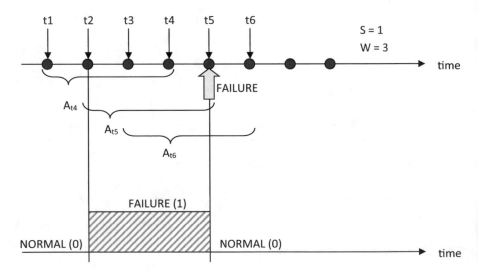

Fig. 8.23 Data labeling: classification

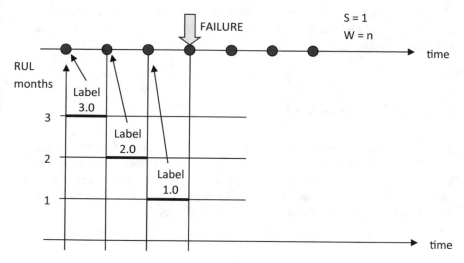

Fig. 8.24 Data labeling: regression

We may, of course, also decide to encode the root cause of the failure in the failure label and use different values depending on the root cause instead of just 1 (FAILURE). We will then have multiple failure classes and the machine learning model will not only predict the machine failure but it will also give the root cause! More information about root cause analysis with machine learning can be found in case study/Sect. 8.5.

In the case of a regression model (see Fig. 8.24) we usually calculate the remaining useful life (RUL) of the machine and simply increment it in the past. For example, if the time step is 1 month (S), then each time step in the past has a label with +1 month. Each label is the remaining useful life of the machine. Remember that in the case of regression the label is a decimal number!

8.7.4   Splitting the Final Dataset into Training and Test Sets

After aggregating the data into a final dataset the order of the records is less important than before but since it is still in a time-based order it may affect the machine learning model. For this reason it is often more useful to split the data into a training and test sets in a time-based manner. Usually we choose the first, larger part of the aggregated dataset as the training set and the second, smaller part as the test set and make sure that there is no failure label on the border of the two datasets (within W distance). In case we need a dataset for optimization or validation purposes then we take a chunk out of the training dataset.

Another way of splitting the data is by machine ID if there are several machines included in the dataset.

If for some reason you are not able to make a time-based manner or machine ID split, then it is not a huge problem since the original time series data is aggregated and labeled appropriately.

8.7.5   Imbalance in the Data

There is usually a huge imbalance in predictive maintenance datasets because we typically have much fewer failure records than normal records! Remember that we have two options to handle imbalance in the data, by using performance measures which are not sensitive to imbalance, for example, recall (see Chap. 3), and/or by using resampling techniques (see Sect. 4.3.2.6 in Chap. 4).

8.7.6   Example: Predictive Maintenance of Hydropower Turbines

Hydropower turbines are used in the energy industry for making electricity. They first convert the kinetic and potential energy of streaming water into mechanical work (by rotating a runner) and then into electricity by using a generator. Hydropower turbines are complex machines with many moving parts and therefore need regular maintenance. It is expensive to stop the machines and perform the maintenance work (no electricity production and maintenance costs), and therefore the correct scheduling of such maintenance is important.

In this example we will use data collected from 100 hydropower turbines [54] and we will train a machine learning model by using the AI-TOOLKIT. The model will be able to predict when the next failure may occur. Such a model can then be deployed in a real environment to make continuous predictions about future failures. The example is specific to hydropower turbines but very similar to other cases in different industries; therefore, if you understand this example then you will also be able to solve problems in other situations/industries.

8.7.6.1   The Input Data: Feature Engineering & Labeling

The *Error list*, *Failure list*, *Telemetry data* and *Machine properties* tables contain an excerpt from the four data sources in this example. Please refer to the previous sections for more information about these types of data sources.

Error list

Time	Machine ID	Error ID
7/9/2015 9:00:00 PM	3	4
7/18/2015 7:00:00 PM	3	1
7/20/2015 6:00:00 AM	3	2
7/20/2015 6:00:00 AM	3	3
...	...	...

Failure list

Time	Machine ID
12/14/2015 6:00:00 AM	9
1/19/2015 6:00:00 AM	10
4/4/2015 6:00:00 AM	10
5/19/2015 6:00:00 AM	10
...	...

Telemetry data: Machine operating conditions

Time	Machine ID	Voltage	Rotation	Pressure	Vibration
6/14/2015 9:00:00 AM	24	188.503	436.688	104.203	33.474
6/14/2015 10:00:00 AM	24	183.643	438.019	109.105	38.487
6/14/2015 11:00:00 AM	24	201.739	440.172	95.493	39.854
6/14/2015 12:00:00 PM	24	155.041	442.763	97.887	32.126
...	...	...	...	...	...

Machine properties

Machine ID	Model ID	Age
43	3	14
44	4	7
45	3	14
46	4	10
...	...	...

The data was collected from 100 machines in 2015. The telemetry data is real time data averaged per hour [54]. The error list comes from error logs and contains different types of non-breaking errors which were collected while the machines were running (no failure). The times of error occurrences are rounded to the closest hour in order to be in sync with the telemetry data. The failure list contains the times when the machines had a failure (stopping), and there is also a machine properties list with the model number and age of the machines.

A preview of the telemetry data for machine 1 can be seen on Fig. 8.25 (only the first 1000 points are shown).

The aim of predictive maintenance is that the machine learning model learns when (after which conditions of the machine) a failure is expected. Every machine

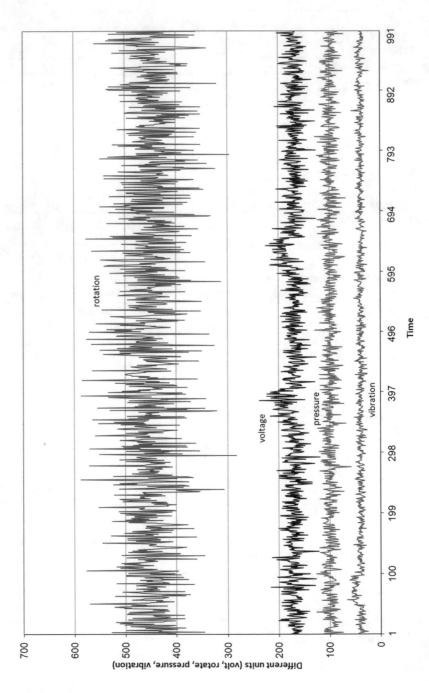

Fig. 8.25 Telemetry machine 1

and its components degrade with time and this degradation creates trends in the data which can be discovered by the machine learning model. The telemetry data, in the case of the turbines, is hourly data and we could combine it with the other data to have a final hourly dataset. But the question is whether hourly data would be appropriate in this case or not? This is an important decision which depends on the business case (see Sect. 8.7.2.1)!

In the case of the hydropower turbines we need to take a longer period than one hour because one hour does not contain enough information and the hourly data is very noise. Let us take for the time window length 24 h ($W = 24$) and for the step size 3 hours ($S = 3$). This means that every 3 h we will aggregate 24 h of (overlapping) historical data. If we train a machine learning model with this aggregated dataset it will be able to predict whether a machine failure is expected the next 24 h.

After the time window and the step size have been chosen the next step is to choose which type of aggregate we want to add (average, standard deviation, count...). This decision is also business case dependent and often a process of trial and error. It is possible that the set of aggregated features we have chosen might need to be modified if the performance of the machine learning model is not satisfactory. In the case of the hydropower turbines we take the average and the standard deviation of the telemetry data and the count of the errors in the chosen time window (W).

The final set of features can be seen in Table 8.43.

Note that we count the occurrence of each error in the given time window and this count is added to each record. The failure (decision) column is added already to the above aggregated dataset but we still need to decide how to set its values for each record every 3 h (step size). Since we are predicting whether the machine will fail in the next 24 h we set the failure (decision value) to 1 (TRUE) in the 24 h window just before a failure occurs, and to 0 (FALSE) for all other values. Since we have chosen a time step of 3 h several records will be marked with 1 in the failure column in any given time window.

You may automate the above aggregating and combining steps with a script or use a data mining tool; however, it is not the aim of this book to explain this process in detail.

8.7.6.2 The Machine Learning Model & Performance Evaluation

After the important steps of data preparation and feature engineering have been completed, the next step is to choose a machine learning model and to train it. As usual you should divide the dataset into training and test sets and evaluate the performance of the model on both sets. In this example we will keep it simple and train and test the model on the whole dataset.

Table 8.43 Final features

Failure	Machine ID	Voltage avg	Rotation avg	Pressure avg	Vibration avg	Voltage std	Rotation std	Pressure std	Vibration std	Error 1	Error 2	Error 3	Error 4	Error 5	Model	Age
...	...	...	...	...	...	...	...	...	...	...	...	...	...	...	...	...

Table 8.44 RF model

```
RF model - hydropower turbine
model:
  id: 'ID--eXGOKIIAs'
  type: RF
  path: 'pm.sl3'
  params:
    - min_leaf_size: 5
    - num_trees: 5
    - k_fold: 3
  training:
    - data_id: 'pm'
    - dec_id: 'decision'
  test:
    - data_id: 'pm'
    - dec_id: 'decision'
  input:
    - data_id: 'pm_input_data'
    - dec_id: 'decision'
  output:
    - data_id: 'pm_output_data'
    - col_id: 'decision'
```

We will build, train and test two machine learning models using the AI-TOOLKIT, a random forest classification model and an SVM classification model, and we will compare the results.

As usual the first step is to create an AI-TOOLKIT database with the "Create New AI-TOOLKIT Database" command on the Database tab on the left taskbar. Save the database in a directory of your choice. The second step is to import the data in the database created in the previous step with the "Import Data into Database" command. Do not forget to indicate the number of header rows (if any) and the correct zero based index of the decision column (0 in this example)! Next we must create the AI-TOOLKIT project file. Use the "Open AI-TOOLKIT Editor" command and then insert the chosen model template with the "Insert ML Template" button.

Random Forest (RF) Model
There are three parameters which can be adjusted for an RF model: the minimum number of leaves, the number of trees which must be trained and the number of folds if cross validation is desired. For more information about this model read Sect. 2.2.4 in Chap. 2. The optimal parameters are filled in already in the RF model in Table 8.44. It is usually a trial and error process to find the optimal parameters.

The extended performance results after training the RF model can be seen in Table 8.45. Remember that these results are for the whole training dataset and that in practice you should divide the dataset into training and test sets and evaluate the model on both datasets! The global accuracy is very high even with three fold cross

Table 8.45 RF model extended performance results

```
RF model performance evaluation results - hydropower turbine


Training Accuracy: 99.97 %
KFoldCV Accuracy:  99.92 %

Extended Performance Evaluation Results:
Confusion Matrix [predicted x original] (number of classes: 2):


                                    (0)                      (1)
                    (0)           285649                      37
                    (1)               56                    5558


        Accuracy: 99.97%
           Error: 0.03%
        C.Kappa: 99.15%

                                    (0)                      (1)
            Precision:            99.99%                   99.00%
               Recall:            99.98%                   99.34%
                  FNR:             0.02%                    0.66%
                   F1:            99.98%                   99.17%
                  TNR:            99.34%                   99.98%
                  FPR:             0.66%                    0.02%

```

validation. Cross validation divides the dataset into three random parts and trains and tests the model with each dataset and finally averages the accuracies. Remember also that because of the imbalance in the data the global accuracy is not very important in this case, but we should concentrate on performance measures which are not sensitive to imbalance in the data, i.e., recall, Cohen's kappa (abbreviated to C. Kappa in the AI-TOOLKIT), FNR, TNR and FPR.

Cohen's kappa is a kind of accuracy measure corrected for chance, and therefore we should say that the accuracy of this model is 99.15% and not 99.97% because of the high imbalance! The recall value also indicates very good results. The model wrongly predicts a machine failure in 0.02% of the cases (these are false alarms and should not be too high) and it misses a failure in 0.66% of the cases only (this should be as low as possible because missing a failure means extra costs).

Table 8.46 Hydropower turbine SVM model

```
SVM model - hydropower turbine
model:
  id: 'ID-wkxvQdAjuo'
  type: SVM
  path: 'pm.sl3'
  params:
    - svm_type: C_SVC
    - kernel_type: RBF
    - gamma: 0.60822
    - C: 79.45974
    - degree: 3
    - coef0: 0
    - p: 0.1
    - cache_size: 1000
    - max_iterations: -1
  training:
    - data_id: 'pm'
    - dec_id: 'decision'
  test:
    - data_id: 'pm'
    - dec_id: 'decision'
  input:
    - data_id: 'pm_input_data'
    - dec_id: 'decision'
  output:
    - data_id: 'pm_output_data'
    - col_id: 'decision'
```

In the next section we will compare the RF results with the SVM results.

SVM Model and Comparison of the RF & SVM Results
There are several parameters which must be adjusted for an SVM model. We can use the built-in SVM parameter optimization module in the AI-TOOLKIT for determining the optimal values of these parameters (see Sect. 9.2.3.1 in Chap. 9). For more information about this model refer to Sect. 2.2.1 in Chap. 2. The optimal parameters are filled in already in the SVM model (see Table 8.46).

The extended performance results after training the SVM model can be seen in Table 8.47. Remember that these results are for the whole training dataset and that in

Table 8.47 SVM model extended performance results

```
SVM model performance evaluation results - hydropower turbine


Training Accuracy = 99.97%

Extended Performance Evaluation Results:
Confusion Matrix [predicted x original] (number of classes: 2):

                                        (0)                    (1)
                   (0)                285622                     18
                   (1)                    83                   5577


   Accuracy: 99.97%
      Error: 0.03%
   C.Kappa: 99.08%

                                        (0)                    (1)
          Precision:                  99.99%                 98.53%
             Recall:                  99.97%                 99.68%
                FNR:                   0.03%                  0.32%
                 F1:                  99.98%                 99.10%
                TNR:                  99.68%                 99.97%
                FPR:                   0.32%                  0.03%

```

practice you should divide the data into training and test sets and evaluate the model on both datasets! The global accuracy is very high at 99.97%. Remember also that because of the imbalance in the data the global accuracy is not very important in this case, but we should concentrate on performance measures which are not sensitive to imbalance in the data, i.e., recall, Cohen's kappa (C.Kappa), FNR, TNR and FPR.

As mentioned previously Cohen's kappa is a kind of accuracy corrected for chance, and therefore we should say that the accuracy of this model is 99.08% and not 99.97% because of the imbalance! The recall value also indicates very good results. The model wrongly predicts a machine failure in 0.03% of the cases (these are false alarms and should not be too high) and it misses a failure in 0.32% of the cases only (this should be as low as possible because missing a failure means extra costs).

Both RF and SVM models have similar results; the SVM model is slightly better in missing less failures but it gives a few more false alarms.

It always depends on the business case as to which performance measure(s) we focus on and which model(s) we prefer. In the manufacturing industry it is usually the costs and safety factors that are driving these decisions whereas in, for example, healthcare it is usually the issues which are more important to patients.

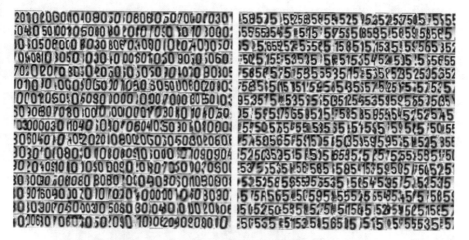

Fig. 8.26 Image collage with numbers for number recognition [62]

CASE STUDY 7

8.8 Image Recognition with Machine Learning

It may be useful in many industries to be able to automatically recognize specific forms, numbers, letters, etc. For example, erroneous products on an industrial production line may have a different shape than normal products. This section will introduce image recognition with machine learning and by using the AI-TOOLKIT. Usually we use feedforward convolutional neural networks for image learning because of their high performance in this field. More information about this subject can be found in Sect. 2.2.3 of Chap. 2.

We will go, step by step, through an example which trains a machine learning model to recognize numbers between 0 and 9. We will not just feed nice, clear numbers to the model, we will make it more difficult and more real by using many different types of numbers and noise. In this way, the example will be very close to what would happen in a real problem.

8.8.1 The Input Data

As usual the first step is to create an AI-TOOLKIT database with the "Create New AI-TOOLKIT Database" command on the Database tab on the left taskbar. Save the

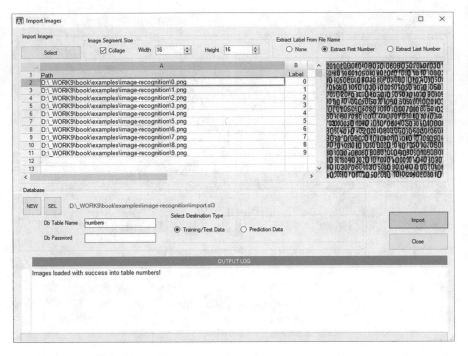

Fig. 8.27 AI-TOOLKIT import image module

database in a directory of your choice. The second step is to import the data, images in this case, into the database created in the previous step with the "Import Images into Database" command. The AI-TOOLKIT can import single images and also a collage of images in many different formats. A collage is an image which contains several equal sized images. By using a collage we can group the same classes together, in this case the same numbers. The following two images show the collage used for image class 0 and for image class 5 (Fig. 8.26):

The other classes (numbers) have similar collages.

As you can see the numbers contain many different types of numbers (skewed, distorted, etc.) and also noise in the form of partial numbers. The reason for doing this is so that the machine learning model will recognize the number even in difficult circumstances.

On Fig. 8.27 you can see the import image module. You can import single images and/or a collage of images. If you import a collage it is important to check the 'Collage' check box and set the width and height of the sub-images (16x16 in this example) before you select the files with the 'Select' button. The software can also extract the class labels automatically from the file names in the form of a number at the beginning or at the end of the file name. For this you have to select one of the 'Extract First Number' or 'Extract Last Number' options (only numbers are supported).

Table 8.48   Project file for image recognition

```
model:
  id: 'ID-6'
  type: FFNN2_C
  path: 'image-recognition.sl3'
  params:
    - layers:
      - Convolution:
          inSize: 1        # in depth
          outSize: 12      # out depth
          kW: 4            # filter/kernel width
          kH: 4            # filter/kernel height
          dW: 1            # stride w
          dH: 1            # stride h
          padW: 0          # padding w
          padH: 0          # padding h
          inputWidth: 16   # input width
          inputHeight: 16  # input height
      - MaxPooling:
          kW: 3            # pooling width
          kH: 3            # pooling height
          dW: 2            # stride w
          dH: 2            # stride h
          floor: true      # floor (true) or ceil (false)
                           # rounding operator
      - Convolution:
          inSize: 12
          outSize: 18
          kW: 2
          kH: 2
          dW: 1
          dH: 1
          padW: 0
          padH: 0
          inputWidth: 6
          inputHeight: 6
      - Convolution:
          inSize: 18
          outSize: 24
          kW: 2
          kH: 2
          dW: 1
          dH: 1
          padW: 0
          padH: 0
          inputWidth: 5
          inputHeight: 5
      - Linear:
      - Dropout:
          ratio: 0.9
          nodes: 96
      - Linear:
    - num_cycles: 200
    - batch_size: 32
    - learning_rate: 0.01
  training:
    - data_id: 'training_data'
```

(continued)

Table 8.48 (continued)

```
     - dec_id: 'label'
   test:
     - data_id: 'training_data'
     - dec_id: 'label'
   input:
     - data_id: 'input_data'
     - dec_id: 'label'
   output:
     - data_id: 'input_data'
     - col_id: 'label'
```

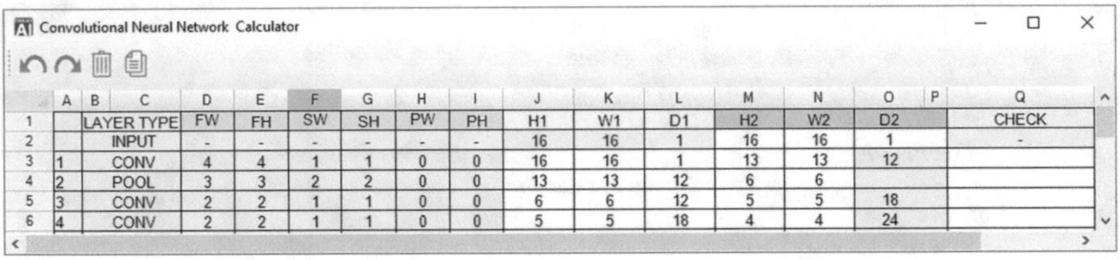

	A	B	C	D	E	F	G	H	I	J	K	L	M	N	O	P	Q
			LAYER TYPE	FW	FH	SW	SH	PW	PH	H1	W1	D1	H2	W2	D2		CHECK
1																	
2			INPUT	-	-	-	-	-	-	16	16	1	16	16	1		-
3	1		CONV	4	4	1	1	0	0	16	16	1	13	13	12		
4	2		POOL	3	3	2	2	0	0	13	13	12	6	6			
5	3		CONV	2	2	1	1	0	0	6	6	12	5	5	18		
6	4		CONV	2	2	1	1	0	0	5	5	18	4	4	24		

Fig. 8.28 Convolutional neural network calculator

The rest of the import options are similar to the import delimited data file options seen several times in previous sections of the book (database name and the selection of training or prediction).

The import module will not only import the images into the database but it will also convert them into the appropriate format for the machine learning model.

8.8.2   The Machine Learning Model

Next we must create the AI-TOOLKIT project file. Use the "Open AI-TOOLKIT Editor" command and then insert the chosen model template with the "Insert ML Template" button. Insert the Convolutional Neural Network Classification template from the Supervised Learning group and then save the project file in the same folder as the database.

Table 8.48 shows the final project file with the optimal parameters already filled in (as will be explained later).

As you can see a convolutional neural network is a bit more complex than the other models, with many more parameters. More information about the parameters and possible layer types can be found in Sect. 2.2.3 of Chap. 2!

In order to make the calculation of the parameter values easier there is a built-in convolutional neural network calculator in the AI-TOOLKIT (see Fig. 8.28).

In the first convolutional layer the input width and input height are 16, the size of our images (the sub-image size in the collage). The filter size, stride and the padding must be chosen in each convolutional and pooling layer (green input cells). It is not easy to choose these numbers for the first time but as you will get more experience it will become easier and the calculation module will warn you if you choose the wrong numbers. See also Sect. 2.2.3 for directions about how to choose these values!

The input width and height and the inSize (input depth) for each subsequent layer are calculated by the calculator (Fig. 8.28; white cells) and can be added to the project file (the calculation is explained in Chap. 2). Notice that for the first convolution layer the input depth is always 1 because we use grayscale images! Color images would take too much memory and too long to train, and usually there is no need for using color.

The last dropout layer is added to improve generalization (see Sect. 2.2.2.2 in Chap. 2). The number of nodes (96) is taken as a value between the output nodes count (10 in this example because there are 10 classes!) and the output nodes count from the preceding convolutional layer which is 4 x 4 x 24 = 384. Choosing the value 96 is a trial and error process and a question of experience. There are no scientific rules for these kinds of decisions. The same is true for the number of iterations (cycles), batch size and optimal learning rate.

8.8.3   Training & Evaluation

The AI-TOOLKIT is accelerated in several ways with parallel processing, by using special processor instructions etc., and because of this the training of this example takes only 23 seconds! Usually image learning takes a long time due to the huge number of input nodes and internal nodes in the network and therefore all kinds of accelerations are used. One of the fastest existing learning algorithms today uses GPU acceleration but you need special video card(s) (a graphics accelerator card) in order to use it. High end video algorithms usually work with GPU acceleration for training and GPU or CPU acceleration for inference (prediction). The AI-TOOLKIT uses special CPU instruction set acceleration and parallel processing for both training and prediction.

The extended performance evaluation results on the whole training dataset can be seen in Table 8.49.

We get the above results by testing the model on the training dataset. In order to make the test more realistic (generalization error) let us create some modified images of numbers which are not seen during the training as shown on Figs. 8.29, 8.30 and 8.31.

Table 8.49 Extended performance evaluation results

Iteration 200: Training Error (mse = 0.025859; on the predicted label = 0.00%)

Confusion Matrix [predicted x original] (number of classes: 10):

Confusion Matrix	0	1	2	3	4	5	6	7	8	9
0	256	0	0	0	0	0	0	0	0	0
1	0	256	0	0	0	0	0	0	0	0
2	0	0	256	0	0	0	0	0	0	0
3	0	0	0	256	0	0	0	0	0	0
4	0	0	0	0	256	0	0	0	0	0
5	0	0	0	0	0	256	0	0	0	0
6	0	0	0	0	0	0	256	0	0	0
7	0	0	0	0	0	0	0	256	0	0
8	0	0	0	0	0	0	0	0	256	0
9	0	0	0	0	0	0	0	0	0	256

Accuracy:	100.00%									
Error:	0.00%									
C.Kappa:	100.00%									

	0	1	2	3	4	5	6	7	8	9
Precision:	100%	100%	100%	100%	100%	100%	100%	100%	100%	100%
Recall:	100%	100%	100%	100%	100%	100%	100%	100%	100%	100%
FNR:	0%	0%	0%	0%	0%	0%	0%	0%	0%	0%
F1:	100%	100%	100%	100%	100%	100%	100%	100%	100%	100%
TNR:	100%	100%	100%	100%	100%	100%	100%	100%	100%	100%
FPR:	0%	0%	0%	0%	0%	0%	0%	0%	0%	0%

Fig. 8.29 Modified 'eight'
image

Fig. 8.30 Modified 'zero'
image

Fig. 8.31 Modified 'five'
image

Fig. 8.32 Rotated 'eight'
image

Fig. 8.33 Rotated 'zero'
image

Fig. 8.34 Rotated 'five'
image

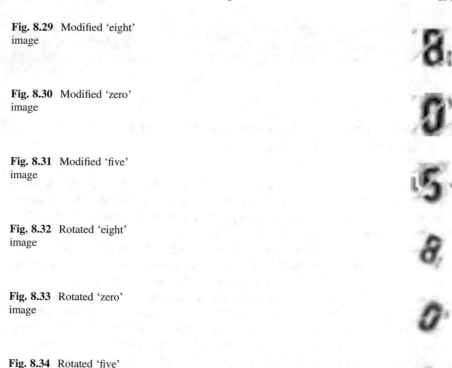

When importing single images (not collages) you should not select the collage option! Make sure also that you select prediction instead of training as the destination data type! After importing the images into the database change the data_id in the 'input' and 'output' sections of the project file to the database table name of the imported images! Then simply by using the 'Predict with AI model' command on the left toolbar, all input images will be processed by the model automatically.

The model can still recognize all three numbers perfectly! You can see the results by opening the database in the built-in database editor and looking at the database table used to store the input images. The label column is filled in by the model. Note that we use the same table for input and output.

Let us try now something different and rotate the numbers clockwise by 25 degrees as it is shown on Figs. 8.32, 8.33 and 8.34.

When feeding these images to the model it identifies the following numbers: 8, 7 and 3. Only the number 8 is recognized correctly. In order to make rotated numbers work well, we would need to add rotated numbers to the training data!

8.9   Detecting Future Cardiovascular Disease (CVD) with Machine Learning

Cardiovascular diseases (CVDs) are a group of disorders of the heart and blood vessels. The World Health Organization (WHO) reported in [53] that "CVDs are the number 1 cause of death globally: more people die annually from CVDs than from any other cause. An estimated 17.9 million people died from CVDs in 2016, representing 31% of all global deaths. Of these deaths, 85% are due to heart attack and stroke. Over three quarters of CVD deaths take place in low- and middle-income countries. Out of the 17 million premature deaths (under the age of 70) due to non-communicable diseases in 2015, 82% are in low- and middle-income countries, and 37% are caused by CVDs. Most cardiovascular diseases can be prevented by addressing behavioral risk factors such as tobacco use, unhealthy diet and obesity, physical inactivity and harmful use of alcohol using population-wide strategies. People with cardiovascular disease or who are at high cardiovascular risk (due to the presence of one or more risk factors such as hypertension, diabetes, hyperlipidemia or already established disease) need early detection and management using counseling and medicines, as appropriate."

"The most important behavioral risk factors of heart disease and stroke are unhealthy diet, physical inactivity, tobacco use and harmful use of alcohol. The effects of behavioral risk factors may show up in individuals as raised blood pressure, raised blood glucose, raised blood lipids, and overweight and obesity. These "intermediate risks factors" can be measured in primary care facilities and indicate an increased risk of developing a heart attack, stroke, heart failure and other complications. Cessation of tobacco use, reduction of salt in the diet, consuming fruits and vegetables, regular physical activity and avoiding harmful use of alcohol have been shown to reduce the risk of cardiovascular disease. In addition, drug treatment of diabetes, hypertension and high blood lipids may be necessary to reduce cardiovascular risk and prevent heart attacks and strokes. Health policies that create conducive environments for making healthy choices affordable and available are essential for motivating people to adopt and sustain healthy behavior." [53]

8.9.1   The Dataset

The Framingham Heart Study is a long-term cardiovascular cohort study of residents of the city of Framingham, Massachusetts, in the USA. The study began in 1948

Table 8.50 Cardiovascular disease (CVD) dataset description

Variable	Description	Values/Ranges/Units
SEX	Participant sex	1 = men 0 = women
AGE	Age at exam (years)	Continuous
EDUCATION	Education level	1,2,3 or 4
CURSMOKE	Current cigarette smoking at exam	0 = not current smoker 1 = current smoker
CIGPDAY	Number of cigarettes smoked each day	0 = not current smoker 1–90 cigarettes per day
BPMEDS	Use of anti-hypertensive medication at exam	0 = not currently used 1 = current use
PREVHYP	Prevalent hypertensive. Subject was defined as hypertensive if treated or if second exam showed mean systolic was > = 140 mmHg or mean diastolic > = 90 mmHg	0 = free of disease 1 = prevalent disease
PREVSTRK	Prevalent stroke	0 = free of disease 1 = prevalent disease
DIABETES	Diabetic according to criteria of first exam treated or first exam with casual glucose of 200 mg/dL or more	0 = not a diabetic 1 = diabetic
TOTCHOL	Serum total cholesterol (mg/dL)	107.0–696.0
SYSBP	Systolic blood pressure (mean of the last two of three measurements) (mm/hg)	82.0 300.0
DIABP	Diastolic blood pressure (mean of the last two of three measurements) (mm/hg)	30.0–160.0
BMI	Body mass index (weight in kilograms/height meters squared)	14.0–57.0
HEARTRTE	Heart rate (ventricular rate) in beats/min	37–220
GLUCOSE	Casual serum glucose (mg/dL)	39.0–478.0
TENYEARCVD	Within 10 years high risk of having CVD	1 = high chance of disease 0 = no disease expected *Decision variable!* Classification!

with more than 5000 adult subjects from Framingham, and it is still an ongoing study with third generation of participants. The study generates a lot of relevant data which can be used to train a machine learning model. The trained model is then able to predict if a person is at high risk of developing a cardiovascular disease within the next 10 years. This can, of course, be very useful because the life style of the person can be adapted and a serious disease (and possible death) can be prevented!

The description of the Framingham dataset can be seen in Table 8.50.

Before using any dataset it is important to review it and clean it if required. For more information about all of the tasks concerning the input data see Chap. 4. This dataset had several records with missing data which have been removed! No other data cleaning actions were needed. The two classes of the decision variable (TEN YEARCVD) are not completely in balance but this causes no problem.

Table 8.51 The CVD SVM model

```
model:
  id: ID-xER-OjnxqS'
  type: SVM
  path: 'chd.sl3'
  params:
    - svm_type: C_SVC
    - kernel_type: RBF
    - gamma: 15.0
    - C: 1000
    - cache_size: 100
    - max_iterations: -1
  training:
    - data_id: 'chd'
    - dec_id: 'decision'
  test:
    - data_id: 'chd'
    - dec_id: 'decision'
  input:
    - data_id: 'chd_input_data'
    - dec_id: 'decision'
  output:
    - data_id: 'chd_output_data'
    - col_id: 'decision'
```

As usual the first step is to create an AI-TOOLKIT database with the "Create New AI-TOOLKIT Database" command on the Database tab on the left taskbar. Save the database in a directory of your choice. The second step is to import all data into the database created in the previous step with the "Import Data into Database" command. Do not forget to indicate the number of header rows (if any) and the correct zero based index of the decision column! Next we must create the AI-TOOLKIT project file. Use the "Open AI-TOOLKIT Editor" command and then insert the chosen model template with the "Insert ML Template" button. In this example we will choose a supervised SVM model.

8.9.2   Modeling & Parameter Optimization

Each ML model has its own parameters. For SVM models there is a built-in parameter optimization module in the AI-TOOLKIT. For a classification SVM model (CVD within 10 years is the decision variable) the svm_type is 'C_SVC'. This model uses the kernel_type, gamma, C, p, cache_size and max_iterations parameters. The degree and coef0 parameters are not used in this model because the kernel_type is RBF. More information about this model type can be found in

Table 8.52 The CVD SVM
model performance results

Confusion matrix		Original	
		class 0	class 1
Predicted	Class 0	2034	3
	Class 1	0	1619
Accuracy	99.92%		
Error	0.08%		
C.Kappa	99.83%		
Precision		99.85%	100.00%
Recall		100.00%	99.82%
FNR		0.00%	0.18%
F1		99.93%	99.91%
TNR		99.82%	100.00%
FPR		0.18%	0.00%

Sect. 2.2.1 of Chap. 2! All algorithm and parameter options are explained there in detail.

The ML model for this example with the optimal parameters can be seen in Table 8.51.

Before starting the optimization set the max_iterations parameter to a lower value (e.g., 100 or 1000) because the model will be trained many times and this will reduce the time needed for the optimization. An unlimited number of iterations can be set by max_iterations $= -1$! The cache_size (in MB's) is important in the case of large datasets because it may speed up the calculations significantly. This dataset contains only 3656 records, and therefore the default value of 100 is sufficient.

The parameter optimization can be initiated on the left sidebar with "Svm Parameter Optimizer". The module will search for the optimal parameters and at the end it will train the model with the optimal parameters and with an unlimited number of iterations. You will have to adjust the project file manually with these optimal parameters (see output log)!

You may, of course, also train a neural network model for the same purpose and check if the accuracy is better. The next section will explain the results.

8.9.3   Analysis of the Results

The performance results can be seen in Table 8.52. The SVM model is able to learn with 99.92% global accuracy and there were only three mistakes of *not* reporting a possible risk of CVD disease from 3656 cases (recall of 99.82%), and no mistakes at all in terms of falsely reporting a risk of CVD disease (recall of 100%).

It is always important in the case of healthcare problems to look at the per-class performance measures and evaluate the performance of the model depending on what important is for the patients! In this case potential false reporting of a risk of

CVD disease is important, but it is more important to reduce (if possible) the three instances of not reporting a risk of CVD disease! The patients who are falsely warned about CVD disease can have a checkup in a hospital to test the results, but the patients who are not warned may continue with their unhealthy lifestyle and may be in severe danger!

These results were measured on the whole training dataset. In practice you should use more data (if possible) and divide the input data into separate training and test sets. The test set should not exist in the training dataset (unseen during the training). More information about the division of training and test sets and about the generalization error can be found in Sect. 1.2 of Chap. 1 and in Sect. 4.3.2.9 of Chap. 4.

CASE STUDY 9

8.10 Business Process Improvement with Machine Learning

The main aim of this very simple example is to demonstrate the applications of machine learning in business process improvement in the healthcare sector. Many of the principles and ideas applied in this case study are also applicable in many other sectors!

We will try to improve the post-operative patient care process in a hospital. After an operation, according to the current post-operative patient care process, patients need to be examined by a medical doctor in order to determine where the patients should be sent from the post-operative recovery area. The possibilities are the following:

- The patient may go home
- The patient needs to go to the general care hospital floor (GC)
- The patient needs to be transferred to intensive care (IC)

In order to improve this process (make the process much faster and more reliable) the hospital needs to collect all necessary data which is needed for decision making for multiple patients and then use this data to train a machine learning model. After the machine learning model is successfully trained, a hospital employee (e.g., a nurse) can simply feed the specific patient data to the model and the machine learning model will instantly determine (inference) what should happen with the patient. This improved process is much faster because the waiting time for the medical doctor is eliminated, and in many cases the decision is more reliable because the machine learning model does not get tired or confused by external factors, the medical doctor or specialist can do other important things and last, but not least, the patient will be more satisfied with the faster process! Several important reasons to implement such a process improvement!

8.10.1   The Dataset

A subset of the data which will be used to train the machine learning model can be seen in Table 8.53. The data is real patient data collected in a hospital [55]. The different symbols in the data are explained below.

Most of the numerical attributes, such as temperature, are grouped and converted to textual categories. This is one of the tricks which can be used while training a machine learning model. You may of course also use numerical values but then you will give more freedom to the model! In many situations it is sufficient to group the data into well-chosen textual categories. The AI-TOOLKIT can handle both categorical and numerical attributes.

The collected attributes and groupings are the following:

- **L-CORE**: the patient's internal temperature: high >37°, 37° ≥ mid ≥ 36°, low <36°.
- **L-SURF**: the patient's surface temperature: high >36.5°, 35° ≥ mid ≥ 36.5°, low <35°.
- **L-O2**: the oxygen saturation: excellent ≥98%, 98% > good ≥90%, 90% > fair ≥80%, poor <80%.
- **L-BP**: the last measurement of blood pressure: high >130/90, 130/90 ≥ mid ≥ 90/70, low <90/70.
- **SURF-STBL**: the stability of patient's surface temperature: stable, mod-stable and unstable.
- **CORE-STBL**: the stability of patient's core temperature: stable, mod-stable and unstable.
- BP-STBL: the stability of patient's blood pressure: stable, mod-stable and unstable.
- **COMFORT**: the patient's perceived comfort at discharge, measured as an integer between 0 and 20.
- **DECISION**: the discharge decision: Home, the patient needs to be prepared to go home; GC; the patient must be sent to the general care hospital floor; IC, the patient must be sent to intensive care.

There is a huge imbalance in the data. There are 64 cases with a 'Home' decision, 24 cases with a 'GC' decision and only 2 cases with an 'IC' decision. The dataset is also very small with only 90 records. We will need to take the imbalance into account while evaluating the machine learning model, or we could also resample the data and remove the imbalance. We will try both in this example.

Table 8.53 The patient care dataset

L-CORE	L-SURF	L-O2	L-BP	SURF-STBL	CORE-STBL	BP-STBL	COMFORT	DECISION
mid	low	excellent	mid	stable	stable	stable	15	GC
mid	high	excellent	high	stable	stable	stable	10	Home
high	low	excellent	high	stable	stable	mod-stable	10	GC
mid	low	good	high	stable	unstable	mod-stable	15	GC
mid	mid	excellent	high	stable	stable	stable	10	GC
high	low	good	mid	stable	stable	unstable	15	Home
mid	low	excellent	high	stable	stable	mod-stable	5	Home
...	...	...	...	...	...	...	...	...

As usual the first step is to create an AI-TOOLKIT database with the "Create New AI-TOOLKIT Database" command on the Database tab on the left taskbar. Save the database in a directory of your choice. The second step is to import all data into the database created in the previous step with the "Import Data into Database" command. Do not forget to indicate the number of header rows (if any) and the correct zero based index of the decision column! We also have to select the two 'Conversion/Resampling' options for the automatic conversion of the categorical values and, if we want to resample the dataset, the resample for imbalance reduction option.

In this example we will import the dataset twice, first without resampling and then with resampling in order to test both cases. The dataset(s) can be imported into the same AI-TOOLKIT database but with a different name. For the categorical conversion we use the default option which is integer encoding. When resampling the dataset we must also set the 'Majority Limit' option to 64 (the number of cases with the majority class) in order to have 64 resampled cases for all three classes (Home, GC and IC).

More information about categorical values conversion can be found in Sect. 4.3.2.2 of Chap. 4, and more information about resampling for imbalance removal can be found in Sect. 4.3.2.6 of Chap. 4.

8.10.2   Training the Machine Learning Model

Next we must create the AI-TOOLKIT project file. Use the "Open AI-TOOLKIT Editor" command and then insert the chosen model template with the "Insert ML Template" button. We will train a simple neural network classification model with three internal layers. The project file and the optimal parameters can be seen in Table 8.54.

More information about neural networks and their parameters can be found in Sect. 2.2.2 of Chap. 2.

As mentioned previously, we will train the model on the original dataset and also on the resampled version. The results are shown in the next section.

8.10.3   Analysis of the Results

The training process and the extended evaluation results with the original (not resampled) dataset can be seen in Table 8.55.

Table 8.54 Patient care project file

```
model:
  id: 'ID-BBEMTNTNEY'
  type: FFNN1_C
  path: 'postoperative.sl3'
  params:
    - layers:
      - Linear:
      - TanHLayer:
          nodes: 100
      - Linear:
      - TanHLayer:
          nodes: 80
      - Linear:
      - TanHLayer:
          nodes: 40
      - Linear:
    - iterations_per_cycle: 1000
    - num_cycles: 10
    - step_size: 5e-5
    - batch_size: 1
    - optimizer: SGD_ADAM
    - stop_tolerance: 1e-5
    - sarah_gamma: 0.125
  training:
    - data_id: 'postoperative'
    - dec_id: 'decision'
  test:
    - data_id: 'postoperative'
    - dec_id: 'decision'
  input:
    - data_id: 'po_input_data'
    - dec_id: 'decision'
  output:
    - data_id: 'po_output_data'
    - col_id: 'decision'
```

The training process and the extended evaluation results with the resampled dataset can be seen in Table 8.56.

Both results, trained with the original dataset and trained with the resampled dataset, are good but there is a slight difference between them. The resampled dataset has better results at first sight but we should think about what happens to the patients, and which kind of errors are 'better' to make and which errors should not be made!

When using the resampled dataset the model sends three patients home but they should go to general care (see confusion matrix). This is the kind of error which should not happen and therefore we should prefer the model trained with the original dataset because it only makes this mistake once! And this is despite the fact that the model trained with the resampled dataset has higher accuracy.

Sending patients to general care when they could go home is less dangerous for the patients and the doctor can still send them home from general care; therefore, this kind of error is preferred compared to the other errors.

None of the models make the worst mistake and send a patient home or to general care when he/she should go to intensive care. This is very important in this case!

Table 8.55 Patient care training process and extended evaluation results (original dataset)

```
AI Training... (Model ID: ID-BBEMTNTNEY).
 0 - training accuracy = 71.11 % (Model ID: ID-BBEMTNTNEY) (training time: 0m 2s ).
 1 - training accuracy = 81.11 % (Model ID: ID-BBEMTNTNEY) (training time: 0m 0s ).
 2 - training accuracy = 90.00 % (Model ID: ID-BBEMTNTNEY) (training time: 0m 0s ).
 3 - training accuracy = 91.11 % (Model ID: ID-BBEMTNTNEY) (training time: 0m 0s ).
 4 - training accuracy = 92.22 % (Model ID: ID-BBEMTNTNEY) (training time: 0m 0s ).
 5 - training accuracy = 92.22 % (Model ID: ID-BBEMTNTNEY) (training time: 0m 0s ).
 6 - training accuracy = 92.22 % (Model ID: ID-BBEMTNTNEY) (training time: 0m 0s ).
 7 - training accuracy = 93.33 % (Model ID: ID-BBEMTNTNEY) (training time: 0m 0s ).
 8 - training accuracy = 93.33 % (Model ID: ID-BBEMTNTNEY) (training time: 0m 0s ).
 9 - training accuracy = 93.33 % (Model ID: ID-BBEMTNTNEY) (training time: 0m 0s ).
10 - training accuracy = 93.33 % (Model ID: ID-BBEMTNTNEY) (training time: 0m 0s ).

The best model is chosen with training accuracy = 93.33 % (Model ID: ID-BBEMTNTNEY).

Performance Evaluation Results:
Confusion Matrix [predicted x original] (number of classes: 3):
                          GC              Home                IC
            GC            63                 5                 0
          Home             1                19                 0
            IC             0                 0                 2

    Accuracy: 93.33%
       Error: 6.67%
    C.Kappa: 83.46%

                          GC              Home                IC
    Precision:          92.65%           95.00%           100.00%
       Recall:          98.44%           79.17%           100.00%
          FNR:           1.56%           20.83%             0.00%
           F1:          95.45%           86.36%           100.00%
          TNR:          80.77%           98.48%           100.00%
          FPR:          19.23%            1.52%             0.00%
```

(continued)

Table 8.55 (continued)

```
Category mapping [table 'postoperative']:

Column col1:
        _cat_type => integer
        high => 1
        low => 2
        mid => 0
Column col2:
        _cat_type => integer
        high => 1
        low => 0
        mid => 2
Column col3:
        _cat_type => integer
        excellent => 0
        good => 1
Column col4:
        _cat_type => integer
        high => 1
        low => 2
        mid => 0
Column col5:
        _cat_type => integer
        stable => 0
        unstable => 1
Column col6:
        _cat_type => integer
        mod-stable => 2
        stable => 0
        unstable => 1
Column col7:
        _cat_type => integer
        mod-stable => 1
        stable => 0
        unstable => 2
Column decision:
        GC => 0
        Home => 1
        IC => 2
        _cat_type => integer
```

Table 8.56 Patient care training process and extended evaluation results (resampled dataset)

```
AI Training... (Model ID: ID-BBEMTNTNEY_RS).
 0 - training accuracy = 68.23 % (Model ID: ID-BBEMTNTNEY_RS) (training time: 0m 0s ).
 1 - training accuracy = 79.17 % (Model ID: ID-BBEMTNTNEY_RS) (training time: 0m 0s ).
 2 - training accuracy = 85.94 % (Model ID: ID-BBEMTNTNEY_RS) (training time: 0m 0s ).
 3 - training accuracy = 89.06 % (Model ID: ID-BBEMTNTNEY_RS) (training time: 0m 0s ).
 4 - training accuracy = 93.23 % (Model ID: ID-BBEMTNTNEY_RS) (training time: 0m 0s ).
 5 - training accuracy = 95.31 % (Model ID: ID-BBEMTNTNEY_RS) (training time: 0m 0s ).
 6 - training accuracy = 97.40 % (Model ID: ID-BBEMTNTNEY_RS) (training time: 0m 0s ).
 7 - training accuracy = 98.44 % (Model ID: ID-BBEMTNTNEY_RS) (training time: 0m 0s ).
 8 - training accuracy = 98.44 % (Model ID: ID-BBEMTNTNEY_RS) (training time: 0m 0s ).
 9 - training accuracy = 98.44 % (Model ID: ID-BBEMTNTNEY_RS) (training time: 0m 0s ).
10 - training accuracy = 98.44 % (Model ID: ID-BBEMTNTNEY_RS) (training time: 0m 0s ).

The best model is chosen with training accuracy = 98.44 % (Model ID: ID-BBEMTNTNEY_RS).

Performance Evaluation Results:
Confusion Matrix [predicted x original] (number of classes: 3):
                        GC              Home            IC
        GC              61              0               0
        Home            3               64              0
        IC              0               0               64

   Accuracy: 98.44%
      Error: 1.56%
   C.Kappa: 97.66%

                        GC              Home            IC
     Precision:         100.00%         95.52%          100.00%
        Recall:         95.31%          100.00%         100.00%
           FNR:         4.69%           0.00%           0.00%
            F1:         97.60%          97.71%          100.00%
           TNR:         100.00%         97.66%          100.00%
           FPR:         0.00%           2.34%           0.00%
```

IMPORTANT: This way of thinking about the results is very typical to sectors which are more sensitive to specific errors than to other, more general, errors (e.g., healthcare)! This is also the reason why analyzing the extended evaluation results and the per-class metrics is important! More information about the performance evaluation of machine learning models can be found in Chap. 3.

8.11 Replacing Measurements with Machine Learning

The aim of this very simple example is to demonstrate how to replace expensive and/or inconvenient measurements with machine learning.

This case is about estimating the amount of body fat for a person. The amount of body fat is important because it is partly an indication of the health of the individual. "Percentage of body fat for an individual can be estimated once body density has been determined", body density "can be accurately measured a variety of ways. The technique of underwater weighing 'computes body volume as the difference between body weight measured in air and weight measured during water submersion... with the appropriate temperature correction for the water's density" [65]. Body density determined from underwater weighing along with a person's age, weight, height and several circumference measurements are used in predictive equations. The methodology is described in [66].

This example is about replacing body fat measurements, but the same principle can be applied in order to replace many other types of measurements with machine learning!

8.11.1 The Body Fat Dataset

A part of the body fat dataset [65, 66] can be seen in Table 8.57. The columns from left to right are the following:

- C1: Percentage body fat
- C2: Age (years)
- C3: Weight (lbs)
- C4: Height (inches)
- C5: Neck circumference (cm)
- C6: Chest circumference (cm)
- C7: Abdomen circumference (cm)
- C8: Hip circumference (cm)
- C9: Thigh circumference (cm)
- C10: Knee circumference (cm)
- C11: Ankle circumference (cm)
- C12: Biceps (extended) circumference (cm)
- C13: Forearm circumference (cm)

Table 8.57 Body fat dataset preview

C1	C2	C3	C4	C5	C6	C7	C8	C9	C10	C11	C12	C13	C14	C15
12.3	23	154.25	67.75	36.2	93.1	85.2	94.5	59	37.3	21.9	32	27.4	17.1	1.0708
6.1	22	173.25	72.25	38.5	93.6	83	98.7	58.7	37.3	23.4	30.5	28.9	18.2	1.0853
25.3	22	154	66.25	34	95.8	87.9	99.2	59.6	38.9	24	28.8	25.2	16.6	1.0414
10.4	26	184.75	72.25	37.4	101.8	86.4	101.2	60.1	37.3	22.8	32.4	29.4	18.2	1.0751
28.7	24	184.25	71.25	34.4	97.3	100	101.9	63.2	42.2	24	32.2	27.7	17.7	1.034
20.9	24	210.25	74.75	39	104.5	94.4	107.8	66	42	25.6	35.7	30.6	18.8	1.0502
19.2	26	181	69.75	36.4	105.1	90.7	100.3	58.4	38.3	22.9	31.9	27.8	17.7	1.0549
12.4	25	176	72.5	37.8	99.6	88.5	97.1	60	39.4	23.2	30.5	29	18.8	1.0704
4.1	25	191	74	38.1	100.9	82.5	99.9	62.9	38.3	23.8	35.9	31.1	18.2	1.09
11.7	23	198.25	73.5	42.1	99.6	88.6	104.1	63.1	41.7	25	35.6	30	19.2	1.0722
...	...	...	...	...	...	...	...	...	...	...	...	...	...	...

- C14: Wrist circumference (cm)
- C15: Density determined from underwater weighing

The data contains the percentage of body fat determined by underwater weighing, various body circumference measurements, age, weight and height for 252 men (252 data records). The decision variable (which will be learnt by the machine learning model) is in the first column and it is called "C1: Percentage body fat". The principle is very simple, we take a number of measurements and we let the machine learning model learn the relationship in the data in order to replace future body density measurements with just a few simple circumferences and other data from an individual.

If you are going to replace other types of measurements in a similar manner, then try to collect as much measurement data as possible (also taking into account the costs, time, etc. for taking the measurements) in order to provide enough information to the machine learning model!

8.11.2   Training the Body Fat Machine Learning Model

Next we must create the AI-TOOLKIT project file. Use the "Open AI-TOOLKIT Editor" command and then insert the chosen model template with the "Insert ML Template" button. We will train a simple SVM regression model in which the decision variable will be the percentage of body fat in the first column of the dataset. The project file and the optimal parameters can be seen in Table 8.58.

More information about the SVM regression model and parameter values can be found in Sects. 2.2.1 and 9.2.3.1. The creation of the database, importing of the data and training of the machine learning model have already been explained several times in previous examples and will not be explained again here. The optimal parameters were automatically determined by the SVM parameter optimization module in the AI-TOOLKIT.

8.11.3   Analysis of the Results

The training process and the evaluation results can be seen in Table 8.59.

Despite the fact that there is only a small dataset (252 records) the mean squared error (MSQE) (performance) of the model is quite good and amounts to only 0.009623.

Table 8.58 The body fat project file

```
--- # Body fat SVM Regression Example {
model:
 id: 'ID-2'
 type: SVM
 path: 'project.sl3'
 params:
  - svm_type: EPSILON_SVR
  - kernel_type: RBF
  - gamma: 15.0
  - C: 79.4
  - p: 0.1
 training:
  - data_id: 'training_data_2'
  - dec_id: 'decision'
 test:
  - data_id: 'training_data_2'
  - dec_id: 'decision'
 input:
  - data_id: 'idd_2'
  - dec_id: 'decision'
 output:
  - data_id: 'idd_2'
  - col_id: 'decision'
#}
```

Table 8.59 Body fat training process & extended evaluation results

```
[TRAINING START]
SQLite database file project.sl3 opened successfully.
Data is read.
Start training...
Model optimization step: support vectors = 233.
Calculating accuracy on whole training dataset...
MSQE = 0.00962341
The model is trained and saved in the database with success. Model ID: ID-2, Database:
project.sl3
[TRAINING END]
```

The trained machine learning model can be used to estimate the body fat without expensive and inconvenient measurements, for example, underwater weighting! The only data you have to provide to the model for the prediction (inference) are columns 2 to 15 in the input dataset.

Part VI
The AI-TOOLKIT
Machine Learning Made Simple

Chapter 9
The AI-TOOLKIT: Machine Learning Made Simple

Abstract In this chapter you will learn how the AI-TOOLKIT works and how to use it. The full version of the AI-TOOLKIT is free, for non-commercial purposes, and therefore you can experiment with all of the examples in this book and apply the knowledge you have learned about machine learning in the previous chapters without any programming! The AI-TOOLKIT supports all three major forms of machine learning: supervised, unsupervised and reinforcement learning! No programming skills are needed at all to build and use state-of-the-art machine learning models!

9.1 Introduction

The AI-TOOLKIT contains several easy to use machine learning tools which are considered to be the building blocks of Artificial Intelligence. At the time of writing this book the following tools are included (which may be extended later):

- AI-TOOLKIT Professional
- DeepAI Educational
- VoiceBridge
- VoiceData
- Document Summary

The aim of the AI-TOOLKIT is to make the use of machine learning very easy and accessible to everyone. No programming skills are needed for most of the tools, except for the fully open source VoiceBridge, which is a ready to use C++ source code for very easy high performance speech recognition on MS Windows. VoiceBridge contains a DLL and also an application, which are both covered by the AI-TOOLKIT open source license (Apache 2.0 based permissive license).

The AI-TOOLKIT, together with this book, will help beginners in machine learning to quickly acquire the necessary knowledge to use machine learning in their work. The flagship product AI-TOOLKIT Professional may help experts in the application of machine learning across many industries.

© Springer Nature Switzerland AG 2021
Z. Somogyi, *The Application of Artificial Intelligence*,
https://doi.org/10.1007/978-3-030-60032-7_9

All of the available tools will be explained in the following sections in more detail.

9.2   AI-TOOLKIT Professional

9.2.1   Introduction

AI-TOOLKIT Professional (AI-TOOLKIT-Pro) is a machine learning software toolkit which can be used for easy training, testing and inference (prediction) with several types of machine learning (ML) models and for creating machine learning flow (ML flow). ML flow means that several ML models are connected (input-output) and working together or working in parallel to each other. No programming skills are needed at all for building and using sophisticated machine learning models!

There is also a very fast built-in SQL database to make the storage of your machine learning data compact and easy. The database supports several GB's of data storage and several databases can be used, even in one project. You can use the database as a dedicated ML database per ML project. You can import any delimited text data file into the database. You can also import images which will be converted automatically to machine learning data and saved in the database. The data can be automatically resampled.

There is also an easy to use database editor which can be used to view and edit all AI-TOOLKIT-Pro databases. AI-TOOLKIT-Pro and the database editor both support encrypted databases in case you need to secure your data.

AI-TOOLKIT-Pro supports all three major categories of machine learning: supervised, unsupervised and reinforcement learning, with several types of algorithms per category, and also several hybrid machine learning (ML) applications:

1. Supervised learning

    (a) Support Vector Machine (SVM) Model
    (b) Random Forest (RF) Classification Model
    (c) Feedforward Neural Network (FFNN) Regression Model
    (d) Feedforward Neural Network (FFNN) Classification Model
    (e) Convolutional Feedforward Neural Network (CFFNN) Classification Model

2. Unsupervised Learning

    (a) K-Means Classification Model
    (b) MeanShift Classification Model
    (c) DBScan Classification Model
    (d) Hierarchical Classification Model

3. Reinforcement Learning

    (a) Deep Q-Learning (Neural Network)

4. Applications

   (a) Dimensionality Reduction with Principal Component Analyzes (PCA)
   (b) Recommendation with Explicit Feedback (Collaborative Filtering)
   (c) Recommendation with Implicit Feedback (Collaborative Filtering)

Other algorithms and applications may be added later!

There are several built-in and easy to apply machine learning model templates which means you build a complex AI model by just using a simple selection process. You can train, test and use your models easily without the need for programming or scripting. Most of the tasks are automated and the necessary parameters are minimized.

> NOTE: The AI-TOOLKIT also includes several built-in easy to use applications/tools such as a professional face recognition, speaker recognition and fingerprint recognition application, image editor, large text file editor, etc. More information about these tools can be found in later sections.

9.2.2   AI Definition Syntax

AI-TOOLKIT-Pro uses a simple and easy to learn machine learning model definition syntax. This section will explain this syntax in detail.

There is a user friendly syntax colored text editor (specially developed for AI-TOOLKIT-Pro) in which you can define your machine learning models and ML flow (one project file may contain several connected or independent machine learning models).

In order to make it simple let us look at one of the supervised learning example projects (see Table 9.1 below).

The following are the basic model definition syntax rules:

1. Each model must start with three hyphens '- - -', which indicates the start of a new model (there can be several models in a project file).

```
---
```

2. Comments can be added anywhere by starting a line with a '#'. The rest of the line will be considered as part of the comment and will also be colored green.

```
--- # SVM Classification Example {
```

3. If you add several models you can make each model foldable (open/close) by adding '# {' at the start of the model and '#}' at the end of the model. Between the two curly brackets the text will be foldable. Note that the brackets must come after '#' (the comment symbol)!

Table 9.1 Supervised learning model syntax example

```
--- # SVM Classification Example {
model:
  id: 'ID-1'
  type: SVM
  path: 'project.sl3'
  params:
    - svm_type: C_SVC
    - kernel_type: RBF
    - gamma: 15.0
    - C: 1.78
  training:
    - data_id: 'training_data_1'
    - dec_id: 'decision'
  test:
    - data_id: 'training_data_1'
    - dec_id: 'decision'
  input:
    - data_id: 'idd_1'
    - dec_id: 'decision'
  output:
    - data_id: 'idd_1'
    - col_id: 'decision'
#}
```

```
--- # SVM Classification Example {
...
#}
```

4. IMPORTANT: Tabs are prohibited and all indentation must be made with spaces!
5. Each line in the model must start with a keyword and then a colon ':'. We could call these keywords nodes because the whole model has a kind of tree structure.

```
id: 'ID-1'
```

6. Each model must start with the keyword (or node) 'model:'. Nothing is allowed behind it except comments.

```
model:
```

7. The next line must be indented by two spaces and then indicate the unique ID of the model (all IDs in a project file must be different/unique). Each property starts with a keyword (here 'id'), then a colon and the value of the property. *The hierarchy of the properties (a kind of tree structure) is defined by the indentation!* Because the ID is the property of the model it is indented to the right by two spaces (it could also be by one or more spaces but for simplicity and clarity let us use two). The model ID must be between apostrophes (text property).

```
id: 'ID-1'
```

8. The next line must be indented by two spaces and then indicate the type of model. Model types are special keywords and are colored accordingly. Please read the section below about the available model types and their keywords. *Note that 'id' and 'type' are both parameters of the model and therefore indented by the same amount to the right!*

```
type: SVM
```

9. The next line must be indented by two spaces (model parameter) and then indicate the path to the AI-TOOLKIT database. You may indicate the full path to the database or use only the file name. *If you only use the file name then the database must be located in the same folder as the project file!*

```
path: 'project.sl3'
```

   IMPORTANT: Each AI-TOOLKIT database has a special table called 'trained_models' which must be setup correctly; therefore, always make a new database with the 'Create New AI-TOOLKIT Database' button on the toolbar of the DATABASE tab!

10. The next line must be indented by two spaces and then indicate the start of the model type parameters with the keyword 'params'. Each parameter will be defined below this line!

```
params:
```

11. The 'type' parameter must be indented by two more spaces (second level deep) and be defined by starting with a dash '-', then the parameter name, a colon and the parameter value. Parameter values may be special keywords (e.g., C_SVC), strings between apostrophes or numbers. *Note that each parameter is indented the same amount to the right!*

```
- svm_type: C_SVC
```

12. After all parameters are defined (which may be several levels deep as designated by indentation to the right), there are four sections to indicate the database tables to be used for training, testing, input (prediction) and saving the output of the AI model after prediction.

13. Each data section starts with the data definition name, 'training:', 'test:', 'input:' or 'output:', and then on a new line the column definitions by data_id, dec_id and, optionally, several col_id's defining the column names (Table 9.2).

   'data_id' defines the database table name where the data is located, and 'dec_id' defines the decision/label column name in the database; there must be one and only one 'dec_id'. You may open the example database and check the table names to fully appreciate this point.

Table 9.2 AI-TOOLKIT project file data section

```
training:
  - data_id: 'training_data_1'
  - dec_id: 'decision'
test:
  - data_id: 'training_data_1'
  - dec_id: 'decision'
input:
  - data_id: 'idd_1'
  - dec_id: 'decision'
output:
  - data_id: 'idd_1'
  - col_id: 'decision'
```

IMPORTANT: If you do not define col_id's then the whole data table will be used like in this example! But you may choose to use only a part of the table by defining col_id's. In this way you can easily select the feature you want to use.

EXPERT INFO: The data may come from several data tables. You can indicate this by starting with 'data_id:' followed by one or more column definitions with 'col_id:', and then the next 'data_id' and one or more col_id's, etc. There must be only one dec_id! All tables must have the same number of columns!

14. IMPORTANT: All four data definitions ('training', 'test', 'input', 'output') are defined similarly. The data in the 'input' and 'output' data tables in the database will be modified by the AI-TOOLKIT; therefore, make sure that you do not use the same data table as you used for the 'training' data definition!

15. The prediction result(s) (decision/label) will be written to both the 'input' and 'output' data tables. These two may be the same tables.

These are the only rules to be followed. With the 'Insert Template' button you can insert into the editor a pre-defined model for each model type. You only have to change the values of the parameters or insert some more elements (like neural network layers). This makes it very easy to build your AI models. Most of the above rules are guided by the templates.

TIP: The allowed parameter value ranges are indicated in the model templates by comments with the following syntax: range [0, 16] or range [0,]. The first range indicates that the parameter value may be between 0 and 16; the second range indicates that the parameter value may be between 0 and any sensible number. The type of number permitted is also indicated with 'double' (decimal number) or 'int' (integer). An example can be seen in the next section!

The following section will explain all model types and their parameters in detail.

Table 9.3 SVM model parameters

Parameter Name	Description
svm_type:	Choose *C_SVC* for classification or *EPSILON_SVR* for regression.
kernel_type:	*RBF* (radial basis function), *LINEAR*, *POLY* (polynomial) and *SIGMOID*. The RBF kernel is recommended in most cases.
C:	All SVM models use this parameter. It is important to set this value correctly because the model has a high sensitivity to it!
gamma:	The RBF, polynomial and sigmoid kernels use this parameter. It must be optimized because the model has a high sensitivity to it!
degree:	Only the polynomial kernel uses this parameter.
coef0:	The polynomial and sigmoid kernels use this parameter.
p:	All models use this parameter in the case of regression (EPSILON_SVR).
cache_size:	The memory used during the calculations in order to speed up processing. The default value is 100 MB. If you do not define this parameter then the default value will be used (this is also true for many other parameters).
max_iterations:	When this value is set to −1 there is no iteration limit (the model does not stop until there is an improvement), otherwise the indicated number of iterations is performed. Set this to a lower value, e.g., 100 while optimizing the SVM model in order to speed up the process.

9.2.3 Model Types and Parameters

9.2.3.1 Supervised Learning: Support Vector Machine (SVM) Model

The support vector machine model (type: SVM) parameters can be seen in Table 9.3. Please note that the parameters C, γ (gamma), d (degree), coef0 and p are parameters of the functions which are being optimized during machine learning. C is the penalty parameter of the error term and the other parameters are parameters of the so called kernel function(s). It is not the aim of this book to explain the solution of the SVM optimization problem in more detail. For more information please refer to the literature at the end of the book.

The SVM parameter syntax and the permitted parameter value ranges are as shown in Table 9.4.

Table 9.4 SVM parameter syntax

```
params:
    - svm_type: C_SVC # options: C_SVC, EPSILON_SVR
    - kernel_type: RBF # options: RBF, LINEAR, POLY, SIGMOID
    - gamma: 15.0 # double range[0,16], Used for: RBF, POLY, SIGMOID
    - C: 1.78 # double range[0,4000], Used for: RBF, LINEAR, POLY, SIGMOID
    - degree: 3 # int range[0,10], Used for: POLY
    - coef0: 0 # double range[0,100], Used for: POLY, SIGMOID
    - p: 0.1 # double range[0.0,1.0], Used for: EPSILON_SVR only
    - cache_size: 100 # int range[10,] (MB)
    - max_iterations: -1 # No limit if set to -1.
```

With the built-in parameter optimization module you can automatically determine the optimal combination of the parameters influencing the SVM model. Depending on the chosen SVM model (classification or regression) and the kernel type there are different parameters which are optimized and used. The optimized parameters are C, Gamma and p. The other parameters are not optimized but you can change their value in the project file and make several optimization runs to check the sensitivity to the accuracy.

You can access the SVM parameter optimizer on the AI-TOOLKIT tab with the toolbar button symbol shown on Fig. 9.1.

More information about this model can be found in Sect. 2.2.1.

9.2.3.2   Supervised Learning: Random Forest Classification Model (RF)

The random forest model (type: RF) parameters can be seen in Table 9.5.

The RF syntax in the syntax highlighting editor is as shown in Table 9.6.

More information about this model can be found in Sect. 2.2.5 of Chap. 2.

Fig. 9.1 SVM parameter optimizer toolbar button

Table 9.5 Random forest model parameters

Parameter name	Description
min_leaf_size:	The minimum number of leaves used while building the decision tree.
num_trees:	The number of decision trees used.
k_fold:	The number of folds used in cross validation. If this value is set to -1 then no cross validation will be performed. TIP: Do not use cross validation if the dataset is very small (it may not be feasible) or very big (it may take too long).

Table 9.6 RF model syntax

```
params:
    - min_leaf_size: 5 # int range[1,]
    - num_trees: 10 # int range[1,]
    - k_fold: -1 # int range[2,], -1 means no cross-validation!
```

9.2.3.3   Supervised Learning: Feedforward Neural Network Regression Model (FFNN1_R)

More information about this model and the available layers can be found in Sect. 2.2.2, and the parameters for this model (type: FFNN1_R) can be seen in Table 9.7.

Table 9.7 The feedforward neural network regression model parameters

Parameter name	Description
layers:	Under this keyword the neural network layers are defined. NOTE: Each layer, and each parameter within these layers, are indented to the right accordingly! This indentation is very important because it tells the software which parameter belongs to which keyword! Note also that there is a hyphen in front of each layer type BUT no hyphen in front of the layer parameters (e.g., 'nodes:')! Look at the built-in templates for several examples. There are also some examples in the main text.
Intermediate connection layers:	
– Linear	No parameters. Most of the time this is the first layer in non-convolutional neural networks.
– LinearNoBias	No parameters. This is a linear layer but bias is not added.
– DropConnect	There is one optional parameter: 'ratio' (default value = 0.5). It randomly regularizes the input with the 'ratio' probability by setting the connection values to zero and by scaling the remaining elements by a factor of $1/(1 - \text{ratio})$.
Activations layers:	
– IdentityLayer	No parameters.
– TanHLayer	There is one parameter: number of nodes ('nodes:').
– SigmoidLayer	There is one parameter: number of nodes ('nodes:').
– ReLULayer	There is one parameter: number of nodes ('nodes:').
– LeakyReLU	There are two parameters: number of nodes ('nodes:') and the non-zero gradient, which can be adjusted by specifying the parameter 'alpha:' in the range 0 to 1. The default value of alpha = 0.03.
– PReLU	There are two parameters: number of nodes ('nodes:') and the non-zero gradient, which can be adjusted by specifying the parameter 'alpha:' in the range 0 to 1. The default value of alpha = 0.03.
– FlexibleReLU	There are two parameters: number of nodes ('nodes:') and the non-zero gradient, which can be adjusted by specifying the parameter 'alpha:' in the range 0 to 1. The default value of alpha = 0.0.
– SoftPlusLayer	There is one parameter: number of nodes ('nodes:').
Special intermediate layers:	
– Dropout	There is one optional parameter: 'ratio' (default value = 0.5). It randomly regularizes the input with the 'ratio' probability by setting the connection values to zero and by scaling the remaining elements by a factor of $1/(1 - \text{ratio})$.
– BatchNorm	No parameters. This transforms the input data into zero mean and unit variance and then scales and shifts the data. Normalization is done for individual training cases, and the mean and standard deviations are computed across the batch.

(continued)

Table 9.7   (continued)

Parameter name	Description
– LayerNorm	No parameters. This transforms the input data into zero mean and unit variance and then scales and shifts the data. Normalization is done for individual training cases, and the mean and standard deviations are computed across the layer dimensions.
iterations_per_cycle:	The total number of iterations per cycle. *– For SGD_ADAM and SGD_ADADELTA: the normal value for iterations_per_cycle is the number of records in data file (n)!* *– For SARAHPLUS: the iterations_per_cycle are extra external iterations for each n / batch_size inner iterations! The total number of iterations will thus be: (iterations_per_cycle * n / batch_size)!*
num_cycles:	The number of optimization cycles. *Each cycle has iterations_per_cycle number of inner iterations!*
step_size:	The step size of the optimizer. Depending on the chosen optimizer and on the input data this value may have to be adjusted and may have an important effect on the optimization! Permitted range of [1e-10,].
batch_size:	The batch size (number of grouped data points) used per step.
optimizer:	The name of the optimizer to use. There are three options: *SGD_ADAM, SARAHPLUS and SGD_ADADELTA.* SGD_ADAM is the recommended optimizer.
stop_tolerance:	The tolerance to stop the optimization. If it is a very small value the optimization will not be stopped until the maximum number of iterations is reached.
sarah_gamma:	The parameter for the SARAHPLUS optimizer. The default value = 0.125.

TIP: Look at the built-in templates for the acceptable ranges for numerical parameters. For example, range [1e-10,] shown by the comment in the template means that 1e-10 is the minimum value and there is no maximum (see below). Range [1e-10, 1] means a value between 1e-10 and 1 must be entered. For example:

```
- step_size: 5e-5 # double range[1e-10,]
```

An example of the activation function parameters is shown in Table 9.8.
An example with all model parameters can be seen in Tables 9.7 and 9.9.

NOTE: Please note that the input and output nodes are not defined in the layers section of a neural network because they are determined automatically. Feedforward neural networks start and end with a 'linear' connection most of the time. Convolutional feedforward neural networks normally start with a convolutional layer and end with a 'linear' layer.

Table 9.8   Activation function parameters

```
params:
    - layers:
       # ... etc some layers
       - LeakyReLU:
           nodes: 150
           alpha: 0.03
       # ... etc. the model is not complete!
```

Table 9.9   Feedforward neural network regression model example

```
params:
    - layers:
       - Linear: # input connection; no parameters
       - TanHLayer: # layer1; 1 parameter: number of nodes
           nodes: 200
       - Linear: #intermediate connection
       - TanHLayer: # layer2; 1 parameter: number of nodes
           nodes: 150
       - Linear: #intermediate connection
       - TanHLayer: # layer3; 1 parameter: number of nodes
           nodes: 80
       - Linear:   #connection to output
         # add more layers hereunder if needed
    - iterations_per_cycle: 10000
    # for SGD_ADAM and SGD_ADADELTA: normal value is the number
    # of records in data file (n)! For SARAHPLUS: these are extra
    # external iterations for each n/batch_size inner iterations.
    # Total number of iterations:
    # (iterations_per_cycle * n / batch_size)!
    - num_cycles: 50 # int range[1,]
                     # (each cycle has the above defined inner iterations!)
    - step_size: 5e-5 # double range[1e-10,]
    - batch_size: 10   # int range[1,] (Power of 2)
    - optimizer: SGD_ADAM   # Options: SGD_ADAM, SARAHPLUS, SGD_ADADELTA
    - stop_tolerance: 1e-5 # double range[1e-10,1]
    - sarah_gamma: 0.125 # double range[0,]
```

More information about this model can be found in Sect. 2.2.2.

Neural Network Examples

A selection of examples for several neural network layer types can be seen in Table 9.10.

Table 9.10 Neural network examples

Layer Type	Architecture	Notes

TanHLayer	- Linear: # connection to input
	- TanHLayer: # layer 1
	    nodes: 200
	- Linear: # intermediate connection
	- TanHLayer: # layer 2
	    nodes: 150
	- Linear: # intermediate connection
	- TanHLayer: # layer 3
	    nodes: 80
	- Linear: # connection to output
		TanH activation function.
		
		The input and output nodes are automatically configured and need not be defined.
		
		There must be an intermediate connection layer between activation layers.

LinearNoBias	- LinearNoBias: # connection to input
	- TanHLayer: # layer 1
	    nodes: 200
	- LinearNoBias: # intermediate connection
	- TanHLayer: # layer 2
	    nodes: 150
	- LinearNoBias: # intermediate connection
	- TanHLayer: # layer 3
	    nodes: 80
	- LinearNoBias: # connection to output
		Connection between two layers without bias.

SigmoidLayer	- Linear: # connection to input
	- SigmoidLayer: # layer 1
	    nodes: 200
	- Linear: # intermediate connection
	- SigmoidLayer: # layer 2
	    nodes: 150
	- Linear: # intermediate connection
	- SigmoidLayer: # layer 3
	    nodes: 80
	- Linear: # connection to output
		Sigmoid activation function.

SoftPlusLayer	- Linear: # connection to input
	- SoftPlusLayer: # layer 1
	    nodes: 200
	- Linear: # intermediate connection
	- SoftPlusLayer: # layer 2
	    nodes: 150
	- Linear: # intermediate connection
	- SoftPlusLayer: # layer 3
	    nodes: 80
	- Linear: # connection to output
		SoftPlus activation function.

IdentityLayer	- Linear: # connection to input
	- IdentityLayer: # layer 1
	    nodes: 200
	- Linear: # intermediate connection
	- IdentityLayer: # layer 2
	    nodes: 150
	- Linear: # intermediate connection
	- IdentityLayer: # layer 3
	    nodes: 80
	- Linear: # connection to output
		No activation function.

(continued)

Table 9.10 (continued)

| ReLULayer | ```
- Linear: # connection to input
- ReLULayer: # layer 1
    nodes: 200
- Linear: # intermediate connection
- ReLULayer: # layer 2
    nodes: 150
- Linear: # intermediate connection
- ReLULayer: # layer 3
    nodes: 80
- Linear: # connection to output
``` | ReLU activation function. |
| LeakyReLU | ```
- Linear: # connection to input
- LeakyReLU: # layer 1
    nodes: 200
    alpha: 0.03
- Linear: # intermediate connection
- LeakyReLU: # layer 2
    nodes: 150
    alpha: 0.03
- Linear: # intermediate connection
- LeakyReLU: # layer 3
    nodes: 80
    alpha: 0.03
- Linear: # connection to output
``` | LeakyReLU activation function. |
| PReLU | ```
- Linear: # connection to input
- PReLU: # layer 1
    nodes: 200
    alpha: 0.03
- Linear: # intermediate connection
- PReLU: # layer 2
    nodes: 150
    alpha: 0.03
- Linear: # intermediate connection
- PReLU: # layer 3
    nodes: 80
    alpha: 0.03
- Linear: # connection to output
``` | PReLU activation function. |
| FlexibleReLU | ```
- Linear: # connection to input
- FlexibleReLU: # layer 1
    nodes: 200
    alpha: 0.03
- Linear: # intermediate connection
- FlexibleReLU: # layer 2
    nodes: 150
    alpha: 0.03
- Linear: # intermediate connection
- FlexibleReLU: # layer 3
    nodes: 80
    alpha: 0.03
- Linear: # connection to output
``` | FlexibleReLU activation function. |
| Dropout | ```
- Linear: # connection to input
- TanHLayer: # layer 1
    nodes: 200
- Dropout: # special intermediate layer
    ratio: 0.5
- Linear: # intermediate connection
- TanHLayer: # layer 2
    nodes: 150
- Linear: # intermediate connection
- TanHLayer: # layer 3
    nodes: 80
- Linear: # connection to output
``` | This special intermediate dropout layer may be placed between the input and output layers as required. |

(continued)
```

Table 9.10 (continued)

| DropConnect | ```
- Linear: # connection to input
- TanHLayer: # layer 1
    nodes: 200
- DropConnect: # special intermediate
connection
    ratio: 0.5
- TanHLayer: # layer 2
    nodes: 150
- Linear: # intermediate connection
- TanHLayer: # layer 3
    nodes: 80
- Linear: # connection to output
``` | This special intermediate connection layer may be used to replace the linear intermediate connection layer. |
| BatchNorm | ```
- Linear: # connection to input
- BatchNorm: # special intermediate layer
- TanHLayer: # layer 1
    nodes: 200
- Linear: # intermediate connection
- BatchNorm: # special intermediate layer
- TanHLayer: # layer 2
    nodes: 150
- Linear: # intermediate connection
- TanHLayer: # layer 3
    nodes: 80
- Linear: # connection to output
``` | This special intermediate BatchNorm layer may be placed between the input and output layers as required. |
| LayerNorm | ```
- Linear: # connection to input
- LayerNorm: # special intermediate layer
- TanHLayer: # layer 1
    nodes: 200
- Linear: # intermediate connection
- LayerNorm: # special intermediate layer
- TanHLayer: # layer 2
    nodes: 150
- Linear: # intermediate connection
- TanHLayer: # layer 3
    nodes: 80
- Linear: # connection to output
``` | This special intermediate LayerNorm layer may be placed between the input and output layers as required. |

9.2.3.4 Supervised Learning Feedforward Neural Network Classification Model (FFNN1_C)

This model (type: FFNN1_C) has the same parameters as the previous regression model (Sect. 9.2.3.3).

9.2.3.5 Supervised Learning Convolutional Feedforward Neural Network Classification Model Type 2 (FFNN2_C)

The parameters of this model (type: FFNN2_C) can be seen in Table 9.11.

> NOTE: look at the built-in templates for how to set these parameters! Use the built-in *Convolutional Network Calculator* (AI-TOOLKIT tab, toolbar button) to determine the right values for all of these parameters automatically!
```

Table 9.11 Convolutional feedforward neural network classification model parameters

Parameter name	Description
Layers:	Under this keyword the neural network layers are defined. All previously explained layer types are available (Sect. 9.2.3.3) (but they may not all be applicable in this model!) and the following additional layer types are available for convolutional networks:
Convolution layers:	
– Convolution	
inSize:	The number of input filters (depth).
outsize:	The number of output filters (depth).
kW:	Width of the filter/kernel.
kH:	Height of the filter/kernel.
dW:	Stride of filter application in the x direction. Default: 1.
dH:	Stride of filter application in the y direction. Default: 1.
padW:	Padding width of the input. Default: 0.
padH:	Padding height of the input. Default: 0.
inputWidth:	The width of the input data. Default: 0.
inputHeight:	The height of the input data. Default: 0.
– MaxPooling – MeanPooling	
kW:	Width of the pooling window.
kH:	Height of the pooling window.
dW:	Width of the stride operation. Default: 1
dH:	Width of the stride operation. Default: 1
floor:	Rounding operator (floor or ceil). Default: True
num_cycles:	The number of iterations.
batch_size:	The data batch size used per step.
learning_rate:	The learning rate of the optimizer.

An example of a convolution and pooling layer definition is shown in Table 9.12. More information about this model can be found in Sect. 2.2.3.

Convolutional Network Calculator Application

This application (Fig. 9.2) can be initiated from the AI-TOOLKIT toolbar. It is a simple spreadsheet where you fill in the green input cells and automatically receive the results in the form of the convolutional network parameters. You will need a basic understanding of convolutional neural networks in order to build one; therefore, make sure that you have read Sect. 2.2.3 in Chap. 2 first!

In the LAYER_TYPE column you must enter *CONV* (the abbreviation for a convolutional layer) or *POOL* (the abbreviation for a pooling layer) and build up you network. The column names are the following:

Table 9.12 Convolution and pooling layer definition

```
params:
    - layers:
      - Convolution:
          inSize: 1           # in depth
          outSize: 6          # out depth
          kW: 5               # filter/kernel width
          kH: 5               # filter/kernel height
          dW: 1               # stride w
          dH: 1               # stride h
          padW: 0             # padding w
          padH: 0             # padding h
          inputWidth: 30      # input width
          inputHeight: 30     # input height
      - MaxPooling:           # and also MeanPooling
          kW: 2               # pooling width
          kH: 2               # pooling height
          dW: 2               # stride w
          dH: 2               # stride h
          floor: true         # floor (true) or ceil (false) rounding operator
```

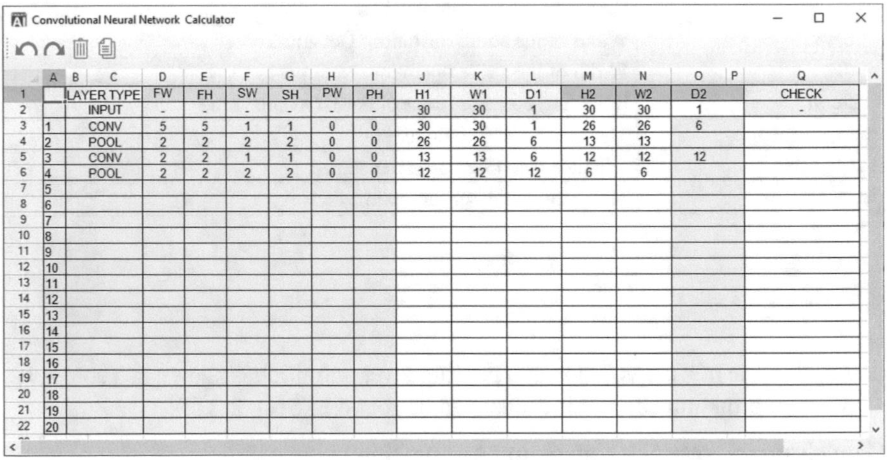

	A	C	D	E	F	G	H	I	J	K	L	M	N	O	Q
1		LAYER TYPE	FW	FH	SW	SH	PW	PH	H1	W1	D1	H2	W2	D2	CHECK
2		INPUT	-	-	-	-	-	-	30	30	1	30	30	1	-
3	1	CONV	5	5	1	1	0	0	30	30	1	26	26	6	
4	2	POOL	2	2	2	2	0	0	26	26	6	13	13		
5	3	CONV	2	2	1	1	0	0	13	13	6	12	12	12	
6	4	POOL	2	2	2	2	0	0	12	12	12	6	6		
7	5														
8	6														
9	7														
10	8														
11	9														
12	10														
13	11														
14	12														
15	13														
16	14														
17	15														
18	16														
19	17														
20	18														
21	19														
22	20														

Fig. 9.2 Convolutional neural network calculator

(a) **FW**: kW, filter/kernel width (CONV) or pooling width (POOL).
(b) **FH**: kH, filter/kernel height (CONV) or pooling height (POOL).
(c) **SW**: dW, stride width.
(d) **SH**: dH, stride height.
(e) **PW**: padW, padding width.
(f) **PH**: padH, padding height.
(g) **W1**: inputWidth, input width.

(h) **H1**: inputHeight, input height.
(i) **W2**: output width.
(j) **H2**: output height.
(k) **D1**: inSize, in depth.
(l) **D2**: outsize, out depth.

You will receive feedback on the spreadsheet if you fill in something incorrectly, in the form of a red colored message or question mark.

When you have finished setting up your network then push the 'Copy to Clipboard as AI-TOOLKIT configuration script' button to get the script which you can paste into the project file directly.

9.2.3.6 Unsupervised Learning K-Means Classification Model

Information about the K-Means unsupervised learning model can be found in Sect. 2.3.1 of Chap. 2.

This model has the parameters shown in Table 9.13.

IMPORTANT: In the case of all unsupervised learning models, a decision column (a column for holding cluster/group assignments or class labels) must exist in the database tables (training, test and input) and must also be defined in the model (project file) in the training, test and input tables. And this is even if no data is provided (empty columns)! If the decision values in the training data table are empty or NULL, the predicted values will be filled in after training the model. If the decision values are not empty in the training data table, then the accuracy will be calculated with the trained model by comparing the predicted decision values (cluster numbers) to the existing original values.

Table 9.13 k-Means model parameters

```
params:
  - clusters: 2 # the number of clusters to distribute the data into
  - iterations: 1000 # the number of iteration
  - projections: 0 # if = 1 then the calculation is speed up with
                   # projections (data dimensionality reduction)
```

Type definition:

```
type: CKMEANS
```

Table 9.14   MeanShift model parameters

```
params:
  - UseKernel: true # Use kernel or mean to calculate new centroid.
                    # If false, KernelType will be ignored.
  - Radius: 0.0 # If distance of two centroids is less than this value,
                # one will be removed. If <= 0 it will be estimated.
  - MaxIterations: 1000  # Maximum number of iterations
  - KernelBandwidth: 1.0 # Kernel parameter
  - KernelType: GaussianKernel # kernel type: GaussianKernel,
                               # SphericalKernel, TriangularKernel,
                               # LaplacianKernel, EpanechnikovKernel

Type definition:

    type: CMEANSHIFT
```

9.2.3.7   Unsupervised Learning MeanShift Classification Model

Information about the MeanShift unsupervised learning model can be found in Sect. 2.3.2 of Chap. 2.

This model has the parameters shown in Table 9.14.

9.2.3.8   Unsupervised Learning DBScan Classification Model

Information about the DBScan unsupervised learning model can be found in Sect. 2.3.3 of Chap. 2.

This model has the parameters shown in Table 9.15.

9.2.3.9   Unsupervised Learning Hierarchical Classification Model

Information about the hierarchical clustering unsupervised learning model can be found in Sect. 2.3.4 of Chap. 2.

This model has the parameters shown in Table 9.16.

Table 9.15   DBScan model parameters

```
params:
  - Epsilon: 0.9  # Size of the circular window
  - MinPoints: 20 # Minimum number of points for each cluster

Type definition:

    type: CDBSCAN
```

Table 9.16 Hierarchical clustering model parameters

```
params:
     - clusters: 2 # the number of clusters to distribute the data into
     - ModelType: PairwiseMaximumLinkage # PairwiseMaximumLinkage,
                                         # PairwiseAverageLinkage,
                                         # PairwiseCentroidLinkage

Type definition:

     type: CHIERARCHICAL
```

9.2.3.10   Reinforcement Learning (RL): Deep Q-Learning

There are three RL templates and examples included with the AI-TOOLKIT:

(a) *Cart-pole*
(b) *Acrobot*
(c) *Grid World*

If you look at these examples it will be clear how to define your own reinforcement learning (RL) problems. The Cart-pole and Grid World examples are explained in Sect. 2.4.5 (Example 2: Cart-pole (inverted pendulum)) and Sect. 2.4.4 (Example 1: Simple Business Process Automation in Chap. 2) respectively. In addition, you will be able to find many articles about the Acrobot problem on the internet.

In order to view and use these templates, create a new AI-TOOLKIT project and insert the appropriate template with the 'Insert Template' button. As usual you will have to create a database for the project. There is no need for training data but you will have to define two tables, one for input and one for output. These are used by a trained agent while taking an action/step based on an input state. Both tables are defined in the project file. The easiest way to create these tables is to create a tab delimited text file with as many columns of data as there are states in the problem plus a decision column. In the case of the Cart-pole example there are four states: position, velocity, angle and angular velocity. Therefore, there should be five columns of data in the imported data file. Use the 'prediction' target option when importing the data and not the 'training' data option! You may import just one line of data (with the values all as 0) and create the two tables. In the input table the decision column should be empty or NULL (which signals that it needs to be evaluated).

First we will look at the Cart-pole RL model in Table 9.17 as an example and explain the syntax. Then the special case of the Grid World example will be explained (for problems which can be modeled in a rectangular grid/table form).

There are two types of nodes in RL templates. The dark blue (bold) nodes (e.g., 'model', 'sample'—see Table 9.17 in the case of the Cart-pole template) indicate the fixed built in nodes and the light blue (normal) nodes indicate the parameters or variables we define in the specific RL problem (e.g., gravity, massCart, etc.). In all of your problems the dark blue nodes are always present and the light blue nodes will

Table 9.17   Project file Cart-pole model

```
--- # Cart-pole Reinforcement Learning Example {
model:
  id: 'ID-UNIQUE1'  # must be unique!
  type: RL
  path: 'D:\mypath\MyDB.sl3' # database path containing all data (input, output,
                             # model)
  sample:
    - params:
      - gravity: 9.81
      - massCart: 1.0
      - massPole: 0.1
      - length: 0.5
      - forceMag: 10.0
      - tau: 0.02
      - xThreshold: 2.4
      - doneReward: 0.0
      - notDoneReward: 1.0
      - thetaThresholdRadians: 0.20944 # 12 * 2 * 3.1416 / 360
    - actions:
      - backward: 0
      - forward: 1
    - states:
      - action: 'action1' # action symbol in the equations
      - equations:
        - force: 'action1 ? forceMag : -forceMag'
        - totalMass: 'massCart + massPole'
        - poleMassLength: 'massPole * length'
        - temp: '(force + poleMassLength * angularvelocity * angularvelocity
                 * sin(angle)) / totalMass'
        - thetaAcc: '(gravity * sin(angle) - cos(angle) * temp) / (length
                    * (4.0 / 3.0 - massPole * cos(angle) * cos(angle) / totalMass))'
        - xAcc: 'temp - poleMassLength * thetaAcc * cos(angle) / totalMass'
      - initial:
        - position: '(randu() - 0.5) / 10'
        - velocity: '(randu() - 0.5) / 10'
        - angle: '(randu() - 0.5) / 10'
        - angularvelocity: '(randu() - 0.5) / 10'
      - next:
        - position: 'position + tau * velocity'
        - velocity: 'velocity + tau * xAcc'
        - angle: 'angle + tau * angularvelocity'
        - angularvelocity: 'angularvelocity + tau * thetaAcc'
      - isterminal:'abs(position)>xThreshold or abs(angle)>thetaThresholdRadians'
      - reward: 'isterminal ? doneReward : notDoneReward'
  network:
    - layers:
      - Linear:
      - ReLULayer:
          nodes: 20
      - Linear:
      - ReLULayer:
          nodes: 20
      - Linear:
    - StepSize: 0.01
    - Discount: 0.9
```

(continued)

Table 9.17 (continued)

```
        - TargetNetworkSyncInterval: 100
        - ExplorationSteps: 100
        - DoubleQLearning: true
        - StepLimit: 200
        - Episodes: 1000
        - MinAverageReward: 70 # if you do not know this number then use trial and
                               # error or set it to a value that it does not stop
                               # the iterations
        - SampleBatchSize: 10 # Number of examples returned at each sample.
        - SampleCapacity: 10000 # Total memory size in terms of number of examples.
    input: # input data for Agent prediction.
        - data_id: 'my_input_data' # input data table name
        - dec_id: 'decision' # the value in this column must be NULL or empty
                             # ( = signals that it should be evaluated)!
        # NOTE: Without col_id's defined all columns are used! The number of
        # columns must correspond to the number of states!
        # - col_id: 'decision'
        # - col_id: 'col1'
    output: # output data for Agent prediction.
        - data_id: 'my_output_data' # output data table name (for the prediction)
        - dec_id: 'decision' # the column where the action taken by the Agent
                             # will be written.
        # NOTE: Without col_id's defined all columns are used! The number of
        # columns must correspond to the number of states!
        # - col_id: 'decision'
        # - col_id: 'col1'
#}
```

be different depending on the problem. If you define parameters in the 'params' section then it is important to use the same name later in the equations describing the problem (i.e., the 'equations' section). Equations are evaluated sequentially and must be defined from top to bottom in the correct order!

> IMPORTANT: In the initial state definition we use the randu() (uniform random number generator in the range [0,1]) function, which will generate random initial states each time you run the training of the agent! For this reason we run the training several times and choose the best result.

The 'Params' Section (Model → Sample > Params)
In this section you can define global parameters which can be used in any equation as e.g. the constant for gravity (which is 9.81). As usual the left side gives the name of the parameter followed by a colon and the right side gives the parameter's value (Table 9.18).

Table 9.18   Global parameters

```
- params:
        - gravity: 9.81
        - massCart: 1.0
        - massPole: 0.1
        - length: 0.5
        - forceMag: 10.0
        - tau: 0.02
        - xThreshold: 2.4
        - doneReward: 0.0
        - notDoneReward: 1.0
        - thetaThresholdRadians: 0.20944 # 12 * 2 * 3.1416 / 360
```

IMPORTANT: You may not use reserved names as parameter name. For a list of reserved names please see later in Section 'Equation syntax and available built-in functions'. There are also some globally reserved parameter names ('pi', 'epsilon' and 'inf') which may also not be used. The easiest way of making sure that you have unique names is, for example, to add a prefix to your variables such as 'myvar_paramname'.

The 'Actions' Section (Model → Sample → Actions)

In this section you have to define the different actions the agent can take. All actions have an integer value starting from 0! An action is, e.g., moving the object forward or backward (Table 9.19).

The 'States' Section (Model → Sample → States)

This section defines all necessary information about the states in the problem. States are special properties which we keep track of.

(a) *The 'states → action' section.*

     This contains the action variable definition. Choosing the correct action to take in a given state is what the agent learns!

(b) *The 'states → equations' section.*

     These are the equations which will be evaluated sequentially for each step of the agent. These generate the sample data (states) which are used for learning.

Table 9.19   Agent's actions

```
- actions:
    - backward: 0
    - forward: 1
```

Table 9.20   Initial state

```
- initial:
  - position: '(randu() - 0.5) / 10'
  - velocity: '(randu() - 0.5) / 10'
  - angle: '(randu() - 0.5) / 10'
  - angularvelocity: '(randu() - 0.5) / 10'
```

(c) *The 'states → initial' section.*

This is the initial state from which the agent starts at the beginning of each episode (Table 9.20).

(d) *The 'states → next' section.*

This is how the agent calculates the next state after an action is taken.

(e) *The 'states → isterminal' section.*

This is how the agent decides if the problem is in the terminal (completed) state. Reaching the terminal state is the objective of the problem.

The 'Network' Section (Model → Network)

This is where the deep neural network is defined which is used for learning. The format is the same as that explained before in the supervised learning sections.

The RL network parameters can be seen in Table 9.21.

Table 9.21   RL network model parameters

Parameter name	Default	Description
StepSize	0.01	The step size of the neural network optimizer. You will have to experiment to find the best value.
Discount	0.9	Future reward discount factor.
TargetNetworkSyncInterval	100	The number of steps after which the target neural network will be updated with the learning network. Note: the learning network is used to estimate the action in the case of double Q-Learning.
ExplorationSteps	100	The agent will try unseen action/state combinations until this number of steps reached. After this the exploration probability will anneal at each step and the agent will take less and less exploration (unseen) steps.
DoubleQLearning	True	When set to true the agent uses a learning network to select the best action at each step. It is called double Q-Learning because the subsequent action values are estimated with a neural network and the action is also selected with the network. When set to false the action will be chosen according to the list of past actions and the rewards by maximizing the reward.

(continued)

Table 9.21 (continued)

Parameter name	Default	Description
StepLimit	200	The maximum number of steps taken by the agent in one episode. An episode normally ends when the problem reaches the terminal state or at this number.
Episodes	1000	The maximum number of episodes used by the agent while learning the problem.
MinAverageReward	70	When the average reward reaches this value the iteration is stopped. You will have to experiment to find the right value for this parameter.
SampleBatchSize	10	The number of examples returned at each sample.
SampleCapacity	10,000	The total memory size in terms of number of examples.

The 'Input' and 'Output' Sections (Model → Input/Output)

These are the input and output table definitions. If you import the data with the AI-TOOLKIT then the dec_id will always be 'decision'! The number of columns must correspond to the number of states plus a decision column! The format of these sections is the same as explained in Sect. 9.2.2.

Functions

You can also define functions. A function is a block of equations which may have a maximum of six parameters and can be called from any other equation or function. Functions may also call other functions. The example in Table 9.22 is from the

Table 9.22 Example function definitions in RL models

```
- functions: # functions are very similar to the functions in Excel
  - Dsdt:
    - params: # only maximum 6 parameters are allowed of the type
                # double (real) number
      - th1 # Theta1
      - th2 # Theta2
      - av1 # AngularVelocity1
      - av2 # AngularVelocity2
      - tr  # torque
      - ret # if 0 return val2; if 1 return val3 !
    - equations:
      - d1: 'linkMass1 * pow(linkCom1, 2) + linkMass2 *
            (pow(linkLength1, 2) + pow(linkCom2, 2) + 2 * linkLength1 *
            linkCom2 * cos(th2)) + linkMoi + linkMoi'
      - d2: 'linkMass2 * (pow(linkCom2, 2) + linkLength1 * linkCom2 *
            cos(th2)) + linkMoi'
      - phi2: 'linkMass2 * linkCom2 * gravity * cos(th1 + th2 - pi / 2)'
      - phi1: '- linkMass2 * linkLength1 * linkCom2 * pow(av2, 2) *
            sin(th2) - 2 * linkMass2 * linkLength1 * linkCom2 * av2 *
            av1 * sin(th2) + (linkMass1 * linkCom1 +  linkMass2 *
            linkLength1) * gravity * cos(th1 - pi / 2) + phi2'
      - val3: '(tr + d2 / d1 * phi1 - linkMass2 * linkLength1 * linkCom2 *
            pow(av1, 2) * sin(th2) - phi2) / (linkMass2 *
            pow(linkCom2, 2) + linkMoi - pow(d2, 2) / d1)'
      - val2: '-(d2 * val3 + phi1) / d1'
      - val: 'ret == 1 ? val3 : val2' # the last equation's result will be
                                      # returned by the function
```

Acrobot template/example. The dark keywords/nodes (bold font) are fixed and the light blue ones (normal font) are problem dependent, e.g., 'Dsdt' is the name of the function and 'd1' is the first equation in the function. The equations are evaluated sequentially and thus the order of the equations is important. Functions may also use the global variables defined before.

> IMPORTANT: The result of the last equation will be returned by the function!

Equation Syntax and Available Built-in Functions

> IMPORTANT: Please keep in mind while using the information in this section that each equation must evaluate to a single value! As long as this rule is maintained you can use any of the following built-in operators and functions!

1. **Arithmetic and Assignment Operators** can be seen in Table 9.23.

2. **The operators for equalities and inequalities** can be seen in Table 9.24.

Table 9.23 Arithmetic and assignment operators

Operator	Definition
+	Addition between x and y. (i.e., x + y)
−	Subtraction between x and y. (i.e., x - y)
*	Multiplication between x and y. (i.e., x * y)
/	Division between x and y. (i.e., x / y)
%	Modulus of x with respect to y. (i.e., x % y)
^	x to the power of y. (i.e., x ^ y)
:=	Assign the value of x to y. Where y is either a variable or vector type. (i.e., y: = x)
+=	Increment x by the value of the expression on the right-hand side. Where x is either a variable or vector type. (i.e., x + = abs(y - z))
-=	Decrement x by the value of the expression on the right-hand side. Where x is either a variable or vector type. (i.e., x[i] - = abs(y + z))
*=	Assign the multiplication of x by the value of the expression on the right-hand side to x. Where x is either a variable or vector type. (i.e., x * = abs(y / z))
/=	Assign the division of x by the value of the expression on the right-hand side to x. Where x is either a variable or vector type. (i.e., x[i + j] /= abs(y * z))
%=	Assign x modulo the value of the expression on the right-hand side to x. Where x is either a variable or vector type. (i.e., x [2] % = y ^ 2)

Table 9.24 Equalities and inequalities

Operator	Definition
== or =	True only if x is strictly equal to y. (i.e., x == y)
<> or !=	True only if x does not equal y. (i.e., x <> y or x != y)
<	True only if x is less than y. (i.e., x < y)
<=	True only if x is less than or equal to y. (i.e., x <= y)
>	True only if x is greater than y. (i.e., x > y)
>=	True only if x greater than or equal to y. (i.e., x >= y)

3. **Boolean Operations** can be seen in Table 9.25.

4. **General Purpose Functions** can be seen in Table 9.26.

5. **Trigonometry Functions** can be seen in Table 9.27.

6. **Control Structures** can be seen in Table 9.28.

Table 9.25 Boolean operations

Operator	Definition
true	True state or any value other than zero (typically 1).
false	False state, value of exactly zero.
and	Logical AND, True only if x and y are both true. (i.e., x and y)
mand	Multi-input logical AND, True only if all inputs are true. Left to right short-circuiting of expressions. (i.e., mand(x > y, z < w, u or v, w and x))
mor	Multi-input logical OR, True if at least one of the inputs is true. Left to right short-circuiting of expressions. (i.e., mor(x > y, z < w, u or v, w and x))
nand	Logical NAND, True only if either x or y is false. (i.e., x nand y)
nor	Logical NOR, True only if the result of x or y is false (i.e., x nor y)
not	Logical NOT, Negate the logical sense of the input. (i.e., not (x and y) == x nand y)
or	Logical OR, True if either x or y is true. (i.e., x or y)
xor	Logical XOR, True only if the logical states of x and y differ. (i.e., x xor y)
xnor	Logical XNOR, True if the biconditional of x and y is satisfied. (i.e., x xnor y)
&	Similar to AND but with left to right expression short circuiting optimization. (i.e., (x & y) == (y and x))
|	Similar to OR but with left to right expression short circuiting optimization. (i.e., (x | y) == (y or x))

Table 9.26 General purpose functions

Function	Definition
abs	Absolute value of x. (i.e., abs(x))
avg	Average of all the inputs. (i.e., avg(x,y,z,w,u,v) == (x + y + z + w + u + v) / 6)
ceil	Smallest integer that is greater than or equal to x.
clamp	Clamp x in range between r0 and r1, where r0 < r1. (i.e., clamp(r0,x,r1))
equal	Equality test between x and y using normalized epsilon
erf	Error function of x. (i.e., erf(x))
erfc	Complementary error function of x. (i.e., erfc(x))
exp	e to the power of x. (i.e., exp(x))
expm1	e to the power of x minus 1, where x is very small. (i.e., expm1(x))
floor	Largest integer that is less than or equal to x. (i.e., floor(x))
frac	Fractional portion of x. (i.e., frac(x))
hypot	Hypotenuse of x and y (i.e., hypot(x,y) = sqrt(x*x + y*y))
iclamp	Inverse-clamp x outside of the range r0 and r1. Where r0 < r1. If x is within the range it will snap to the closest bound. (i.e., iclamp(r0,x,r1)
inrange	In-range returns 'true' when x is within the range r0 and r1. Where r0 < r1. (i.e., inrange(r0,x,r1)
log	Natural logarithm of x. (i.e., log(x))
log10	Base 10 logarithm of x. (i.e., log10(x))
log1p	Natural logarithm of 1 + x, where x is very small. (i.e., log1p(x))
log2	Base 2 logarithm of x. (i.e., log2(x))
logn	Base N logarithm of x, where n is a positive integer. (i.e., logn(x,8))
max	Largest value of all the inputs. (i.e., max(x,y,z,w,u,v))
min	Smallest value of all the inputs. (i.e., min(x,y,z,w,u))
mul	Product of all the inputs. (i.e., mul(x,y,z,w,u,v,t) == (x * y * z * w * u * v * t))
ncdf	Normal cumulative distribution function. (i.e., ncdf(x))
nequal	Not-equal test between x and y using normalized epsilon
pow	x to the power of y. (i.e., pow(x,y) == x ^ y)
root	The nth-root of x, where n is a positive integer. (i.e., root(x,3) == x^(1/3))
round	Round x to the nearest integer. (i.e., round(x))
roundn	Round x to n decimal places (i.e., roundn(x,3)) where n > 0 and is an integer. (i.e., roundn(1.2345678,4) == 1.2346)
sgn	Sign of x, -1 where x < 0, +1 where x > 0, else zero. (i.e., sgn(x))
sqrt	Square root of x, where x >= 0. (i.e., sqrt(x))

(continued)

Table 9.26 (continued)

Function	Definition
sum	Sum of all the inputs. (i.e., sum(x,y,z,w,u,v,t) == (x + y + z + w + u + v + t)).
swap	Swap the values of the variables x and y and return the current value of y. (i.e., swap(x,y) or x <=> y)
trunc	Integer portion of x. (i.e., trunc(x))

Table 9.27 Trigonometry functions

Function	Definition
acos	Arc cosine of x expressed in radians. Interval [-1,+1] (i.e., acos(x))
acosh	Inverse hyperbolic cosine of x expressed in radians. (i.e., acosh(x))
asin	Arc sine of x expressed in radians. Interval [-1,+1] (i.e., asin(x))
asinh	Inverse hyperbolic sine of x expressed in radians. (i.e., asinh(x))
atan	Arc tangent of x expressed in radians. Interval [-1,+1] (i.e., atan(x))
atan2	Arc tangent of (x / y) expressed in radians. [-pi,+pi] (i.e., atan2(x,y))
atanh	Inverse hyperbolic tangent of x expressed in radians. (i.e., atanh(x))
cos	Cosine of x. (i.e., cos(x))
cosh	Hyperbolic cosine of x. (i.e., cosh(x))
cot	Cotangent of x. (i.e., cot(x))
csc	Cosecant of x. (i.e., csc(x))
sec	Secant of x. (i.e., sec(x))
sin	Sine of x. (i.e., sin(x))
sinc	Sine cardinal of x. (i.e., sinc(x))
sinh	Hyperbolic sine of x. (i.e., sinh(x))
tan	Tangent of x. (i.e., tan(x))
tanh	Hyperbolic tangent of x. (i.e., tanh(x))
deg2rad	Convert x from degrees to radians. (i.e., deg2rad(x))
deg2grad	Convert x from degrees to gradians. (i.e., deg2grad(x))
rad2deg	Convert x from radians to degrees. (i.e., rad2deg(x))
grad2deg	Convert x from gradians to degrees. (i.e., grad2deg(x))

Table 9.28 Control structures

Structure	Definition
if	If x is true then return y, else return z. For example, 1. if (x, y, z) 2. if ((x + 1) > 2y, z + 1, w / v) 3. if (x > y) z; 4. if (x <= 2*y) { z + w };
if-else	The if-else/else-if statement. Subject to the condition branch the statement will return either the value of the consequent or the alternative branch. For example, 1. if (x > y) z; else w; 2. if (x > y) z; else if (w != u) v; 3. if (x < y) { z; w + 1; } else u; 4. if ((x != y) and (z > w)) { y := sin(x) / u; z := w + 1; } else if (x > (z + 1)) { w := abs (x - y) + z; u := (x + 1) > 2y ? 2u : 3u; }

switch	The first true case condition that is encountered will determine the result of the switch. If none of the case conditions hold true, the default action is assumed as the final return value. This is sometimes also known as a multi-way branch mechanism. e.g.,
	switch
	{
	case x > (y + z) : 2 * x / abs (y - z);
	case x < 3 : sin (x + y);
	default : 1 + x;
	}

while	The structure will repeatedly evaluate the internal statement(s) 'while' the condition is true. The final statement in the final iteration will be used as the return value of the loop, e.g.,
	while ((x -= 1) > 0) { y := x + z;
	w := u + y;
	}

repeat/
until
	The structure will repeatedly evaluate the internal statement(s) 'until' the condition is true. The final statement in the final iteration will be used as the return value of the loop, e.g.,
	repeat
	y := x + z;
	w := u + y;
	until ((x += 1) > 100)

for	The structure will repeatedly evaluate the internal statement(s) while the condition is true. On each loop/iteration, an 'incrementing' expression is evaluated. The conditional is mandatory whereas the initialiser and incrementing expressions are optional, e.g.,
	for (var x := 0; (x < n) and (x != y); x += 1)
	{
	y := y + x / 2 - z;

(continued)

Table 9.28 (continued)

Structure	Definition
| | ```
w := u + y;
}
``` |
| break<br>break[] | Break terminates the execution of the nearest enclosed loop, allowing for the execution to continue external to the loop. The default break statement will set the return value of the loop to NaN, where as the return based form will set the value to that of the break expression, e.g.,<br>```
while ((i += 1) < 10)
{
if (i < 5)
j -= i + 2;
else if (i % 2 == 0)
break;
else
break[2i + 3];
}
``` |
| continue | Continue results in the remaining portion of the nearest enclosing loop body being skipped, e.g.,<br>```
for (var i := 0; i < 10; i += 1)
{
if (i < 5)
continue;
j -= i + 2;
}
``` |
| return | Return immediately from within the current expression, with the option of passing back a variable number of values (scalar, vector or string), e.g.,<br>```
1. return [1];
2. return [x, 'abx'];
3. return [x, x + y, 'abx'];
4. return [];
5. if (x < y)
return [x, x - y, 'result-set1', 123.456];
else
return [y, x + y, 'result-set2'];
``` |
| ?: | Ternary conditional statement, similar to that of the above denoted if-statement, e.g.,<br>```
1. x ? y : z
2. x + 1 > 2y ? z + 1 : (w / v)
3. min(x,y) > z ? (x < y + 1) ? x : y : (w * v)
``` |
| ~ | Evaluate each sub-expression, then return as the result the value of the last sub-expression. This is sometimes known as multiple sequence point evaluation, e.g.,<br>```
~(i := x + 1, j := y / z, k := sin(w/u)) == (sin(w/u)))
~{i := x + 1; j := y / z; k := sin(w/u)} == (sin(w/u)))
``` |
| [*] | Evaluate any consequent for which its case statement is true. The return value will be either zero or the result of the last consequent to have been evaluated, e.g.,<br>```
[*]
``` |

(continued)
```

Table 9.28   (continued)

Structure	Definition
|  | ```
{
case (x + 1) > (y - 2) : x := z / 2 + sin (y / pi);
case (x + 2) < abs (y + 3) : w / 4 + min (5y, 9);
case (x + 3) == (y * 4) : y := abs (z / 6) + 7y;
}
``` |
| [] | The vector size operator returns the size of the vector being actioned, e.g.,
1. v[]
2. max_size := max (v0 [] , v1 [] , v2 [] , v3 []) |

Source: [92]

The Grid World Template

Many real world problems can be modeled in a rectangular grid/table form (logistics, management, etc.). This template and example shows how to do this in the AI-TOOLKIT. Most of the work is done behind the scenes and you only have to define a limited number of parameters and use two built-in functions. More information about this example can be found in Sect. 2.4.4 (Example 1: Simple Business Process Automation). The Grid World template is shown in Table 9.29 (Fig. 9.3).

The comment in the beginning of the template explains how the states (cells in the grid) are numbered between 0 and n-1, where n is the number of cells in the grid (nrows * ncols). In this example there are three rows and four columns (12 cells). The grid parameters (availability and reward) are stored in one vector which is indexed column wise (see the first 'States' table). Each cell in the grid can be active and non-active. Non-active (not available) cells are marked with 0 and active cells are marked with 1 in the cell availability vector (second 'Cell availability' table). Non-active means that the agent may not move into the cell. The vector must be defined in the 'sample' section with any name (in the example 'ca' is used). The same is true for the reward vector which is defined in the example with 'rw'. A separate reward is defined (negative or positive) for moving into each cell. The terminal states have of course a specific reward (in this example +1 and −1). The 'isterminal' function shows which cells (states) are terminal states, in this example (3, 7).

There are two special built-in 'grid world' functions for calculating the next state (nextstate_gridw) and for calculating the reward (reward_gridw). The first parameter in both functions is the state variable, called 'state' in this example. This name must also be the same in the 'initial' and 'next' sections! The rest of the input parameters are very similar to the ones in the other RL example discussed before!

More information about this example can be found in Sect. 2.4.4 (Example 1: Simple Business Process Automation).
```

Table 9.29   Grid World reinforcement learning example

```
--- # GridWorld Reinforcement Learning Example {
# Expected result: Average reward in deterministic test: 0.84 in 10 test
episodes. The weights of the neural network
# are initialized with random weights, therefore it is not guaranteed that every
training will be successful. Make several runs!
# States:                Cell availability:   Reward:
# ----------------        ----------------     --------------------------------
# | 0 | 1 | 2 | 3 |       | 1 | 1 | 1 | 1 |    | -0.04 | -0.04 | -0.04 |   1  |
# ----------------        ----------------     --------------------------------
# | 4 | 5 | 6 | 7 |       | 1 | 0 | 1 | 1 |    | -0.04 | -0.04 | -0.04 |  -1  |
# ----------------        ----------------     --------------------------------
# | 8 | 9 | 10| 11|       | 1 | 1 | 1 | 1 |    | -0.04 | -0.04 | -0.04 | -0.04|
# ----------------        ----------------     --------------------------------
# Terminal states: 3, 7
model:
  id: 'ID-UNIQUE1'  # must be unique!
  type: RL
  path: 'D:\mypath\MyDB.sl3' # database path containing all data
                            # (input, output, model)
  sample:
    - params:
      - nrows: 3
      - ncols: 4
      - ca: '1,1,1,1,
             1,0,1,1,
             1,1,1,1'
      - rw: '-0.04, -0.04, -0.04,  1,
             -0.04, -0.04, -0.04, -1,
             -0.04, -0.04, -0.04, -0.04'
    - actions:
      - up: 0
      - down: 1
      - left: 2
      - right: 3
    - states:
      - action: 'action1' # action symbol in the equations
      - equations:  # NOTE: no equations.
      - initial:
        - state: 8
      - next:
        - state: 'nextstate_gridw(state, nrows, ncols, action1, ca)'
                  # NOTE: special built-in function!
      - isterminal: 'state == 3 or state == 7'
      - reward: 'reward_gridw(state, nrows, ncols, rw)'
                  # NOTE: special built-in function!
  network:
    - layers:
      - Linear:
      - ReLULayer:
          nodes: 20
      - Linear:
      - ReLULayer:
          nodes: 20
      - Linear:
    - StepSize: 0.01
    - Discount: 0.9
    - TargetNetworkSyncInterval: 50
```

(continued)

Table 9.29 (continued)

```
            - ExplorationSteps: 50
            - DoubleQLearning: true
            - StepLimit: 50
            - Episodes: 100
            - MinAverageReward: 0.2 # if you do not know this number then use trial
                                    # and error or set it to a value that it does not
                                    # stop the iterations
            - SampleBatchSize: 10 # Number of examples returned at each sample.
            - SampleCapacity: 10000 # Total memory size in terms of number of examples.
        input: # input data for Agent prediction.
            - data_id: 'my_input_data' # input data table name
            - dec_id: 'decision' # the value in this column must be NULL or empty
                                 # ( = signals that it should be evaluated)!
          # NOTE: Without col_id's defined all columns are used! The number of columns
          #  must correspond to the number of states!
          # - col_id: 'decision'
          # - col_id: 'col1'
        output: # output data for Agent prediction.
            - data_id: 'my_output_data' # output data table name (for the prediction)
            - dec_id: 'decision' # the column where the action taken by the Agent will
                                 # be written.
          # NOTE: Without col_id's defined all columns are used! The number of columns
          #  must correspond to the number of states!
          # - col_id: 'decision'
          # - col_id: 'col1'
    #}
```

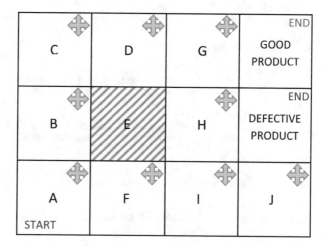

Fig. 9.3 Example 1: Business process Grid World

NOTE: The 'ca' and 'rw' vectors are **comma** separated numbers (with a **point** as the decimal separator) and they can be structured with new lines in a way that makes it more clear what the grid looks like. The **apostrophe** in the beginning and at the end of the sequence is important!

9.2.3.11   Dimensionality Reduction with Principal Component Analyzes (PCA)

If you have a dataset with a lot of columns (features) then you may use this machine learning model reduce the number of columns in the dataset (dimensionality reduction). This may speed up the training of your machine learning model if there is a lot of data, and it may also improve the quality (accuracy) of the machine learning model.

More information about principal component analysis (PCA) can be found in Sect. 4.3.2.2.

> IMPORTANT: The decision column must be defined in the training data table because it will be separated from the rest of the data and added back in after the dimensionality reduction. The decision column will be the first column in the output data table.

The parameters of the PCA model can be seen in Table 9.30.

The PCAType parameter has four possible values which determine which singular value decomposition (SVD) method is used:

Table 9.30 PCA model parameters

```
params:
  - NewDimension: 2
    # The requested new dimension of the dataset (must be < (number of
    # columns - 1) in the original dataset (-1 is for the decision
    # column)) if VarianceToRetain = -1 then NewDimension is used,
    # otherwise VarianceToRetain is used and must be NewDimension = -1!

  - VarianceToRetain: -1
    # Variance retained must be between 0 and 1 (1 = retain all
    # variance, 0 = retain only 1 dimension/column).

  - Scale: 0
    # If set to 1, the data will be scaled before running PCA, such
    # that the variance of each feature is 1.

  - PCAType: PCA_EXACT
    # Options: PCA_EXACT, PCA_RANDOMIZED, PCA_RANDOMIZED_BLOCK_KRYLOV,
    #          PCA_QUIC

Type specifier:

  type: PCA
```

- Exact SVD
- Randomized SVD
- Randomized Block Krylov SVD
- QUIC-SVD

The exact SVD method performs an exact calculation of the singular value decomposition (linear algebraic operation) which is very time consuming in the case of large datasets. For this reason several other methods have been developed for approximating SVD with much faster algorithms, which are the three other methods available. It is not the aim of this book to explain these algorithms in detail. More information can be found in [31–33].

There is more information about the PCA project file parameters (database tables and columns) in the comments of the AI TOOLKIT PCA template!

This model cannot be tested or used for further input processing because it directly transforms a dataset. For this reason no machine learning model is saved in the database after training the model. The quality of the model is determined by the retained variance in the resultant dataset.

9.2.3.12 Recommendation with Explicit Feedback (Collaborative Filtering: CFE)

More information about collaborative filtering with explicit feedback (CFE) can be found in Sect. 8.4.1.6 in Chap. 8.

The built-in CFE model template can be seen in Table 9.31.

All parameters are explained in the comments (green text).

There are some important requirements concerning the data (training, test, input and output) for collaborative filtering models which will be explained hereunder!

This model requires that the data is in a specific order in the database tables (training, test and input): the first column must be the rating (decision), the second column must be the user ID and the third column must be the item ID (movie ID in this example)! The decision (rating) column will be automatically the first column if you import the data; therefore, when you import your data into the database make sure that you select the correct decision column in the import module! After importing the data, the second and third columns must be the user ID and item ID in this order! Therefore, always make sure that you have the user ID first and then the item ID in the original CSV file and that there are no other columns in between or before (Figs. 9.4 and 9.5)!

The effective user IDs and item IDs (the values) in the test and input datasets must all exist in the training dataset! Only the users and/or items present in the original training dataset can be tested or predicted/recommended!

The output data table may be the same as the input data table, but you must make sure that the recommendation columns are defined if recommendations are requested! The contents of the recommendation columns are not important because they will be overwritten. The user ID and item ID columns in the output definition

Table 9.31 CFE model template

```
--- # Recommendation with Explicit Feedback (Collaborative Filtering) Example {
# Explicit Feedback means that the users provided the rating of items.
# IMPORTANT: This model requires that the data is in a specific order in the DB #
data # tables (training, test, input): 1st column Rating (decision), 2nd
# column UserID, 3rd column ItemID! When you import your data into the DB make
# sure that you select the right decision column in your data! The decision
# column will be automatically the first column after importing the data and
# the 2nd and 3rd columns must be the UserID and ItemID! The UserID's and
# ItemID's in the test and input datasets must all exist in the training
# dataset!
model:
  id: 'ID-pfwgflPaBd' # Must be unique!
  type: CFE
  path: 'D:\mypath\MyDB.sl3' # Database path containing all data
  params:
    - Neighborhood: 5
      # Size of the neighborhood of similar users to consider for each
      # query user.
    - MaxIterations: 1000
      # Maximum number of iterations. If set to zero, there is no limit on the
      # number of iterations.
    - MinResidue: 1e-5
      # Residue required for terminating the factorization (lower values
      # generally
      # mean better fits).
    - Rank: 0
      # Rank of decomposed matrices (if 0, a heuristic is used to estimate the
      # rank).
    - Recommendations: 1
      # The number of recommendations to generate for each User.
  training:
      # Training data - the data may come from several tables!
    - data_id: 'my_training_data'
      # Training data table name in the database defined by 'model > path'
      # above.
    - dec_id: 'decision'
      # The decision column ID. There must be ONE defined in one of the tables!
      # (Ratings!)
      # NOTE: you can select 3 columns or without col_id's the first 3 columns
      # are used! If you select the columns then it is important that the
      # columns appear in the DB table in the following order:
      #    - col_id: 'decision' # 'decision' must also be defined as a col_id!
      #    - col_id: 'UserID'
      #    - col_id: 'ItemID'.
  test:
      # Test data - the data may come from several tables.
    - data_id: 'my_test_data'
```

(continued)

Table 9.31 (continued)

```
          # Test data table name.
     - dec_id: 'decision'
          # Only 1 dec_id is allowed!
          # NOTE: you can select 3 columns or without col_id's the first 3
          # columns are used! If you select the columns then it is important
          # that the columns appear in the DB table in the following order:
          #    - col_id: 'decision' # 'decision' must also be defined as a col_id!
          #    - col_id: 'UserID'
          #    - col_id: 'ItemID'.
   input:
          # ML Flow input data - the data may come from several tables
     - data_id: 'my_input_data'
          # Input data table name
     - dec_id: 'decision'
          # The value in the DB must be empty or NULL otherwise it will not be
          # evaluated! Only 1 dec_id is allowed!
          # NOTE: you can select 3 columns or without col_id's the first 3
          # columns are used! If you select the columns then it is important
          # that the columns appear in the DB table in the following order:
          #    - col_id: 'decision' # 'decision' must also be defined as a col_id!
          #              Must be empty or NULL!
          #    - col_id: 'UserID'
          #    - col_id: 'ItemID'.
   output:
     - data_id: 'my_output_data'
          # Output data table name (for the prediction).
     - col_id: 'userid'
          # The column where the userID will be written; this must be the first
          # col_id here!
          # NOTE: the userid column will not be used if the input data table is
          # chosen as output but must be defined!
     - col_id: 'itemid'
          # The column where the item id will be written; this must be the
          # second col_id here!
     - col_id: 'decision'
          # The column where the output rating will be written; this must be the
          # third col_id here!
     - col_id: 'recom1'
          # The column where the output recommendation will be written (optional)
          # if there are more recommendations then put them hereunder with
          # separate col_id's (different names).
 #}
```

will not be used if the input data table is chosen as output (because there is already a user ID and item ID column in the input) but they must be defined in the output section!

The order of the output columns is very important for this model. The first col_id must be the user ID, then the item ID, then the rating (decision) and finally all of the recommendations in separate columns! The only exception to this rule is when the input data table is the same as the output data table because then the order of the

Database Structure	Browse Data	Edit Pragmas	Execute SQL

Table: input_data

	flow_id	decision	userid	itemid
	Filter	Filter	Filter	Filter
1	1	6.155903	6464	111878

Fig. 9.4 Example input data table for recommendation models

Database Structure	Browse Data	Edit Pragmas	Execute SQL				

Table: output_data   New Record Delete Record

	flow_id	userid	itemid	rating	recom1	recom2	recom3	
	Filter	Filter	Filter	Filter	Filter	Filter	Filter	F
1	1	6464	111878	6.1559	128987	81117	130347	1

Fig. 9.5 Example output data table for recommendation models

columns is as follows: decision column (rating), user ID, item ID and recommendation columns (if recommendations are requested).

In all tables, except the output data table, the 'col_id' definitions (column selection) may be omitted and then all columns are used in the data table.

The data in the input data table's decision (rating) column must be empty or NULL. This signals the program to evaluate that record! The recommendation columns (if they exist in the input because they are also used as output) may contain data but will be replaced.

9.2.3.13 Recommendation with Implicit Feedback (Collaborative Filtering: CFI)

More information about collaborative filtering with implicit feedback (CFI) can be found in Sect. 8.4.1 in Chap. 8.

The built-in CFI model template can be seen in Table 9.32.

This model requires that the data is in a specific order in the database tables (training, test, input): the first column must be the observations (decisions), the second column must be the userID and the third column must be the itemID! Observations are, for example, the count of events (e.g., viewing a webpage, reading an article, etc.). In the case of the BPR model these numbers will be binarized (all numbers greater than 0 will be converted to 1). When you import your data into the database make sure that you select the right decision column in your data! The decision column will be automatically the first column after importing the data and

Table 9.32   CFI model built-in template

```
--- # Recommendation with Implicit Feedback (Collaborative Filtering) Example {
# Implicit Feedback means that there are no ratings of the items but just
# observation counts.
# IMPORTANT: This model requires that the data is in a specific order in the
# DB data tables (training, test, input): 1st column Observations (decisions),
# 2nd column UserID, 3rd column ItemID! Observations are for Example the count
# of events (e.g. viewing a webpage, reading an article, etc.). In case of the
# BPR model these numbers will be binarized (all numbers greater than 0 will
# be converted to 1). When you import your data into the DB make sure that you
# select the right decision column in your data! The decision column will be
# automatically the first column after importing the data and the 2nd and 3rd
# columns must be the UserID and ItemID! The UserID's and ItemID's in the test
# and input datasets must all exist in the training dataset!
#
model:
  id: 'ID-IzUqLbHzxN' # must be unique!
  type: CFI
  path: 'D:\mypath\MyDB.sl3'
      # database path containing all data (input, output, model)
  params:
    - Algorithm: BPR
      # Options: BPR, WALS (see help for more info!)
    - Recommendations: 1
      # The number of recommendations to generate for each User
    - RecommendationType: DEFRECOM
      # DEFRECOM, SIMILAR_USERS_PLUS, SIMILAR_ITEMS_PLUS (see help for info)
      # general parameters:
    - Cutoff: 0.0
      # below this value the observations will be considered as negative and
      # set to 0.0 !
    - Epochs: 50 # number of iterations
    - Factors: 100
      # dimension of learned factors (higher value higher precision but slower
      # model)
  # BPR parameters
    - Init_learning_rate: 0.01 # initial learning rate
    - Bias_lambda: 1.0 # regularization on biases
    - User_lambda: 0.025 # regularization on user factors
    - Item_lambda: 0.0025 # regularization on item factors
    - Decay_rate: 0.9 # decay rate on learning rate
    - Use_biases: 1 # set to 1 to use biases; to 0 to not use
    - Num_negative_samples: 3
      # number of negative items to sample for each positive item
    - Eval_num_neg: 3 # number of negatives generated per positive in
                      # evaluation
  # WALS parameters
    - Boost: 0
      # set to 1 if you want each observation value to be boosted in three
      # ranges (see help for more info!)
    - Regularization_lambda: 0.05 # regularization parameters
    - Confidence_weight: 30 # confidence weight (alpha)
    - Init_distribution_bound: 0.01 # initial weight distribution bound (-b, b)
training: # training data - the data may come from several tables!
  - data_id: 'my_training_data'
```

(continued)

Table 9.32 (continued)

```
        # training data table name in the database defined by 'model > path'
        # above
    - dec_id: 'decision'
        # the decision column ID. There must be ONE defined in one of the tables!
        # (Observations!)
        # NOTE: you can select 3 columns or without col_id's the first 3
        # columns are used! If you select the columns then it is important
        # that the columns appear in the DB table in the following order:
        # - col_id: 'decision' # 'decision' must also be defined as a col_id !
        # - col_id: 'UserID'
        # - col_id: 'ItemID'
  test: # test data - the data may come from several tables
    - data_id: 'my_test_data' # test data table name
    - dec_id: 'decision'
        # NOTE: you can select 3 columns or without col_id's the first 3
        # columns are used! If you select the columns then it is important
        # that the columns appear in the DB table in the following order:
        # - col_id: 'decision' # 'decision' must also be defined as a col_id !
        # - col_id: 'UserID'
        # - col_id: 'ItemID'
  input: # ML Flow input data - the data may come from several tables
    - data_id: 'my_input_data' # input data table name
    - dec_id: 'decision' # must be empty or NULL !
        # NOTE: you can select 3 columns or without col_id's the first 3
        # columns are used! If you select the columns then it is important
        # that the columns appear in the DB table in the following order:
        # - col_id: 'decision' # 'decision' must also be defined as a col_id !
        #              Must be empty or NULL !
        # - col_id: 'UserID'
        # - col_id: 'ItemID'
  output:
    - data_id: 'my_output_data' # output data table name (for the prediction)
    - col_id: 'userid'
        # the column where the userID will be written; this must be the first
        # col_id here! NOTE: the userid column will not be used if the input
        # data table is chosen as output but must be defined!
    - col_id: 'itemid'
        # the column where the item id will be written; this must be the
        # second col_id here!
    - col_id: 'decision'
        # the column where the output rating will be written; this must be
        # the third col_id here!
    - col_id: 'recom1'
        # the column where the output recommendation will be written (optional)
        # if there are more recommendations then put them hereunder with separate
        # col_id's (different names).
#}
```

the second and third columns must be the userID and itemID! The userIDs and itemIDs in the test and input datasets must all exist in the training dataset! Read also the remarks about these columns in the previous section on explicit feedback models!

This is a specialized model for implicit feedback, but it may also be used with explicit feedback data since a rating also means there has been an interaction with an item. One must, however, be careful with the rating values because low rating values may mean a negative preference for the item (dislike) but the implicit model would take these low ratings to be a positive preference (like)! In order to prevent this problem you can use the 'cutoff' parameter and set it to a rating value below which the ratings will not be taken into account!

More information about the two types of models BPR and WALS and about CFI can be found in Sect. 8.4.1 in Chap. 8.

There are three types of recommendation methods in this model:

- *DEFRECOM*—this is the default recommendation method which ranks the items and selects the items with the highest ranks.
- *SIMILAR_USERS_PLUS*—this recommendation method first searches for users similar to the user we are requesting a recommendation for, and then it selects the highest ranked items from several users.
- *SIMILAR_ITEMS_PLUS*—this recommendation method searches for items similar to the item in the recommendation request and then it selects the highest ranked items.

All methods filter out already known items. If there are not enough items to recommend, then some items may be repeated.

The parameter 'factors' determines the number of latent factors used in the CFI model. A higher number of latent factors will result in higher accuracy until a specific limit is reached (it does not make sense to increase it above a specific value) but the model will train slower and will be larger in size.

The parameter 'boost' in the WALS model allows the automatic increase of the feedback values in three ranges (maxIR/3, 2 * maxIR/3 and above, where 'maxIR' is the maximum implicit feedback value). This may be useful if explicit feedback is used as input because it results in a much higher difference between lower and higher ratings (higher ratings mean positive feedback in CFI).

9.2.4 Training, Testing and Prediction

You can train, test and make predictions using individual models defined in the project file. You can select which model you want to use. The results will be saved directly into the database as defined in the project file (trained model, test results and prediction data tables). All operations can be stopped with the 'Stop' button but you may have to wait for a while until the operation is canceled.

9.2.5   AI-TOOLKIT Continuous Operation Module (MLFlow)

You can use the ML Flow module for continuous training and prediction of several (possibly connected) models. After you push the 'START Auto ML Flow' button the software will cycle through all of the models in the project file, train them if necessary, check the database for new prediction requests and evaluate all open predictions. All prediction results will be saved into the database as defined in the project file (input, output). Between each cycle the software will wait during the period (in seconds) defined in the control panel 'timeout' (settings). You can stop the Auto ML Flow operation by pushing the button again (it changes color from green to red to indicate that you can push to stop it).

New prediction requests can be added by the AI models themselves or you can update the SQLite database with new prediction requests using an external application or script.

> NOTE: There is a 'Timeout Database Lock' setting in the control panel which should be set to a high enough value to allow enough time for all necessary updates (the database will be locked (not accessible) while you update it with an external tool and while the AI-TOOLKIT uses the database).
>
> EXPERT INFO: The 'trained_models' table in each AI-TOOLKIT database contains a 're_train' field which is set to 0 after the model is trained. You can force re-training by setting this field to 1. This can be interesting, for example, if you add new data to the training dataset.

9.2.6   The AI-TOOLKIT Database

The AI-TOOLKIT database is a standard SQLite database (a fast serverless database) with a special table called 'trained_models' which must be setup correctly. Therefore, always make a new database with the 'Create New AI-TOOLKIT Database' button on the left toolbar!

The data tables for training/testing and prediction are very similar but each prediction data table (called the 'input' table in the project file) must have a special column called 'flow_id'. Because of these rules (trained_models, flow_id, etc.) it is recommended that the built-in 'Import Data' and 'Import Images' modules are used for creating/importing all data tables. These modules create the right table structure! If you are already an expert in the AI-TOOLKIT then you may of course also create tables manually in the built-in database editor, or in any other SQLite compatible database editor.

Please look at the databases in the examples throughout this book to see the structure and the column types for each data table.

The database can be encrypted. The encryption must be done in the database editor first, before any other operation is performed (see Open Database Editor on the DATABASE tab). Make sure that you provide the correct password in the right control panel of the AI-TOOLKIT tab otherwise all operations will fail! You can remove the encryption and password protection in the database editor any time you want and use the database further without protection.

> EXPERT INFO: In the AI-TOOLKIT it is possible to use more than one database table for each data type (training, testing, and prediction/input) such that the columns come from different tables. It is the responsibility of the user to make sure that all of the tables are synchronized so that specific records in one sub-table match the records in the other sub-tables used together for training, testing or prediction! Also be mindful of this when deleting records from tables in the database editor (delete synchronized records from all relevant tables)!

9.2.6.1   The Database Editor

You can start the built-in database editor (Fig. 9.6) with the Open Database Editor button in the left toolbar. The database editor is an external program called "DB Browser for SQLite" which is embedded into the application for easy use. You will find more information about this program later.

AI-TOOLKIT creates the database and adds the data automatically (importing delimited data files, importing and converting images, etc.); therefore, you only need

Fig. 9.6 AI-TOOLKIT database editor

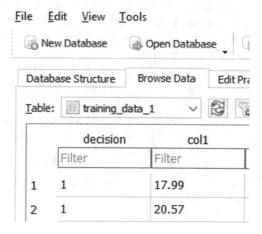

to use the database editor for encrypting the database (if you need this feature), for viewing the data or for deleting obsolete data tables. You can, of course, do a lot more with the database editor but it is not the aim of this book to explain all of its functionality.

The database editor is very easy and intuitive to use. With the Open command in the File menu you can open a database. You can choose the table to view in the Browse Data tab.

You can delete a database table on the 'Database Structure' tab by clicking on the table with the right mouse button and choosing 'Delete Table'.

> IMPORTANT NOTE: Make sure that you do not delete or alter the structure of the table called 'trained_models' because it is needed for the software to function correctly! You can of course delete any record, but the model you delete will not be available for prediction and will need to be retrained.

Database encryption can be performed with the 'Set Encryption' command on the Tools menu. Make sure that you add the database password to the right control panel of the AI-TOOLKIT tab every time you use the database. The password is not saved for security reasons. If you use several databases at once (e.g., because different models use different databases), then all of the databases must be encrypted with the same password or not encrypted at all!

If you need more information about the database editor (has many uses such as, for example, executing complex queries) then please visit https://github.com/sqlitebrowser/sqlitebrowser/wiki.

9.2.6.2   Importing Data (Numerical or Categorical) into the Database

The import data module can be seen on Figs. 9.7 and 9.8. With this module you can import any delimited text data file into the database which can be used for training, testing or prediction of the AI models. The import module is very easy to use. First you have to select the delimited data file, choose the delimiter(s), set some options, and indicate if there are header rows in the file and also the zero based column number of the decision column. Next select the option of whether you want to use the data for training/testing or for prediction. This is important because the data tables have a different structure.

Finally select or create an AI-TOOLKIT database and indicate the table name, the database password (optional) and, if you want to use a single BLOB (binary object) database column for all non-decision columns (i.e., except for the decision column, all columns will be added to a single BLOB column instead of to separate columns). The BLOB column type may be useful in order to decrease the size of the database if you have a lot of columns or some special data. For example, image data is stored in

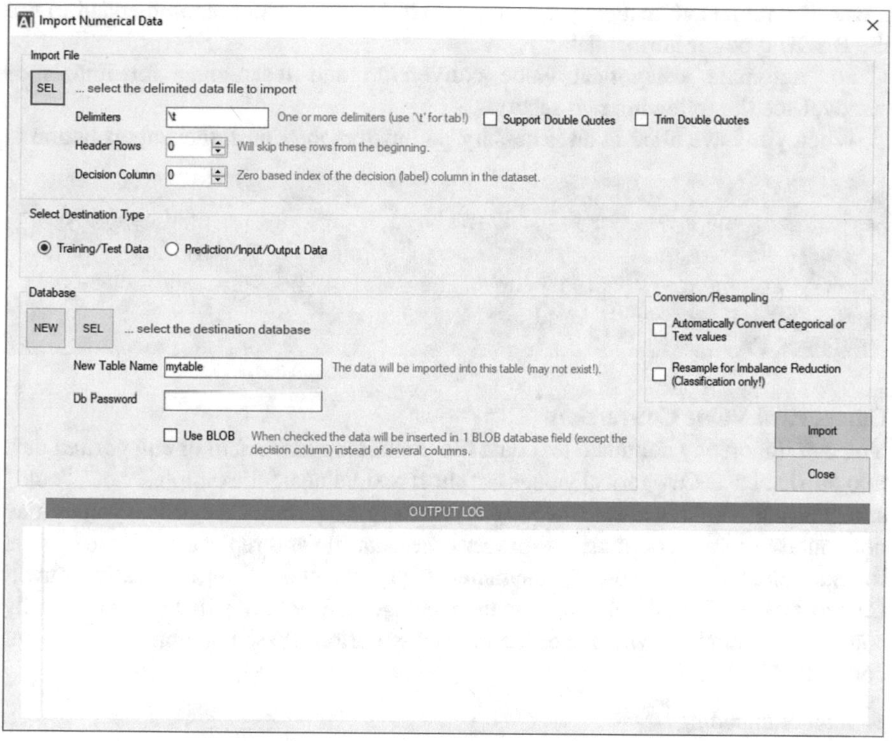

Fig. 9.7 AI-TOOLKIT import data module: part 1

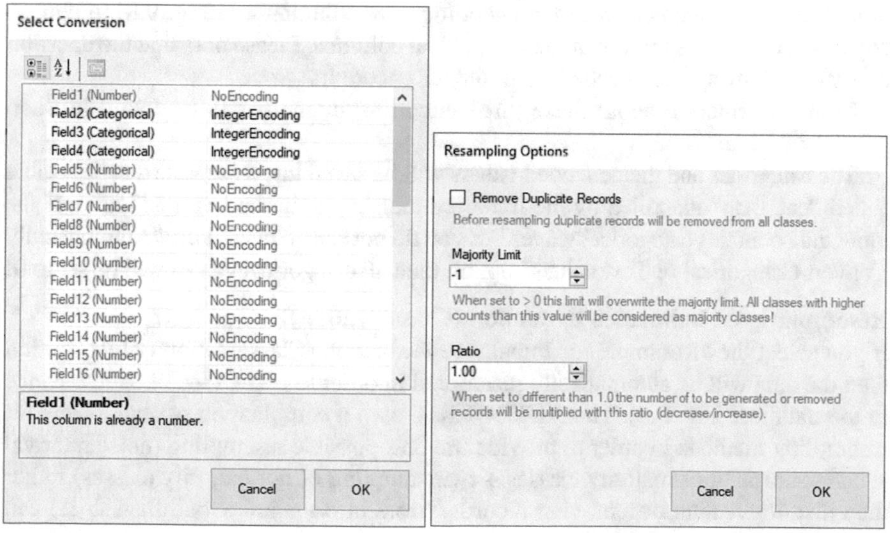

Fig. 9.8 AI-TOOLKIT import data module: part 2 & 3

a BLOB column automatically by the AI-TOOLKIT. It is not recommended to use the BLOB type for normal data.

For automatic categorical value conversion and resampling for imbalance removal see the following two sections.

When you have filled in all necessary parameters then push the Import button.

> NOTE: If you create a new database (NEW button) in this module and you want to encrypt the database then you must first exit this form and add the encryption in the database editor! Then restart this form and select the encrypted database with the SEL (select) button!

Categorical Value Conversion

You can import any delimited text data file containing *numerical* or *categorical* data into the database. Categorical values are short text values; for example, "hot", "cold" and "normal" could be three categorical values in a column. Categorical values may not contain spaces; you must pre-process the data file and replace any spaces with, for example, the underscore "_" character. If you select the "Automatically Convert Categorical or Text values" option then categorical values will be automatically converted to numbers with one of the following options (based on your choice(s) per column):

- Integer encoding
- One-hot encoding (increases the number of features!)
- Binary encoding (increases the number of features!)

The AI-TOOLKIT selects integer encoding as default but you have the option (a dialog box will appear with the encodings per column—see Fig. 9.8) to choose any of the encodings for any of the categorical columns. Please note that the decision column will always be encoded with integer encoding.

More information about these three encoding methods can be found in Sect. 4.3.2.2 in Chap. 4.

The categories and their encoded values will be saved into a separate database table called "cat_map" identified by the imported table name in the "data_id" field. If the input data contains categorical values but you do not select (check) the "Automatically Convert Categorical or Text values" option then all categorical values will be set to 0!

Resampling for Imbalance Reduction (Classification)

If you select the "Resample for Imbalance Reduction (Classification only!)" option then the data will be automatically resampled in order to correct the class imbalance in the data (see Fig. 9.8). The AI-TOOLKIT uses a combination of state-of-the-art resampling methods in order to provide the best possible resampling (noise removal + undersampling of majority classes + oversampling of not majority classes) rather than just duplicating or removing records. More information about this subject can be found in Sect. 4.3.2.6 in Chap. 4.

You have three resampling options:

- **Remove Duplicate Records**—before resampling duplicate records will be removed. This can be useful if the data contains a lot of duplicate records because then the resampling will be more appropriate.
- **Majority Limit**—when it is set to greater than 0, this limit will overwrite the majority limit. All classes with higher counts (the number of records with this class) than this value will be considered as majority classes and not only the class with the highest count! This is very useful if you want to under sample several classes and not only the real majority class, or if you do not want to oversample some classes with high counts. For example, if you have four classes with occurrence counts of 2000, 3000, 15,000 and 25,000 and you set the 'Majority Limit' to 10,000 then the classes with 15,000 and 25,000 will be under sampled to 10,000, and the classes with 2000 and 3000 counts will be oversampled to 10,000. The result will be that all classes are balanced. The oversampling is not duplication but a form of advanced record generation in order to improve the classification boundary.
- **Ratio**—when it is set to a value other than 1.0 then the number of records to be generated or removed will be multiplied by this ratio (decrease/increase). This can be useful if you want to change the oversampling and/or undersampling amount. Be careful because both oversampling and undersampling will use the same ratio but of course undersampling will only under sample majority records and oversampling will only oversample non-majority records!

9.2.6.3   Importing Images into the Database

You can import most image types (png, gif, jpg, bmp, etc.) into the database for training, testing and prediction of your machine learning models (see Fig. 9.9). The images will be converted to an appropriate numerical format and saved into the database.

First you must list the paths to the images in the spreadsheet in the Path column. You can also use the Select button and select several files in the file browser but make sure that you choose the '*Extract Label From File Name*' option first (to extract the integer label from the file name automatically). If the Labels are not filled in automatically then you must indicate all labels in the Label column. The label is the classification of each image in the form of an integer number.

Next set the Collage flag to ON if your images contain several smaller images and indicate the size of these smaller images (segment size) in the Width and Height input boxes. *You must use one collage per class!*

Finally, select or create an AI-TOOLKIT database, and indicate the table name, and the database password if required. When you have filled in all necessary parameters then push the Import button.

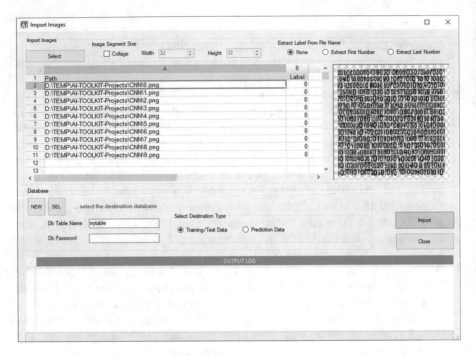

Fig. 9.9   AI-TOOLKIT import images module

NOTE: If you create a new database (NEW button) in this form and you want to encrypt the database then you must first exit this form and add the encryption in the database editor! Then you can restart this form and select with the SEL button the encrypted database!

9.2.7   *Automatic Face Recognition*

AI-TOOLKIT has a built-in application which can be used for automatically recognizing people's faces in images (Fig. 9.10). You can build your own database with the people's faces that you want to recognize and then load an image and search for known and unknown faces. The application will also signal unknown faces which may be useful for security purposes. The face recognition works well even if the face is not a frontal face and it can achieve above 99% accuracy. When you build your reference face database (see later) you may also include several images for the same person in case these images are significantly different. Each face has a corresponding text description in the database which is shown when the face is recognized.

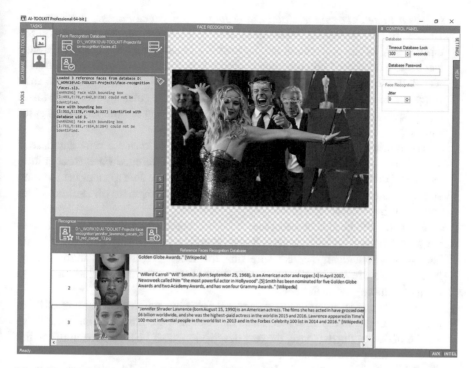

Fig. 9.10 AI-TOOLKIT face recognition application

The database used in speaker recognition (see the following section) is the same as the database used in face recognition and in fingerprint recognition (see Sect. 9.2.9), and therefore you can make a combined face, voice and fingerprint database to identify all three aspects of a person. This will increase the certainty of correctly recognizing a person.

> NOTE: There is an example database (people.sl3) and an example image for face recognition (jennifer_lawrence_oscars_2018_red_carpet_13.jpg) in the folder called 'face-recognition' under the 'Demo' folder installed during the setup of the AI-TOOLKIT!
>
> NOTE: Hover over each command button with the mouse pointer to see a short description!

9.2.7.1   How to Use

The face recognition application in the AI-TOOLKIT is designed to be as user-friendly as possible, and it can be used by simply navigating the commands as follows:

1. *Create a Face Recognition Database & Import Reference Faces.*

The 'Import Reference Faces Into Database | Create Database' button on the top-left side of the application can be used for creating and importing reference faces into a database (with the mouse pointer hover over each command button to see its description). You may create several databases. A single database may contain several hundreds of thousands of people but the software will work faster if you create several smaller databases. Depending on your computer (processor and memory) find out the most optimal database size.

The application will ask you whether you want to create a new database if there is no database selected yet with the 'Select Reference Face Recognition Database' command. If you answer *yes*, then you will be able to enter the file path and file name of the new database in a file browser. Navigate to the desired folder and enter the database file name.

Next you will be able to select one or more images with reference faces. It does not matter if there are other objects in the images because the application will find the faces automatically. Make sure that the face images are of high quality. Frontal faces are the best for this purpose but you may also include a second image (or more) for each person.

NOTE (1): If an image contains several faces but you only want one of them then you can delete the unwanted faces in the database editor later. Only the faces will be added to the database, the text description must be added manually in the database editor (see later).

NOTE (2): You can use the built-in image editor (TOOLS tab) for enhancing and editing your input images.

2. *Select a Face Recognition Database.*

If you have already created a face recognition database then select it with the 'Select Reference Face Recognition Database' button on the top-left. The contents of the database will be shown in the bottom table.

3. *Edit Database.*

The 'Edit Database' button can be used to initiate the database editor. The database editor is explained in the help in the right sidebar of the database editor and in Sect. 9.2.6.1. The database editor can be used for adding a text description for existing records and for removing unwanted records. You may not alter the database table structure or contents in any other way because the application is expecting it in the defined format.

4. *Import Reference Face into Selected Record in Database.*

If you have already created a database then you can import a reference face into a specific record with this command. The first step is to open the database ('Select Reference Face Recognition Database') and then select one of the records in the data table at the bottom of the screen. Next use the 'Import Reference Face into Selected Record in Database' command.

IMPORTANT: If there is already a reference face in the selected record then it will be replaced!

5. *Recognize a Face.*

After the database has been loaded use the 'Load Image' button to load an image in which you want to recognize a face. The image will be shown on the screen. Next use the 'Recognize Faces On Image' button. All recognized (known) and not recognized (unknown) faces will be labeled on the preview image. If the face is recognized then the label will contain the unique identifier (UID) of the person in the database. You can use the database table at the bottom to view the information for the recognized person.

NOTE (1): If a face is not detected on your image then this could be because: a.) the face is too small, in which case enlarge the image in an image editor (the resampling of the image to a higher resolution should be as optimal as possible; you may use the built-in image editor) or b.) the face is not recognizable because of its rotation or image quality, in which case try to get a better image.

NOTE (2): You may select a *region* of the image for recognition (see the following section for more information).

6. *Image Controls*

The image preview has several built-in tools which can be accessed with the small blue buttons next to it. '**S**' is used for switching between zoom and the selection rectangle. '**P**' is used for switching between pan and zoom. '**F**' is used for fitting the image into the whole preview area. The '+/−' signs are used for zooming in/out the image.

The selection rectangle may be useful if you only want to recognize a specific region of the image. Before using the 'Recognize Faces On Image' button select a region (e.g., a face) on the loaded image which you want to recognize.

7. *Settings.*

In the right sidebar you can find the help and also the Settings where you can change some of the parameters.

The database password must only be filled in if you manually encrypted the database in the database editor after it was created in the application.

The 'jitter' parameter can be used for adding some variation to the input images. If jitter >0 then *jitter* number of slightly modified (rotated, deformed in some way, etc.) images will be sent to the machine learning model instead of just one image. This may increase the accuracy of the face recognition slightly but the application will be considerably slower. Normally jitter = 0 will be sufficient for most applications. In case you do decide to use jitter then set this value to at least 20–30 and preferably to 100. The jitter parameter effects how the reference images are imported (learned by the machine learning model) in the database and also the face recognition!

9.2.8 Automatic Speaker Recognition

AI-TOOLKIT has a built-in application which can be used for automatically recognizing people's voices in speech recordings (audio). You can build your own database with the people's voices that you want to recognize and then load a voice recording to identify the people who are speaking in it.

The database used in speaker recognition is the same as the database used in face recognition (see previous section) and in fingerprint recognition (see following section), and therefore you can make a combined face, voice and fingerprint database to identify all three aspects of a person. This will increase the certainty of correctly recognizing a person.

Each voice (and face) has a corresponding text description in the database which is shown when the voice is recognized.

> NOTE (1): You can hover over each command button to see a short description!
>
> NOTE (2): There is an example database (people.sl3) and an example voice recording for speaker recognition in the folder called 'face-recognition' under the 'Demo' folder installed during the setup of the AI-TOOLKIT!

There are three types of speaker recognition models available in the AI-TOOLKIT:

- Supervector
- GMM
- i-vector

A database record may contain one or more types of speaker recognition models. You can choose the model you want to use with the radio buttons in the 'Select Model' group. You may switch to another model any time you want but it is of course better to use the same type of model throughout a project. The supervector model is the easiest to use and provides a good accuracy if clean and unbiased voice recordings are used. The more complex GMM and i-vector models may be useful in case of more biased and large datasets.

> IMPORTANT: The 'Import Reference Voices Into Database' and 'Import Reference Voice into Selected Record' commands depend on the type of the model! Always select the type of the model first before importing the audio recordings because the audio recordings are always imported for the currently selected model type! All three models (supervector, gmm and i-vector) may exist in a single record!

9.2.8.1   How to Use the Supervector Speaker Recognition Model

The speaker recognition application in the AI-TOOLKIT is designed to be as user-friendly as possible, and it can be used by simply navigating the commands as follows:

1. *Create a Speaker Recognition Database*

    The 'Create New Reference Voice Recognition Database' button can be used for creating a new database. You may create several databases. A single database may contain several hundreds of thousands of people but the software will work faster if you create several smaller databases. Depending on your computer (processor and memory) find out the most optimal database size.

2. *Import Reference Voices.*

    The 'Import Reference Voices Into Database' button on the left side of the application can be used for importing reference voices into a database.

    You will be able to select a folder which contains the **subfolders** with the voice recordings of one or more people. Each person is in a separate subfolder which may contain several voice recordings for that person! Make sure that the quality of the voice recordings is very good and contains only the speech of the correct person. The best method is to use short sentences. The reference voices will be added to the database and each subfolder name will be used as speaker ID and will also be added to the information field (e.g., the subfolder name could be the name of the person).

    IMPORTANT: The supervector speaker recognition model needs at least 3 trained individual speaker models for the speaker recognition accuracy to be feasible. The more speakers you add the better the feasibility will be!

    NOTE: You can use the built-in audio editor (TOOLS tab) for editing your recordings.

3. *Select a Speaker Recognition Database.*

    If you have already created a speaker recognition database then select it with the 'Select Reference Speaker Recognition Database' button. The contents of the database will be shown in the bottom table.

4. *Edit Database.*

    The *'Edit Database'* button can be used to initiate the database editor (Fig. 9.11). The database editor is explained in the help in the right sidebar of the database editor and in Sect. 9.2.6.1. The database editor can be used for adding a text description to existing records and for removing unwanted records. You may not alter the database table structure or contents in any other way because the application is expecting it in the defined format.

5. *Import Reference Voice into Selected Record in Database.*

    If you have already created a database then you can import a reference voice into a specific record with this command. The first step is to open the database ('Select Reference Speaker Recognition Database') and then select one of the records in the data table at the bottom of the screen. Next use the 'Import Reference Voice into Selected Record in Database' command. The application

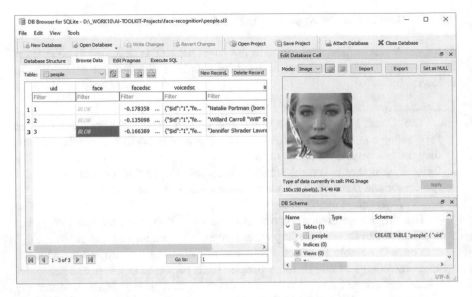

Fig. 9.11 The face and speaker recognition database in the database editor

will ask you to select the folder with the voice recordings for the specific person (only one folder!). The folder name will be appended to the information field.

IMPORTANT: If there is already a voiceprint in the selected record then the new voiceprint will be merged with it! If you want to replace the voiceprint instead, then use the database editor and set the 'voicedsc' field to NULL for that record and save the database (use the 'Set to NULL' and 'Apply' buttons in the database editor).

NOTE: You can see if there is already a voiceprint in any record in the 'Voiceprint' column of the data table. If a '-' is indicated then there is no voiceprint, an 'S' indicates a trained supervector model, a 'G' indicates a GMM model and an 'I' indicates an i-vector model. It is also possible to have several models in a single record which is then indicated with several letters, e.g., 'SGI' means that all three types of models are present.

6. *Recognize a Speaker.*

After the database has been loaded use the 'Load and Recognize Voice' button for loading and recognizing a voice recording. You will be able to select a voice recording (*.wav) and the application will then show the speaker recognition results in a table report.

The first column in the table report is called 'Words' which is the number of spoken words used for recognizing a voice. The input voice recording is split into words when silence is detected. Detecting silence is not an easy task because of the many different noise levels in a recording, and therefore the word splitting may not be perfect but this does not influence the speaker recognition negatively. The silence threshold determines what will be identified as silence and what will be identified as speech. You can set the silence threshold and the 'Recognition

Word Count' in the settings. The 'Recognition Word Count' determines how many words will be taken together for recognizing the speaker. If several people are speaking after each other, then the boundary between speakers may be the mixture of two or more speakers and thus may not be representative.

The report table also contains the unique identifier (UID) of the identified person in the database and the speaker recognition quality measures 'distance' and 'certainty'. A lower value for distance and a higher value for certainty mean a more certain recognition of the person. The recognition quality depends strongly on the quality of the voice recordings (more about this later).

7. *Settings.*

In the right sidebar you can find the help and also the Settings where you can change some of the parameters.

The database password must only be filled in if you encrypted the database in the database editor after it was created in the application.

The other parameters have already been explained above.

IMPORTANT (1): Always use the same sampling rate, *preferably 16,000* sample per second (optimal speaker recognition accuracy). If you are not sure about the sampling rate of a recording then use the built-in audio editor (on the TOOLS tab) to check and convert the audio file to this sampling rate if needed! The minimum sampling rate is 8000 samples per second, below this value the accuracy of speaker recognition cannot be guaranteed. Upsampling an audio to a higher sampling rate will work but it will not have the desired effect; therefore, make sure that the sampling rate of the original recording is at least 8000 and preferably 16,000 samples per second.

IMPORTANT (2): You may use the built-in audio editor for cleaning your voice recordings. You can remove noise, echo, etc. More information about this can be found in the help of the audio editor and in Section 9.2.9. Cleaning up the voice recordings may increase the accuracy of speaker recognition considerably!

NOTE (1): The supervector model for speaker recognition works well if there is not much noise in the audio recording, and if the added audio recordings of a specific person contain the same type of speech (e.g., normal speaking) and were produced under similar recording conditions. The built-in audio editor in the AI-TOOLKIT can be used for removing noise, echo, etc. from your recordings. If the reference recordings in the database (voiceprint) and the input recording for speaker recognition contain a significantly different type of speech (from the same person) or include noise or background sound then the accuracy of speaker recognition may be lower in which case the other models should be used.

NOTE (2): There is an example database (people.sl3) and an example audio file for speaker recognition (jennifer_lawrence.wav) in the folder called 'face-recognition' under the 'Demo' folder installed during the setup of the

---

(continued)

AI-TOOLKIT! Load the database and recognize ('Load and Recognize Voice') the example audio file. The application will indicate that it is Jennifer Lawrence who is speaking

9.2.8.2   How to Use the GMM and i-Vector Speaker Recognition Models

You will need a minimum of basic knowledge of GMM and i-vector speaker recognition in order to be able to use these models (please refer to Chap. 7). It is not possible to explain here everything what you need to know for selecting the right parameters but there is a very simple automatic parameter estimation module called 'Auto Config' on the 'GMM/i-VECTOR SYSTEM' tab. If you enter the number of speakers (people) in your system and then click the 'Propose' button then all parameter values will be adjusted. The proposed parameters are a good starting point and you may still need to adjust them!

The application can be used by simply navigating the commands as follows:

1. Select the desired import audio options on the 'GMM/i-VECTOR SYSTEM' tab. There are two possible scenarios: a.) If the 'Use Folder Name as Speaker ID' is selected then each audio file in a subfolder will get a speaker ID which is equal to the subfolder name (you will have to choose a folder which contains several subfolders with the audio recordings per person). b.) If the 'Use Folder Name as Speaker ID' is *not* selected then each audio file name *must* contain a unique speaker ID in the beginning of the file name until the first underscore character ('_'). For example, in 'myvoice1_recording.wav' 'myvoice1' is the speaker ID! *The audio file names may not contain spaces*!
2. Next select the desired GMM and i-vector options on the 'GMM/i-VECTOR SYSTEM' tab.
3. Next click the 'Train UBM and i-Vector System' command and follow the instructions on the screen which appear. This will train the UBM which will also be the base model for all client GMM-i-vector models!
4. Next open or create a new speaker recognition database.
5. Next Select the GMM or i-vector model option in the 'Select Model' group.
6. Next use the 'Import Reference Voices Into Database' or 'Import Reference Voice into Selected Record' command for importing reference voices into your database. This step is similar to what was explained above in the case of the supervector model. Follow the instructions on the screen.
7. Finally use the 'Load and Recognize Voice' command to recognize someone's voice in an audio recording.

GMM and i-Vector Model Parameters

- 'Use Folder Name as Speaker ID' (for all models)—explained above.

- 'Remove Noise' (for GMM & i-vector)—if selected then the input audio will first be filtered and noise will be removed; it is recommended to use the same setting for the UBM, client models and for the input audio for recognition!
- 'Mixture Distribution Count' (for GMM & i-vector)—the number of mixtures for the UBM and for all individual speaker GMMs (must be the same for both).
- 'Top Distribution Count' (for GMM)—the number of mixtures for the calculation of the log-likelihood; this may be equal or lower to the above mixture distribution count!
- 'TV matrix Rank' (for i-vector)—the rank of the total variability matrix for i-vector calculations.
- 'LDA Rank' (for i-vector)—the rank of the LDA matrix for i-vector calculations.
- 'TV/i Training Iterations' (for i-vector)—the number of iterations while training the TV matrix.
- 'Normalize i-Vectors' (for i-vector)—if selected then the i-vectors will be normalized.
- 'i-Vector Norm. Iterations' (for i-vector)—the number of iterations while normalizing the i-vectors.
- 'Max. Sessions Per Speaker' (for i-vector)—a per class speaker list (speaker separation) is used while normalizing the i-vectors; this parameter will limit the number of input audio files (features) used per speaker; in case this value is higher than the available number of input audio files per speaker then the list will be filled with the last added audio file.
- 'PLDA Iterations' (for i-vector)—the number of PLDA iterations to use.
- 'PLDA Eigen Voice Number' (for i-vector)—the rank of the Eigen voice matrix.
- 'PLDA Eigen Channel Number' (for i-vector)—the rank of the Eigen channel matrix.

9.2.8.3   Output Options

The supervector model has only one type of output depending on the 'Recognition Word Count' defined in the right sidebar. For each 'Recognition Word Count' number of words (the word boundary is automatically detected) the input audio will be tested and the identified speaker reported.

The GMM model has three types of output (only one of them is activated):

- A single speaker recognition output for the whole input audio. This output is activated when neither the per segment output ('Per Segment (GMM)') nor the per window output ('Per Window (GMM)') is activated.
- For each segment defined by the VAD (voice activity detector) a speaker recognition output will be generated. The 'Segment Duration Threshold' parameter may limit the tested segments, i.e., shorter duration segments will not be tested.
- For each window defined by the 'Window Width [sec]' a speaker recognition output will be generated.

The i-vector model has only one type of output which is the speaker recognition output per input audio. The i-vector model does not use segmentation.

> IMPORTANT (1): The audio file names may not contain spaces!
>
> IMPORTANT (2): The GMM and i-vector systems in the AI-TOOLKIT support much more types of audio files than the Supervector system which only supports wav files. You may of course always use the built-in AI-TOOLKIT audio editor for converting any audio to the wav format.
>
> NOTE (1): The audio files used for training the UBM/i-vector system may be different from the reference audio files you will add to the speaker recognition database!
>
> NOTE (2): The quality (accuracy) of speaker recognition depends on the quality of the voice recordings (for training the reference model and also for recognizing a voice), on the background noise and on many other factors! It is not possible (at the time of writing) to have 100% certainty in recognizing a person from a voice recording but we can indicate whether a voice is most probably belongs to a specific person. The better the quality of the voice recording the higher the certainty!

9.2.9   Automatic Fingerprint Recognition

AI-TOOLKIT has a built-in application which can be used to automatically recognize people's fingerprints in images. You can build your own database with the people's fingerprints that you want to recognize and then load an image to identify a person.

The database used in speaker recognition and face recognition (see Sects. 9.2.7 and 9.2.8) is the same as the database used in fingerprint recognition, and therefore you can make a combined face, voice and fingerprint database to identify all three aspects of a person. This will increase the certainty of correctly recognizing a person.

> NOTE (1): Hover over each command button with the mouse pointer to see a short description!
>
> NOTE (2): There is an example database (people.sl3) and an example image for fingerprint recognition (fingerprint.jpg) in the folder called 'biometrics-recognition' under the 'Demo' folder installed during the setup of the AI-TOOLKIT! The fingerprints in the database and the test fingerprint are not the real fingerprints of the actors (the voices and faces are the real ones).

9.2.9.1   How to Use

The application can be used by simply navigating the commands as follows:

1. *Create a Fingerprint Recognition Database*—The 'Create New Reference Fingerprint Recognition Database' button can be used to create and load a new database. You can create several databases. One database may contain several hundreds of thousands of people but the software will work faster if you create several smaller databases. Depending on your computer (processor and memory) find out the most optimal database size!

2. *Import Reference Fingerprints*—The 'Import Reference Fingerprints Into Database' button can be used to import reference fingerprints into the database. Follow the instructions on the 'Select Fingerprints Information Sheet' screen which appears. You will be able to select a folder which contains the subfolders with the fingerprints of one or more people; each person in a separate subfolder which may contain several fingerprints of the person! Each subfolder name will be used as person ID and will also be added to the information field (the subfolder name could be the name of the person). All images will be added (except a reference face image, see later) as fingerprint, therefore make sure that the subfolders only contain fingerprint images which you want to add to the database. Each image may only contain one fingerprint and no other objects, no drawings and no text.

   As next two option dialogs will appear where you will have to indicate your preferred options (for fingerprint ID and for whether to overwrite existing fingerprints). Follow the instructions on the screen. Remember that each person may have several fingerprints in the database and we differentiate these fingerprints by their fingerprint ID which could, for example, simply be the name of the finger. The fingerprint ID is thus not the same as the person ID!

   IMPORTANT (1): If there is already a person in the database with the same person ID that you import then you may replace the fingerprints, add to them or request to not alter them by using the options dialog which appears!

   IMPORTANT (2): The software stores only a small machine learning model (extracted features) of each fingerprint but not the fingerprint images. This makes the database small and the software very fast. The best way of working with fingerprints (and also with other types of biometrics) is that you keep the fingerprint images in a folder structure on the hard drive per database; each person in a separate subdirectory. Like this you will be able to view and visually compare the fingerprints in the software any time you want ('Load Fingerprint' command). You can of course also use a secured cloud storage or network drive.

   You can use the 'Abort Operation' button at the top (x) to stop importing fingerprints. This may be useful if you are importing thousands of fingerprints but you decide to stop the operation.

   After all subfolders are processed the contents of the database will be shown in the bottom table.

3. *Select a Fingerprint Recognition Database*—If you have created a fingerprint recognition database already then select it with the 'Select Reference Fingerprint

Recognition Database' button on the top-left, the contents of the database will be shown in the bottom table.

4. *Edit Database*—The 'Edit Database' button can be used to initiate the database editor. The database editor is explained in the help of the database editor (in the right sidebar). The database editor can be used to add a text description to existing records and to remove unwanted records. You may not alter the database table structure or contents in any other way because the application expects it in the defined format. Always make a backup first before editing the database!

5. *Import Reference Fingerprints into Selected Record in Database*—If you have created a database already, then you can import fingerprints into a specific record with this command. First, open the database ('Select Reference Fingerprint Recognition Database'), then select one of the records in the data table at the bottom of the screen. Next, use the 'Import Reference Fingerprint into Selected Record in Database' command. Follow the instructions on the screen!

   IMPORTANT: If there is already a fingerprint in the selected record, then you can replace it, add to it or request to not alter it by using the options dialog which will appear!

6. *Recognize a Fingerprint*—After the database is loaded, use one of the 'Load Fingerprint' buttons to load a fingerprint image you want to identify. The image will be shown on the screen. Next, use the 'Identify Fingerprint' ('?') button. If the fingerprint is recognized, then the database record will be selected in the preview table at the bottom. All information concerning the fingerprint recognition (recognition score, identified database UID, etc.) is shown in the log screen on the left.

If you load a fingerprint image in one of the four available boxes then, the displayed fingerprint image shows the extracted features with different colors per feature type (called minutiae: ridge endings and ridge bifurcations). These features are used in the machine learning model. It is not the aim of this book to explain the meaning of these fingerprint features. Please consult a reference book about fingerprint recognition for more information.

At the top of the fingerprint image the quality of the fingerprint and the identified number of features are shown. There are five quality levels: 'Excellent', 'Very good', 'Good', 'Fair' and 'Poor'. The better the quality the better the fingerprint recognition! If you build a reference fingerprint database then, try to use excellent or very good quality fingerprint images. You can select the minimum fingerprint quality in the settings, i.e., lower quality fingerprint will not be added to the database.

The fingerprint images together with the features can also be used for visual fingerprint matching. You can compare four fingerprints on the screen.

9.2.9.2 The Image Controls

The four fingerprint boxes (image previews) have several built-in tools which can be accessed with the small blue buttons next to them. 'P' is used for switching between pan and zoom, 'F' is used for fitting the image into the whole preview area, the '+/−' signs are used for zooming in/out the image. The bigger your monitor (screen) the bigger these images will become!

9.2.9.3   Settings

The following parameters are available in the settings in the right sidebar:

- The 'Timeout Database Lock' is used in case of the database is locked to wait the indicated number of seconds.
- The 'Database Password' must only be filled in if you have encrypted the database in the database editor after it was created in the application (you can only encrypt a database in the database editor).
- With the 'Fingerprint Display' options you can adjust how the fingerprint image is displayed and whether the features are shown or not. Each fingerprint feature has also a quality measure (this is not the same as the fingerprint quality) and with the 'Minimum Quality' option you can adjust which features are shown depending on their quality.
- The 'Enhance Image Feature Extraction' option can be used to request the automatic enhancement of fingerprint images before feature extraction. This may sometimes be useful depending on the input fingerprint images. Always try both, with and without enhancement, and only apply the image enhancement when it makes a positive difference. In some rare cases the image enhancement may also degrade the quality of feature extraction (if some ridges disappear from the fingerprint).
- The 'Minimum Fingerprint Quality' option can be used to set the minimum fingerprint quality while importing fingerprints into the database, i.e., lower quality fingerprints will not be imported.
- The 'Score Threshold' is an important fingerprint recognition option because the software will keep searching for fingerprints in the database until it finds a fingerprint with a match score which is at least this value. The optimal value of this parameter depends on your database and on fingerprint quality. The best way of working with the threshold is to start with a high value, e.g., 100, then decrease it step-by-step until a fingerprint is found. The higher the threshold the higher the similarity between two fingerprints, but, of course, the quality of two fingerprints may not be the same and then you will need to decrease the threshold in order to find the right fingerprint!

9.2.10   The Audio Editor

The audio editor (see Fig. 9.12) can be initiated on the TOOLS tab. You can use the audio editor for many tasks such as viewing and editing the audio recording, viewing

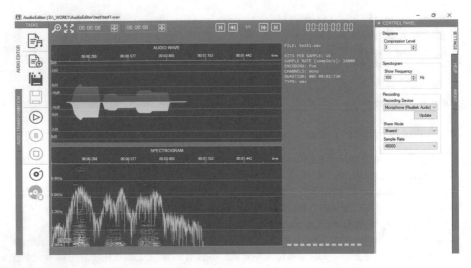

Fig. 9.12 AI-TOOLKIT audio editor

the spectrogram, changing audio properties (e.g., the sample rate, encoding, pitch, ...), removing noise, etc. And all of these tasks are also possible in batch mode applied to several audio files at once.

The audio editor can be used to view (also zoom in/out) the waveform of the audio (top diagram), the spectrogram (bottom diagram) and also to view the properties of the audio (sample rate, encoding, etc.). Many audio file formats are supported such as wav, mp3, etc. The following wav file formats are supported: 16, 24 or 32 bit PCM or IEEE float audio data.

The audio editor is optimized for short audio files, which are typically used for ASR training, but it can also handle longer audio files.

You can zoom in on both diagrams with the left mouse button several levels deep. Select the section you want to enlarge and click on the zoom in (magnifying glass with the + sign) button. Use the fit all button at the top, next to the zoom in button, to reset the zoom to the full diagram.

If you want to save only a specific part of the audio file, then select that part and click the 'Save Audio Selection' button (hover above the buttons to see a tool tip).

You can save the audio in mp3 or wav format.

You can also record an audio file. Select the recording properties in the right control panel.

Longer audio files, or files when zoomed in, are shown on several pages. You can walk through the pages with the left/right and start/end buttons at the top right corner of the waveform diagram. Both diagrams are handled in sync while stepping through the pages or when zooming.

9.2.10.1 Settings

The 'Spectrogram Show Frequency' setting determines which frequency line (the energy content for a specific frequency as a function of time) is shown on the spectrogram; if you set it to 0, then no line will be shown.

In the 'Recording' options you can select the recording device. If you switch the device on after you have started the software, then you must first push the 'Update' button and your device will appear in the list. Please note that not all devices support all sample rates and share mode settings.

9.2.10.2 Transform Audio

You can use the transform audio module to change the properties of wav audio files. The transformation options are shown in the right sidebar. You can select any number of options.

Please note that the following wav file formats are supported: 16, 24 or 32 bit PCM or IEEE float audio data.

> IMPORTANT: The sample rate must be dividable by 100 for several of the transformation options. If your wav file does not comply with this requirement then you must first convert it with the 'Change Sample Rate' transformation option. See the right control panel for the available standard sample rates.

You can apply the transformations to several selected files ('Transform Selected Files') or to a whole folder ('Transform Files in Folder'). The transformations will be done in batch mode and by making use of several processors if available.

9.2.10.3 Suppress Noise and Echo Cancellation

Suppress noise will remove noise from the audio depending on the selected suppression level (right control panel), where 0 means no noise removal and 3 is the highest level of noise removal. Echo removal works similarly but with three strengths (low, moderate and high).

9.2.10.4 Change Pitch

The change pitch transformation can be used to adjust the pitch (fundamental frequency) of the audio to a specific key frequency defined in the right control panel. The speech of a typical adult male has a fundamental frequency ranging from 85 to 180 Hz, and that of a typical adult female from 165 to 255 Hz. Infants show a range of 250 to 650 Hz. By using these values you can change, e.g., an adult voice to an infant one or a male voice to a female-like voice.

9.2.10.5   Removing Audio Without Human Voice

This transformation will remove the parts of the audio in which human speech cannot be detected.

9.2.10.6   Audio File Conversion

You can choose the output file type as wav or mp3 while applying the above transformations.

The 'Convert To WAV Without Encoding (PCM)' command can be used for batch converting several audio files from one audio format to a WAV PCM format without encoding. Several input audio formats are supported depending on your system's configuration (installed encodings). The WAV PCM format without encoding is the audio format which is used by most of the AI-TOOLKIT software (VoiceData, VoiceBridge, etc.).

9.3   DeepAI Educational

DeepAI Educational is a multi-layered neural network machine learning software toolkit for educational purposes (see Fig. 9.13). It helps users to understand the neural network learning process by visualizing it in a heat map and in a data table. DeepAI is very fast (it uses hardware acceleration with multiple processors) but it is not a professional neural network machine learning tool because of the limited options available. If you want state-of-the-art machine learning models and performance then use AI-TOOLKIT Professional instead!

9.3.1   Data Generator

The best way to learn DeepAI is by using the built-in data generator to generate input data. There are several data generators available for both classification and regression problems. You can also add random noise to the data. You can access the data generator on the right sidebar.

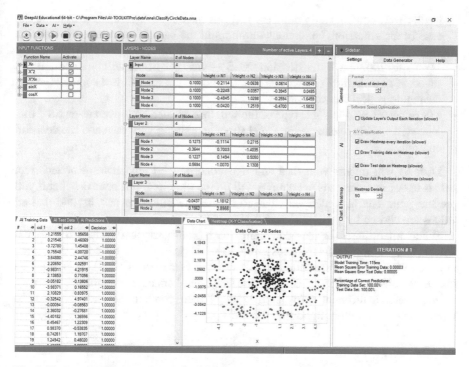

Fig. 9.13 DeepAI Educational

> NOTE: Remember that classification is used when there is a limited number of classes (decisions) assigned to each data record, in the form of integers, and regression is used when a continuous range of values (decisions) is assigned to each data record. Please refer to Chaps. 1 and 2 for more information about supervised learning.

Make sure that you check the "Treat data as X-Y Classification/Regression" in the settings if you use the data generator in order to generate a special heat map of the machine learning model!

9.3.2 Import External Data

If you choose the 'Load Data File' command then you can load different types of files:

- AI-TOOLKIT data file format with extension 'SSV'—the data must have the AI-TOOLKIT SSV data file format (.ssv) which is tab delimited, doesn't have a header row, contains only numbers and has the decision variable (classes in the case of classification, continuous numbers in the case of regression) in the first column.

- MS Excel 2003 file with extension 'XLS'—this type of file will be imported and saved as an SSV file. You will be able to choose the Excel sheet containing your data and the decision column.
- MS Excel XLSX file—this type of file will be imported and saved as an SSV file. You will be able to choose the Excel sheet containing your data and the decision column.

DeepAI will import your data and if needed convert any missing or incorrect values to 0. Empty records or columns will not be imported. It is always better if you prepare your data in a such way that no further cleaning is necessary and that all values are chosen by you!

When importing data files containing text (non-numerical) values the textual values will be automatically transferred to numerical values by converting each unique cell/string to unique numbers per column (see Sect. 4.3.2.2). This is called integer encoding. In different columns the same numbers may occur (this is normal). If you prefer a different conversion strategy then convert the textual values to numbers yourself before importing the file.

9.3.3 Input Functions (Activation)

The chosen (checked) input functions (you can select one or more of them) are used to transform the input training data, and the resulting transformed data will be the input to the machine learning model. This is often called the activation of the input.

There are five input functions available in DeepAI:

- **Xn**—this simply takes the input data points (no transformation). You may choose all or some of the columns (features).
- **X^2**—this takes the square of each data point's value. You can choose all or some of the columns (features).
- **X*Xn**—this function multiplies all the values in a column (X1*X2*X3*...*Xn) and takes the resulting value (one value per column) as input.
- **sinX**—this takes the sin of each data point's value. You can choose all or some of the columns (variables).
- **cosX**—this takes the cos of each data point's value. You can choose all or some of the columns (variables).

9.3.4   Layers and Nodes

You can define as many layers as you need. The first layer (input) and the last layer (output) are fixed and automatically adjusted as you change, e.g., the input functions. For all other layers you can add as many nodes as you need. Each node is automatically connected to the next layer's nodes and in this way the neural network is formed. Each node has a bias parameter which is automatically calculated but can be modified. Each link between two nodes has a weight parameter which is automatically calculated but can be modified. While searching for the right model (learning the input) the machine learning model adjusts the biases and the weights automatically, but if you change a value the machine learning model will take that into account in the further learning process.

9.3.5   Training the Machine Learning Model

You can run the AI training process as many times (number of iterations) as you want. After that you may adjust the training parameters and run the training process further. You may also reset the model and restart the training process. You can also save the machine learning model and all input parameters (not the data) into a project file with extension ".nna".

The following parameters affect the training process (see the built-in help feature by moving the mouse pointer over the labels):

- Learning rate (default value = 0.1).
- Regularization rate (default value = 0.001).
- Batch size (default value = 10).
- Activation function (default = TANH).
- Regularization function (default = NONE).
- Percentage (%) test data.

9.3.6   Predictions

There are two options for asking the AI for a prediction (inference):

- By using the 'Ask Predictions' command and loading an ask data file with the previously described SSV format. The predictions will be shown in the bottom-left 'AI Predictions' tab in a table. The predictions can be exported in MS Excel format.
- By using the 'Ask Predictions Form'. Here you can load an AI model and an ask file in SSV format and ask the AI to make the predictions.

9.3.7 Settings

The Settings tab on the right side of the screen is used in DeepAI to perform several tasks and to adjust the AI parameters.

On the General tab of the Settings you can adjust some optimization parameters which will influence the speed of the software and its look and feel. If DeepAI needs to draw on more things then it will become slower, so choose the settings that best fit your purpose and computer hardware.

A higher 'Heatmap Density' will result in a higher resolution heatmap image but it will be slower to process.

On the 'Charts & Heatmap' tab of the Settings you can choose which data to plot in the data chart and also the colors in the heat map.

On the AI tab of the Settings you can adjust all machine learning training parameters. Move the mouse cursor over the labels to see a short description.

When the 'Treat data as X-Y Classification / Regression' box is checked, and there are only two data columns and one decision column, the data will be presented in a X-Y chart and there will also be a heat map generated. This setting is interesting for 2D data and for educational purposes in order to see how the machine learning model is searching for an optimal solution/model.

9.4 VoiceBridge

EXCLUSIVE CONTENT: Section 9.4.2 contains the exclusive content (only available in this book) for VoiceBridge users. The source code of an inference only application for the Yes-No example. This application makes it possible to decode (inference) an input wav file with a *trained* ASR model. The VoiceBridge open source distribution contains only an example in which the training and decoding are integrated (cannot be separated).

VoiceBridge is a fully open source (AI-TOOLKIT Open Source License— Apache 2.0 based which allows commercial use) speech recognition C++ toolkit optimized for MS Windows 64-bit (it can be modified to compile on other operating systems). VoiceBridge fills the gap for MS Windows speech recognition developers.

VoiceBridge is based on the many years of ASR research put into the KALDI project [63]. VoiceBridge can be considered the MS Windows counterpart of KALDI (a well-known collection of speech recognition software for UNIX like operating systems).

The aim of VoiceBridge is to make writing high quality professional and fast speech recognition software very easy.

The following speech recognition models are currently available in VoiceBridge (see Chap. 5 for more information):

- GMM Mono-phone (see Sect. 5.6)
- GMM Tri-phone (see Sect. 5.6)

These models can be used in combination with the following built-in feature extraction schemes:

- MFCC (see Sect. 5.3)
- MFCC + delta + delta-delta (see Sect. 5.3)
- MFCC + delta + delta-delta + pitch (see Sects. 5.3 and 5.2.1)

And the following feature transformation schemes are supported:

- SAT (see Sect. 5.3.1.3)
- LDA + MLLT (see Sects. 5.3.1.1 and 5.3.1.2)
- LDA + MLLT+SAT (see Sects. 5.3.1.1, 5.3.1.2 and 5.3.1.3)

At the time of writing this book the 'MFCC + delta + delta-delta + SAT' model is the best performing model with the highest accuracy and speed (1/5th of the training time compared to LDA + MLLT+SAT). Due to the automatic tuning of some input (e.g., pronunciation) VoiceBridge achieves a much higher accuracy than other similar software!

VoiceBridge includes the following special modules:

- Automatic language model generation (N-gram; see Sect. 5.5)
- Automatic pronunciation lexicon generation (with pronunciation model training; see Sect. 5.6)
- Semi-automatic speaker group separation (see Sect. 9.4.3.1)

Thanks to these modules VoiceBridge only requires a limited number of inputs:

- Wav files
- Text transcription files for each wav file
- Reference language dictionary (available in the VoiceBridge distribution)

Your speech recognition job has just become much easier!
VoiceBridge is *hardware accelerated* in two ways:

- It uses automatic parallel processing by automatic CPU/core detection and work distribution. More processors/processor cores mean faster processing!
- It makes use of the Intel Math Kernel Library (MKL) which further accelerates processing by drawing on special processor instruction sets.

The VoiceBridge C++ code is organized in a *single dynamic link library (DLL)*. For this reason it is very easy to distribute software built upon VoiceBridge. VoiceBridge aims to be a fast, highly accurate and easy to use professional production-ready system.

VoiceBridge includes two complete *examples* which demonstrate how to use the library. Both examples are also available in Kaldi, to make it much easier for Kaldi users to learn how to use VoiceBridge.

One of the examples is the Yes-No example (see Sect. 9.4.1). This is a very simple speech recognition example in which we train a model to recognize people saying 'yes' or 'no'. The word error rate (WER) for this example in VoiceBridge is 2% (98% accuracy), and the training and testing takes about eight seconds (with four processor cores).

The second example is the so-called *LibriSpeech* example (see Sect. 9.4.2), a real-world speech recognition application which includes several hours of English speech learning and recognition. The word error rate (WER) for this example in VoiceBridge is 5.92% (94% accuracy), and the training and testing takes about 25 minutes (with four processor cores).

Both examples are ready-to-use code templates for your speech recognition projects! The core source code for the DLL and the example application's source codes are heavily documented, which makes them very easy to use.

Everything is included in the VoiceBridge distribution except the Intel MKL library which can be downloaded for free from the Intel website: https://software.intel.com/en-us/mkl.

Please read the VoiceBridge documentation about how to compile the software.

NOTE: The directory 'VoiceBridge\Redistributables' contains all the necessary dll's which need to be redistributed with any software built with the use of VoiceBridge. Most of them are for the Intel MKL library and one is for OpenMP support. You may of course need to distribute some more dll's required by your compiler (MS VS2017), for example, for the C++ runtime.

Please note that the MKL dll's are from the w*mkl*2018.1.156 distribution. You may need to replace these if you download a more recent version!

9.4.1　Yes-No Example

This is a very simple speech recognition example in which we train a simple GMM mono-phone ASR model with MFCC features to recognize people saying 'yes' or 'no'. The word error rate (WER) for this example in VoiceBridge is 2% (98% accuracy), and the training and testing takes about eight seconds (with four processor cores—Intel Core i5–6400 CPU @ 2.70 GHz).

The source code for the example, including explanatory comments, is presented hereunder (the code is slightly modified for the book). With about 300 lines of C++ code we can make a trained automatic speech recognition model thanks to the VoiceBridge DLL which does most of the work automatically.

Please note that we rely heavily on the Boost C++ library file system functions (boost::filesystem) for creating and manipulating files and directories. If you do not know this very popular free C++ library then see its documentation (see [93]).

The VoiceBridge logging mechanism works as follows:

- Use LOGTW_INFO, LOGTW_WARNING, LOGTW_ERROR, LOGTW_FATALERROR and LOGTW_DEBUG to write to the global log file (defined in the beginning of the example) and to the screen in the case of a console application (goes to std.::cout).
- Number formatting must be set for each invocation of the log object or macro, e.g., LOGTW_INFO << "double "<< std.:::fixed << std.:::setprecision(2) << 25.2365874;

NOTE: The Yes-No example first trains the ASR model and then immediately tests it. The input wav files (and transcriptions) are divided into two parts (PrepareData () function), most of them are used for training and the rest for testing the ASR model.

The example code builds the ASR model step by step and then it also tests the model. Please find the summary of the steps hereunder:

1. Initialize all necessary parameters, which mainly setup the directory structure of the project (input data, output, logging, etc.). The software uses different directories for storing the output of data preparation and each model training step. Models are trained hierarchically on top of each other and each sub-model is stored in a separate directory.
2. Initialize the logging and the LOG file.
3. Setup the number of hardware threads used by the software—this is provided by the 'concurentThreadsSupported' parameter. You may decrease this number if you want your computer be more responsive.
4. Define the language model ID which will be appended to the directory name where the trained language model will be created.
5. Check if the ASR model needs to be retrained by comparing the trained model creation time with the creation time of the data files (because we may run the example several times with different parameters). If nothing vital has changed then the model does not need to be retrained.
6. Prepare all necessary training and test data if the ASR model needs to be trained and automatically generate an ARPA N-gram language model if needed. The language model is made from the full text extracted from the transcriptions. A 'vocab.txt' with all unique words used in the text is also saved, which will be used to create the pronunciation lexicon.

The following functions are called in this step:

PrepareData. The `PrepareData` () function has three parameters: the percentage of training data (the rest of the data will be used for testing), the transcription file extension and the *idtype*. The transcription files must have the same name as the wav files except for the extension which is defined in the second parameter. The *idtype* parameter can be 0, 1 or > 1 and used for *speaker separation*. When it is 0 then the directory name in which the wav file (input acoustic signal) is placed will be used as the speaker id (=speaker separation). When it is 1 each wav file name (without extension) will be used as the speaker id (=no speaker separation!). When it is >1 the first *idtype* number of characters from the file name will be used as the speaker id (=speaker separation). The default value is 1!

All necessary data and directory structure will be created automatically for you by this function.

PrepareDict. This function prepares the pronunciation lexicon and all dictionary files automatically. It expects a reference lexicon and from that, if necessary, it trains a pronunciation model which can be used to determine the pronunciation of all words in the project. You can also add silence phones to the dictionary. Silence phones are used to model non-speech sounds (e.g., silence, noise) or even unknown words (UNK or OOV). You can define different types of silence, noise and unknown words. See the LibriSpeech example for how to do this.

PrepareLang. This prepares all language features. It has four parameters, but we only use the first one to define whether position dependent phones must be used. All other parameters must be empty. In the Yes-No example, no position dependent phones are used because of the very simple nature of the words, but in the LibriSpeech example position dependent phones will be used (tri-phone model).

PrepareTestLms. This function prepares the language model for the later testing of the ASR model. It copies all necessary files into a test directory and compiles all finite state transducers (FST) for the language model, which are used in the fast FST searching mechanism discussed in Appendix A. We have separated the training data and the test data with the PrepareData () function, and these will have different language models depending on the individual words in the transcriptions.

MakeMfcc. This function is used to extract all of the MFCC features (no 'delta' features are added!) according to the configuration file "mfcc.conf" for both the test and training acoustic signals (wav files). The configuration file may contain a lot of parameters but the default options are fine in most of the cases. If a parameter is not included in "mfcc.conf" the default parameter value will be used!

All possible parameters in "mfcc.conf" are explained in Table 9.34. There are several similar configuration files for different stages of the ASR process which are all documented in the VoiceBridge Wiki: https://github.com/AI-TOOLKIT/ VoiceBridge/wiki . You can experiment with different parameter values and in case of a problem just revert to the default values (i.e., remove the parameter from the configuration file).

7. **ComputeCmvnStats.** This computes the cepstral mean and variance statistics which will be needed later.

8. **FixDataDir.** This function makes sure that only data which is effectively used in the training of the ASR model is present in the data directory. It checks several input files, for example, "utt2spk" (utterance to speaker conversion). The original contents of the *data-dir* will be copied into *data-dir/backup*.

9. **TrainGmmMono.** This function/step trains the GMM mono-phone model. The "train.conf" file may contain extra training parameters in a similar manner to "mfcc.conf"! See the VoiceBridge Wiki for more information: https://github. com/AI-TOOLKIT/VoiceBridge/wiki. In the Yes-No example only "--totgauss = 400" is present in "train.conf" and thus all other parameters are used with their default values. The "totgauss" parameter determines the target number of Gaussians, and the default value is 1000.

   At this stage the ASR model has been successfully trained. The next step is the decoding phase in which we test the model.

10. Decode with the ASR and language models. Decoding means that we use the above trained model on the test set, or in other words, that we recognize the spoken text and then we determine the word error rate (WER) of the model by comparing the output transcriptions of the decoding to the input transcriptions.

    At this point we have a mono-phone model working properly with a given accuracy. It is still possible to improve the accuracy by 2–5% with a more sophisticated model trained on top of this mono-phone model (e.g., tri-phone, using SAT (speaker adaptive training), etc.).

The test results and accuracy of the trained Yes-No example ASR model can be seen in Table 9.33. The word error rate (WER) is about 2% and the sentence error (SER) rate is about 16%. The 16% sentence error rate may seem a bit high, but it is just because one sentence out of six is not completely correct ($1/6 = 0.16$).

If you look at the output carefully you will see that the first sentence ('yes yes yes no yes no yes yes') is identified nearly correctly, but there is an unknown word ('***') detected near the end of the acoustic recording ('yes yes yes no yes no '***' yes yes'). If we do not take this into account then the accuracy of this simple ASR model is 100%!

Table 9.33 The output of the Yes-No example

```
Scored 6 sentences, 0 not present in hyp.
Found the best WER/SER combination: WER=2.08%
Done 6 sentences, failed for 0


1_1_1_0_1_0_1_1 ref  yes  yes  yes  no  yes  no  '***'  yes  yes
1_1_1_0_1_0_1_1 hyp  yes  yes  yes  no  yes  no  '***'  yes  yes
1_1_1_0_1_0_1_1 op   C    C    C    C    C    C   D    C    C
1_1_1_0_1_0_1_1 #csid 8 0 0 1


1_1_1_1_0_0_1_0 ref  yes  yes  yes  yes  no  no  yes  no
1_1_1_1_0_0_1_0 hyp  yes  yes  yes  yes  no  no  yes  no
1_1_1_1_0_0_1_0 op   C    C    C    C    C    C    C    C
1_1_1_1_0_0_1_0 #csid 8 0 0 0


1_1_1_1_0_1_0_0 ref  yes  yes  yes  yes  no  yes  no  no
1_1_1_1_0_1_0_0 hyp  yes  yes  yes  yes  no  yes  no  no
1_1_1_1_0_1_0_0 op   C    C    C    C    C    C    C    C
1_1_1_1_0_1_0_0 #csid 8 0 0 0


1_1_1_1_1_0_0_0 ref  yes  yes  yes  yes  yes  no  no  no
1_1_1_1_1_0_0_0 hyp  yes  yes  yes  yes  yes  no  no  no
1_1_1_1_1_0_0_0 op   C    C    C    C    C    C    C    C
1_1_1_1_1_0_0_0 #csid 8 0 0 0


1_1_1_1_1_1_0_0 ref  yes  yes  yes  yes  yes  yes  no  no
1_1_1_1_1_1_0_0 hyp  yes  yes  yes  yes  yes  yes  no  no
1_1_1_1_1_1_0_0 op   C    C    C    C    C    C    C    C
1_1_1_1_1_1_0_0 #csid 8 0 0 0


1_1_1_1_1_1_1_1 ref  yes  yes  yes  yes  yes  yes  yes  yes
1_1_1_1_1_1_1_1 hyp  yes  yes  yes  yes  yes  yes  yes  yes
1_1_1_1_1_1_1_1 op   C    C    C    C    C    C    C    C
1_1_1_1_1_1_1_1 #csid 8 0 0 0
```

SPEAKER id	#SENT	#WORD	Corr	Sub	Ins	Del	Err	S.Err
1_1_1_0_1_0_1_1 raw	1	9	8	0	0	1	1	1
1_1_1_0_1_0_1_1 sys	1	9	88.89	0.00	0.00	11.11	11.11	100.00
1_1_1_1_0_0_1_0 raw	1	8	8	0	0	0	0	0
1_1_1_1_0_0_1_0 sys	1	8	100.00	0.00	0.00	0.00	0.00	0.00
1_1_1_1_0_1_0_0 raw	1	8	8	0	0	0	0	0
1_1_1_1_0_1_0_0 sys	1	8	100.00	0.00	0.00	0.00	0.00	0.00
1_1_1_1_1_0_0_0 raw	1	8	8	0	0	0	0	0
1_1_1_1_1_0_0_0 sys	1	8	100.00	0.00	0.00	0.00	0.00	0.00
1_1_1_1_1_1_0_0 raw	1	8	8	0	0	0	0	0
1_1_1_1_1_1_0_0 sys	1	8	100.00	0.00	0.00	0.00	0.00	0.00
1_1_1_1_1_1_1_1 raw	1	8	8	0	0	0	0	0
1_1_1_1_1_1_1_1 sys	1	8	100.00	0.00	0.00	0.00	0.00	0.00
SUM raw	6	49	48	0	0	1	1	1
SUM sys	6	49	97.96	0.00	0.00	2.04	2.04	16.67

(continued)

Table 9.33 (continued)

```
Correct        '***'    '***'   1
Correct         no       no    13
Correct         yes      yes   35
Set1: WER 2.07% +- 3.70%
```

Expert Sect. 9.1 MFCC Features Calculation: Configuration Options (Mfcc.Conf)
The location of this configuration file is passed to the MakeMfcc() function in VoiceBridge. All possible parameters in this configuration file together with their default values are documented in Table 9.34. In most cases the default values will be fine and you will not need to change these parameters. The usage examples show how to define the parameters in the "mfcc.conf" file. *If a parameter is not defined in the conf file then its default value will be used!* If the ASR training process fails then check this file for syntax errors. You will find the theory behind all of these parameters explained in Chap. 5.

IMPORTANT: The configuration file format is as follows:

--param1=value
--param2=value
--param3=value

Notice the name of the parameter is preceded by -- and that there is no space between the equal sign and the parameter name and value! Each parameter is on a new line!

The source code of the Yes-No example is shown in Table 9.35.

9.4.2   Exclusive Content: Inference Only Yes-No Example

This section contains the exclusive content (only available in this book) for VoiceBridge users: the source code of an inference only application for the Yes-No example. This application makes it possible to decode an input wav file with an *already trained* ASR model. The VoiceBridge open source distribution

Table 9.34 MFCC configuration options

Direct MfccOptions
num-ceps : Number of cepstral coefficients in MFCC computation (including C0).
type: int32, default: 13, usage example: --num-ceps=13
use-energy : Use energy (not C0) in MFCC computation.
type: bool, default: true, usage example: --use-energy=false
energy-floor : Floor on energy (absolute, not relative) in MFCC computation. Not in log scale; a small value, e.g., 1.0e-10.
type: BaseFloat, default: 0.0, usage example: --energy-floor=0.0
raw-energy : If true, compute energy before pre-emphasis and windowing.
type: bool, default: true, usage example: --raw-energy=true
MfccOptions for Frame Extraction (FrameExtractionOptions)
sample-frequency : Waveform data sample frequency (must match the waveform file, if specified there).
type BaseFloat, default: 16000, usage example: --sample-frequency=16000
frame-length : Frame length in milliseconds.
type BaseFloat, default: 25.0, usage example: --frame-length=25.0
frame-shift : Frame shift in milliseconds.
type BaseFloat, default: 10.0, usage example: --frame-shift=10.0
preemphasis-coefficient : Coefficient for use in signal preemphasis.
type BaseFloat, default: 0.97, usage example: --preemphasis-coefficient=0.97
remove-dc-offset : Subtract mean from waveform on each frame before FFT.
type bool, default: true, usage example: --remove-dc-offset=true

(continued)

Table 9.34 (continued)

---

dither : Amount of dithering, 0.0 means no dither.

```
type BaseFloat, default: 1.0, usage example: —dither=1.0
```

window-type : Type of window: "hamming"," hanning", " povey", "rectangular", " blackman".

```
type std::string, default: povey, usage example: --window-type=povey
```

NOTE: " povey" is a window developed by Daniel Povey to resemble Hamming but to go to zero at the edges, where pow((0.5 - 0.5$cos(n/N2$*pi)), 0.85). According to Daniel Povey this is the best option to choose [63].

blackman-coeff : Constant coefficient for generalized Blackman window.

```
type BaseFloat, default: 0.42, usage example: --blackman-coeff=0.42
```

round-to-power-of-two : If true, round the window size to power of two by zero-padding input to FFT.

```
type bool, default: true, usage example: --round-to-power-of-two=true
```

snip-edges : If true, end effects will be handled by outputting only frames that completely fit in the file, and the number of frames depends on the frame-length. If false, the number of frames depends only on the frame-shift, and we reflect the data at the ends.

```
type bool, default: true, usage example: --snip-edges=true
```

allow-downsample : If true, allow the input waveform to have a higher frequency than the specified --sample-frequency (and we'll down sample).

```
type bool, default: false, usage example: --allow-downsample=false
```

MfccOptions for Mel Banks (MelBanksOptions)

num-mel-bins : Number of triangular mel-frequency bins".

```
type int32, default: 23, usage example: --num-mel-bins=23
```

NOTE: defaults the mel-banks to 23 for the MFCC computations; this seems to be common for 16khz-sampled data, but for 8khz-sampled data, 15 may be better.

low-freq : Low cutoff frequency for mel bins.

```
type BaseFloat, default: 20.0, usage example: --low-freq=20.0
```

(continued)

Table 9.34 (continued)

high-freq : High cutoff frequency for mel bins (if < 0 -> offset from Nyquist, 0 -> no cutoff, negative). Added to the Nyquist frequency to get the cutoff.
`type BaseFloat, default: 0.0, usage example: --high-freq=0.0`
vtln-low : Low inflection point in piecewise linear VTLN warping function.
`type BaseFloat, default: 100.0, usage example: --vtln-low=100.0`
vtln-high : High inflection point in piecewise linear VTLN warping function (if negative, offset from high-mel-freq). Added to the Nyquist frequency to get the cutoff.
`type BaseFloat, default: -500.0, usage example: --vtln-high=-500.0`
debug-mel : Print out debugging information for mel bin computation.
`type bool, default: false, usage example: --debug-mel=false`
NOTE: this information is based on [63].

contains only an example in which the training and decoding (inference) are integrated (i.e., the training and decoding *cannot* be separated).

How to use

1. Create the necessary directory structure as it is shown on Fig. 9.14.

   First create a directory called 'PredictYesNo' which will hold all project files. Copy the 'conf', 'data/lang/phones' and 'input' directories from the directory structure used during the training of the ASR model into 'PredictYesNo'. Next create a subdirectory called 'predict_yesno' under the 'data' directory. In 'predict_yesno' create a subdirectory called 'mono0a', and copy the file contents (no subdirectories) of the 'train_yesno/mono0a' directory created during the training of the ASR model into 'predict_yesno/mono0a'. Then copy the full contents (with subdirectories) of the directory 'graph_tg' from the 'train_yesno/ mono0a' directory created during the training of the ASR model into 'predict_yesno/mono0a'. Finally create a subdirectory under 'PredictYesNo' called 'waves_yesno' and put the wav (acoustic) file which you want to transcribe (inference) in it.

2. Compile and run the application.
3. Get the transcription.

   After successfully running the application, the transcription can be found in the following location: 'PredictYesNo\data\predict_yesno\mono0a \decode_predict_yesno_tg\transcription.trn'. Each wav file is transcribed on a new line preceded by the file name without extension. If you are creating your own

Table 9.35 Yes-No Example. YesNo.cpp

```
C++ Source Code Yes-No Example. YesNo.cpp


/*
 Copyright 2017-present Zoltan Somogyi (AI-TOOLKIT), All Rights Reserved
 You may use this file only if you agree to the software license:
 AI-TOOLKIT Open Source Software License - Version 2.1 - February 22, 2018:
 https://ai-toolkit.blogspot.com/p/ai-toolkit-open-source-software-license.html.
 Also included with the source code distribution in AI-TOOLKIT-LICENSE.txt.
*/

#include "ExamplesUtil.h"

namespace fs = boost::filesystem;

using string_vec = std::vector<std::string>;

//main program
int TestYesNo()
{
    //Initialize Parameters
    //set project directory
    wchar_t buffer[MAX_PATH];
    GetModuleFileName(NULL, buffer, MAX_PATH);
    fs::path exepath(buffer);
    //use relative path, will work even if the source code is moved
    fs::path project(exepath.branch_path()/"../../../../../VoiceBridgeProjects/YesNo");
    //canonical path will normalize the path and remove "..\"
    project = fs::canonical(project).string();
    bool ret = voicebridgeParams.Init(
                  "train_yesno",
                  "test_yesno",
                  project.string(),
                  (project / "input").string(),
                  (project / "waves_yesno").string());
    if (!ret) {
        LOGTW_ERROR << "Can not find input data.";
        std::getchar();
        return -1;
    }

    fs::path training_dir(voicebridgeParams.pth_data /
                  voicebridgeParams.train_base_name);
    fs::path test_dir(voicebridgeParams.pth_data / voicebridgeParams.test_base_name);

    //in case the data did not change the model will not be retrained unless it is
    // forced with this option
    bool FORCE_RETRAIN_MODEL = false;

    //init general app level log
    fs::path general_log(voicebridgeParams.pth_project_base / "General.log");
    oTwinLog.init(general_log.string());

    LOGTW_INFO << "***************************************";
    LOGTW_INFO << "* WELCOME TO VOICEBRIDGE FOR WINDOWS! *";
    LOGTW_INFO << "***************************************";

    //try to determine the number of hardware threads supported (cores x processors)
    //NOTE: you could also decrease this number by one if you would want to leave
    // one core for an UI thread
    int numthreads = concurrentThreadsSupported;
    if (numthreads < 1) numthreads = 1;
```

(continued)

Table 9.35 (continued)

```
//set which language models we want for the test
    std::vector<std::string> lms;
    //NOTE: there can be several language models. Here we just use one LM designated
    // with 'tg'.
    // Other models could be used and then the directory name would be
    // data\\lang_test_{lm} where {lm} is replaced with the LM id.
    lms.push_back("tg");

    //check if the model needs to be retrained
    //NOTE: We use here model names similar to the names used in Kaldi in order
    //to make it easier for Kaldi users to understand the code. All data structures
    // will also be the same as in Kaldi except that VoiceBridge does not compress
    // files.
    bool needToRetrainModel =
        FORCE_RETRAIN_MODEL ||
        NeedToRetrainModel(
            voicebridgeParams.pth_data / voicebridgeParams.train_base_name /"mono0a",
            voicebridgeParams.pth_project_input,
            voicebridgeParams.waves_dir,
            voicebridgeParams.pth_project_base / "conf" / "train.conf");

    if (needToRetrainModel)
    {
        //Prepare data ------------------------------------------------------------>
        /*
        PrepareData : this steps prepares the training and test data and also generates
        automatically an arpra n-gram language model if needed. The language model is
        made from the full text extracted from the transcriptions. This step saves
        also a 'vocab.txt' with all unique words used in the text which will be used
        for making the lexicon with the pronunciations.
        */
        LOGTW_INFO << "Preparing data...";
        /*
        NOTE: The percentage of training data (the rest of the data will be used for
        testing), the transcription file extension and the idtype are the parameters.
        All necessary data and directory structure will be created automatically for
        you by this function.

        NOTE: idtype parameter to PrepareData(): it can be 0, 1 or >1; when 0 then the
        Directory name in which the wav file is placed will be used as speaker id; when
        1 each wav file name without extension) will be used as speaker id (=no speaker
        separation); when > 1 the first idtype number of characters from the file name
        will be used as speaker id. The default value is 1!
        */
        if (PrepareData(90, ".wav.trn") < 0) {
            LOGTW_ERROR << "***********************************";
            LOGTW_ERROR << "* Error while preparing the data! *";
            LOGTW_ERROR << "***********************************";
            std::getchar();
            return -1;
        }

        /*
        PrepareDict : this step prepares the pronunciation lexicon and all dictionary
        Files automatically. It expects a reference lexicon and from that, if
        necessary, it trains a pronunciation model which can be used to determine all
        pronunciations in the project. You can also add silence phones to the
        dictionary.
        */
        LOGTW_INFO << "Preparing dictionary...";
        fs::path refDict("");
        //NOTE: empty refDict means that we have a lexicon already and we do not
        // need to create it!
        ///fs::path refDict(voicebridgeParams.pth_project_input / "cmudict.dict");
```

(continued)

Table 9.35 (continued)

```
std::map<std::string, std::string> optsilphones = { { "<SIL>","SIL" } };
    std::map<std::string, std::string> silphones = { { "<SIL>","SIL" } };
    if (PrepareDict(refDict, silphones, optsilphones) < 0) {
        LOGTW_ERROR << "*****************************************";
        LOGTW_ERROR << "* Error while preparing the dictionary! *";
        LOGTW_ERROR << "*****************************************";
        std::getchar();
        return -1;
    }

    //Prepare language features
    LOGTW_INFO << "Preparing language features...";

    if (PrepareLang(false, "", "", "") < 0)
    {
        LOGTW_ERROR << "*****************************************";
        LOGTW_ERROR << "*  Error while preparing the language!  *";
        LOGTW_ERROR << "*****************************************";
        std::getchar();
        return -1;
    }

    //Prepare language models for test
    LOGTW_INFO << "Preparing language models for test...";
    if (PrepareTestLms(lms) < 0)
    {
        LOGTW_ERROR << "*****************************************";
        LOGTW_ERROR << "* Error while preparing language models!*";
        LOGTW_ERROR << "*****************************************";
        std::getchar();
        return -1;
    }

    /*
    Feature extraction
    NOTE: mfcc.conf may contain extra parameters!
    */
    LOGTW_INFO << "\n\n";
    LOGTW_INFO << "Starting MFCC features extraction...";

    if (MakeMfcc(training_dir, voicebridgeParams.pth_project_base /
                "conf\\mfcc.conf",
                numthreads) < 0 ||
        MakeMfcc(test_dir, voicebridgeParams.pth_project_base / "conf\\mfcc.conf",
                numthreads) < 0)
    {
        LOGTW_ERROR << "Feature extraction failed.";
        std::getchar();
        return -1;
    }

    /*
    NOTE: The configuration file format:

    --param1=value
    --param2=value
    --param3=value

    Notice the name of the parameter proceeded by -- and then there is no
    space between the equal sign and the parameter name and value! Each parameter
    is on a new line.
    */
```

(continued)

Table 9.35 (continued)

```
/*
            Cepstral mean and variance statistics.
        */
        if (ComputeCmvnStats(training_dir) < 0 ||
            ComputeCmvnStats(test_dir) < 0)
        {
            LOGTW_ERROR << "Feature extraction failed at computing cepstral mean and
                    variance statistics.";
            std::getchar();
            return -1;
        }

        /*
            Fix data directory
        */
        if (FixDataDir(training_dir) < 0 ||
            FixDataDir(test_dir) < 0)
        {
            LOGTW_ERROR << "Feature extraction failed at fixing data directory.";
            std::getchar();
            return -1;
        }

        /*
            Train mono-phone model
            NOTE: train.conf may contain extra training parameters!
        */
        if (TrainGmmMono(training_dir, voicebridgeParams.pth_lang,
            training_dir / "mono0a",
            voicebridgeParams.pth_project_base / "conf\\train.conf", numthreads) < 0)
        {
            LOGTW_ERROR << "Training failed.";
            std::getchar();
            return -1;
        }

        /*
            Graph compilation : compile the graph for each language model
        */
        for (std::string lm : lms) {
            if (MkGraph(voicebridgeParams.pth_data / ("lang_test_" + lm),
                    training_dir / "mono0a",
                    training_dir / "mono0a" / ("graph_" + lm)) < 0)
            {
                LOGTW_ERROR << "Graph compilation failed for language model " << lm;
                std::getchar();
                return -1;
            }
        }

    } ///if (needToRetrainModel
    else {
        LOGTW_INFO
            << "Recently trained model found. Skipping training step. (NOTE: in case
                you want to force to retrain the model then open and save one of the
                input or config files.)";
    }

    //Decode ------------------------------------------------------------------->

    // Decoding of each language model
    UMAPSS wer_ref_filter;
```

(continued)

Table 9.35 (continued)

```
UMAPSS wer_hyp_filter;
    for (std::string lm : lms)
    {
        LOGTW_INFO << "\n\n";
        LOGTW_INFO << "Decoding language model " << lm << "...";
        if (Decode(
            training_dir / "mono0a" / ("graph_" + lm), //graph_dir
            voicebridgeParams.pth_data / voicebridgeParams.test_base_name, //data_dir
            training_dir / "mono0a" / ("decode_" + voicebridgeParams.test_base_name +
            "_" + lm), //decode_dir
            training_dir / "mono0a" / "final.mdl",
                //when not specified "final.mdl" is taken automatically
            "",
            //normally be used, but it can be used if you want to supply existing
            //fMLLR transforms when decoding
            wer_ref_filter,
            wer_hyp_filter,
            "", //iteration of model to test e.g. 'final', if the model is given then
                //this option is not needed
            Numthreads
                //the number of parallel threads to use in the decoding; must
                // be the same as in the data preparation

                //all other parameters are used with threir default values
        ) < 0)
        {
            LOGTW_ERROR << "Decoding failed for language model " << lm;
            std::getchar();
            return -1;
        }
    }
    //-----------------------------------------------------------------------------
    //NOTE: at this point we have a mono-phone model working properly with a given
    //accuracy.
    //It is still possible to improve the accuracy with 2-5% with a more
    //sophisticated model
    //trained on top of the mono-phone model (e.g. using a tri-phone model).
    //-----------------------------------------------------------------------------

    //pause the console application
    LOGTW_INFO << "\n\n";
    LOGTW_INFO << "*****************";
    LOGTW_INFO << "**** ALL OK! ****";
    LOGTW_INFO << "*****************";
    std::getchar();

    return 0;
}
```

application, then you may of course read out the transcription automatically and do with it whatever you choose.

> TIP: You may replace the input wav file with a new file after the transcription has been generated and run the application again. You will then get the new transcription. You may even automate this process and get the transcription of

(continued)

Fig. 9.14 VoiceBridge
directory structure

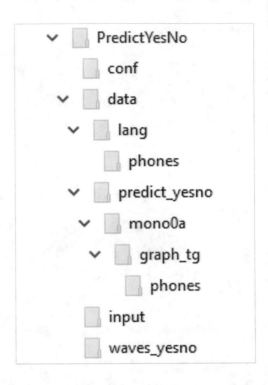

several wav files one after the other. You may also place several wav files into
the input folder at the same time ('waves_yesno') and then all of them will be
decoded at once.

The source code for the inference only Yes-No example is shown in Table 9.36.

9.4.3   LibriSpeech Example

The LibriSpeech example is a real world automatic speech recognition (ASR)
application which includes several hours of English speech learning and recognition.
The WER for this example in VoiceBridge is 5.92% (94% accuracy), and the
training and testing takes about 25 minutes (with four processor cores—Intel Core
i5–6400 CPU @ 2.70 GHz).

Most of the steps in the LibriSpeech example are the same as those in the Yes-No
example with the following extensions:

Table 9.36 Inference Only Yes-No Example. PredictYesNo.cpp

C++ Source Code - Inference Only Yes-No Example. PredictYesNo.cpp

```
/*
Copyright 2017-present Zoltan Somogyi (AI-TOOLKIT), All Rights Reserved
You may use this file only if you agree to the software license:
AI-TOOLKIT Open Source Software License - Version 2.1 - February 22, 2018:
https://ai-toolkit.blogspot.com/p/ai-toolkit-open-source-software-license.html.
Also included with the source code distribution in AI-TOOLKIT-LICENSE.txt.
*/

#include "ExamplesUtil.h"

/*
 The following directory structure is needed for the PredictYesNo application:

 PredictYesNo
    conf                        --> configuration
    data
      lang                      --> language features
         phones
      predict_yesno
         mono0a                 --> trained ASR model
            graph_tg
               phones
    input                       --> input language model
    waves_yesno                 --> prediction input directory

 All directories are the copies of the directories from the model training
 directory structure except two directories (waves_yesno and predict_yesno).
  - In the waves_yesno directory the input for the prediction must be placed
    (wav file).
  - The predict_yesno directory is new but it contains the mono0a directory
    which is a copy of the model directory from the training (the trained model).

 These are the minimum number of files which are needed for the prediction
 (trained model + language features).
*/

//main program
int PredictYesNo()
{
      //Initialize Parameters
      //set project directory
      wchar_t buffer[MAX_PATH];
      GetModuleFileName(NULL, buffer, MAX_PATH);
      fs::path exepath(buffer);
      //use relative path, will work even if the source code is moved
      fs::path project(exepath.branch_path() /
            "../../../../../VoiceBridgeProjects/PredictYesNo");
      //canonical path will normalize the path and remove "..\"
      project = fs::canonical(project).string();
      bool ret = voicebridgeParams.Init("predict_yesno", "predict_yesno",
            project.string(),
            (project / "input").string(),
            (project / "waves_yesno").string());
      if (!ret) {
            LOGTW_ERROR << "Can not find input data.";
            std::getchar();
            return -1;
      }

      fs::path predict_dir(voicebridgeParams.pth_data /
                  voicebridgeParams.test_base_name);
```

(continued)

Table 9.36 (continued)

```
//init general app level log
fs::path general_log(voicebridgeParams.pth_project_base / "General.log");
oTwinLog.init(general_log.string());
LOGTW_INFO << "***************************************";
LOGTW_INFO << "* WELCOME TO VOICEBRIDGE FOR WINDOWS! *";
LOGTW_INFO << "***************************************";

//NOTE: We set the number of threads to 1 because we are predicting
//only one wav file!
int numthreads = 1; // concurentThreadsSupported;
if (numthreads < 1) numthreads = 1;

//set which language models we want for the test
std::vector<std::string> lms;
//NOTE: Must be the same as used during the training of the ASR model!
lms.push_back("tg");

//Clean directories --------------------------------------------------------

/*
Must clean the data directory. All files which were created from the input
wav / trn file(s) have to be removed in order to allow a new and different
input wav file!
*/
// NOTE: the below is hardcoded and could be optimized! Taking only one
//language model into account!
if(fs::exists(voicebridgeParams.pth_data / ("lang_test_" + lms[0])))
  fs::remove_all(voicebridgeParams.pth_data / ("lang_test_" + lms[0]));
if (fs::exists(voicebridgeParams.pth_data / "full_text.txt"))
  fs::remove(voicebridgeParams.pth_data / "full_text.txt");
if (fs::exists(voicebridgeParams.pth_data / "vocab.txt"))
  fs::remove(voicebridgeParams.pth_data / "vocab.txt");
//
if (fs::exists(predict_dir / "backup")) fs::remove_all(predict_dir / "backup");
if (fs::exists(predict_dir / "data")) fs::remove_all(predict_dir / "data");
if (fs::exists(predict_dir / "log")) fs::remove_all(predict_dir / "log");
if (fs::exists(predict_dir / "mfcc")) fs::remove_all(predict_dir / "mfcc");
if (fs::exists(predict_dir / "split1")) fs::remove_all(predict_dir / "split1");
if (fs::exists(predict_dir / "mono0a" / ("decode_" +
              voicebridgeParams.test_base_name + "_" + lms[0])))
  fs::remove_all(predict_dir / "mono0a" / ("decode_" +
      voicebridgeParams.test_base_name + "_" + lms[0]));
//
if (fs::exists(predict_dir / "cmvn.scp")) fs::remove(predict_dir / "cmvn.scp");
if (fs::exists(predict_dir / "feats.scp")) fs::remove(predict_dir / "feats.scp");
if (fs::exists(predict_dir / "spk2utt")) fs::remove(predict_dir / "spk2utt");
if (fs::exists(predict_dir / "text")) fs::remove(predict_dir / "text");
if (fs::exists(predict_dir / "utt2spk")) fs::remove(predict_dir / "utt2spk");
if (fs::exists(predict_dir / "wav.scp")) fs::remove(predict_dir / "wav.scp");

//Prepare data --------------------------------------------------------------

/*
PrepareData : this step prepares the prediction data. The arpra n-gram
language model must exist already (the same as used for the training of
the model).
*/
LOGTW_INFO << "Preparing data...";
/*
All necessary data and directory structure will be created automatically
for you by this function.

    IMPORTANT NOTES:
        - Must be the same as used during the training of the model!
        - The percentage of training data 0% signals that prediction
          will be performed!
*/
```

(continued)

Table 9.36 (continued)

```
if (PrepareData(0, ".wav.trn") < 0) {
        LOGTW_ERROR << "************************************";
        LOGTW_ERROR << "* Error while preparing the data! *";
        LOGTW_ERROR << "************************************";
        std::getchar();
        return -1;
}

//NOTE: the dictionary and pronunciation lexicon must exist already
// (the same as used during the training)!

//NOTE: language features must exist already (the same as used during
// the training)!
//Prepare language models
LOGTW_INFO << "Preparing language models...";
if (PrepareTestLms(lms) < 0)
{
        LOGTW_ERROR << "******************************************";
        LOGTW_ERROR << "* Error while preparing language models!*";
        LOGTW_ERROR << "******************************************";
        std::getchar();
        return -1;
}

/*
Feature extraction
NOTE: mfcc.conf may contain extra parameters!

IMPORTANT NOTE: Must be the same as during the training of the ASR model!
*/
LOGTW_INFO << "\n\n";
LOGTW_INFO << "Starting MFCC features extraction...";
if (MakeMfcc(predict_dir, voicebridgeParams.pth_project_base /
        "conf\\mfcc.conf", numthreads) < 0)
{
        LOGTW_ERROR << "Feature extraction failed.";
        std::getchar();
        return -1;
}

/*
Cepstral mean and variance statistics.
*/
if (ComputeCmvnStats(predict_dir) < 0)
{
        LOGTW_ERROR << "Feature extraction failed at computing cepstral
                         mean and variance statistics.";
        std::getchar();
        return -1;
}

/*
Fix data directory
*/
if (FixDataDir(predict_dir) < 0)
{
        LOGTW_ERROR << "Feature extraction failed at fixing data directory.";
        std::getchar();
        return -1;
}

/*
Graph compilation : compile the graph for each language model
*/
for (std::string lm : lms) {
//NOTE: lang_test_ is used in PrepareTestLms, therefore must stay the
//same here!
```

(continued)

Table 9.36 (continued)

```
        if (MkGraph(voicebridgeParams.pth_data / ("lang_test_" + lm),
                predict_dir / "mono0a",
                predict_dir / "mono0a" / ("graph_" + lm)) < 0)
        {
                LOGTW_ERROR << "Graph compilation failed for language model "
                            << lm;
                std::getchar();
                return -1;
        }
}

//Decode -------------------------------------------------------------

//Decoding of each language model
UMAPSS wer_ref_filter; //ref filter NOTE: must be empty as during the
                       // training of the ASR model!
UMAPSS wer_hyp_filter; //hyp filter NOTE: must be empty as during the
                       // training of the ASR model!
for (std::string lm : lms)
{
        LOGTW_INFO << "\n\n";
        LOGTW_INFO << "Decoding language model " << lm << "...";
        if (Decode(
                predict_dir / "mono0a" / ("graph_" + lm),  //graph_dir
                //NOTE: test_base_name is set to 'predict' in the beginning
                //in the 'init'!
                voicebridgeParams.pth_data /
                        voicebridgeParams.test_base_name,  //data_dir
                predict_dir / "mono0a" / ("decode_" +
                        voicebridgeParams.test_base_name + "_" + lm),
                        //decode_dir
                predict_dir / "mono0a" / "final.mdl", //the trained ASR model
                "", //trans_dir : must be empty as during the training
                    //of the ASR model
                wer_ref_filter, //
                wer_hyp_filter, //
                "", //iteration of model to test e.g. 'final', if the model
                    //is given then this option is not needed
                Numthreads, //the number of parallel threads to use
                0.083333f, //default value; acoustic scale used for lattice
                        // generation;  NOTE: only really affects pruning
                        // (scoring is on lattices).
                0, //default value; stage
                7000, //default value; max_active
                13.0, //default value; beam
                6.0, //default value; lattice beam
                true //skip scoring!
                //all other parameters are used with their default values
        ) < 0)
        {
                LOGTW_ERROR << "Decoding failed for language model " << lm;
                std::getchar();
                return -1;
        }
}

//pause the console application
LOGTW_INFO << "\n\n";
LOGTW_INFO << "******************";
LOGTW_INFO << "**** ALL OK! ****";
LOGTW_INFO << "******************";
std::getchar();

return 0;
}
```

1. After training the mono-phone GMM model, as a first step, an extra phone alignment is performed. A 'delta + delta-delta' GMM tri-phone model is built on top of the mono-phone GMM model, where 'delta + delta-delta' means these extra features are added during feature extraction to the pre-existing MFCC features. The second extension is the use of a tri-phone acoustic model instead of a mono-phone model (the mono-phone models are used to build context dependent models for tri-phones). After the model is built it is also used for decoding the test audio signals in order to test the model. The model is saved into a separate directory.
2. After training the 'delta + delta-delta' GMM tri-phone model, as a first step, an extra phone alignment is performed. A 'delta + delta-delta' GMM SAT or LDA + MLLT+SAT tri-phone model is built on top of the 'delta + delta-delta' GMM tri-phone model. Remember that SAT means feature transformation for speaker adaptive training and LDA + MLLT refers to LDA and MLLT feature transformations. After the model is built it is also used for decoding the test audio signals in order to test the model. You may choose which model you wish to train, the SAT or the LDA + MLLT+SAT model. The SAT model has a better accuracy and is much faster to train in the case of the LibriSpeech example. The model is saved into a separate directory.
3. The LibriSpeech application applies automatic speaker group separation per input wav file subdirectory. Each directory contains speech from different speakers who are identified by the directory name and also by the first number in the file name. The software cannot use the file names because the IDs are not the same length, but it can use the directory names as these are requested by the *idtype* parameter in the `PrepareData()` function (see Sect. 9.4.3.1).

At each stage of model training the trained models and test results are saved in separate directories. You can always use the best performing model from any of the above steps.

The output and accuracy of the LibriSpeech example can be seen in Table 9.37 (only a subset of the 262 test sentences are shown and grouped per speaker). The table shows also the detailed error analysis (why errors occurred) where 'Sub' means substitution error, 'Ins' means insertion error and 'Del' means deletion error. For more information about these errors and other related subjects see Sect. 5.8. All of the errors in the LibriSpeech example are 'deletion errors' (if a word is recognized then it is always correct). The 'Err' column is the sum of all errors. The second line for each speaker ID shows the accuracy as a percentage (for example, the words spoken by speaker 8455 are recognized 98.93% of the time). We can also see that speaker 908 is slightly harder to recognize (96.8% accuracy).

> TIP: You may achieve a higher accuracy by using more input data (more spoken sentences)! The more input the ASR model has during the training the better it can learn to recognize the speech. This is even more true in the case of

(continued)

Table 9.37 The output of the LibriSpeech example

```
Scored 262 sentences, 0 not present in hyp.
Found the best WER/SER combination: WER=5.93%
Done 262 sentences, failed for 0

8455-210777-0002 ref  on  arriving  at  home  at  my  own  residence  i  found  that  our
salon  was  filled  with  a  brilliant  company
8455-210777-0002 hyp  on  arriving  at  home  at  my  own  residence  i  found  that  our
salon  was  filled  with  a  brilliant  company
8455-210777-0002 op   C  C  C  C  C  C  C  C  C  C  C  C  C  C  C  C  C  C
8455-210777-0002 #csid 19 0 0 0
...
SPEAKER id    #SENT  #WORD  Corr   Sub   Ins   Del   Err   S.Err
8455 raw      69     1310   1296   0     0     14    14    12
8455 sys      69     1310   98.93  0.00  0.00  1.07  1.07  17.39

8463 raw      74     1267   1240   0     0     27    27    22
8463 sys      74     1267   97.87  0.00  0.00  2.13  2.13  29.73

8555 raw      62     1360   1346   0     0     14    14    12
8555 sys      62     1360   98.97  0.00  0.00  1.03  1.03  19.35

908 raw       57     1129   1093   0     0     36    36    18
908 sys       57     1129   96.81  0.00  0.00  3.19  3.19  31.58

SUM raw       262    5066   4975   0     0     91    91    64
SUM sys       262    5066   98.20  0.00  0.00  1.80  1.80  24.43

Set1: WER 5.94% +- 0.95%
```

several different speakers, like in the LibriSpeech example, because for good ASR we should include many spoken sentences for each speaker. The above model was trained only for 25 minutes and achieved 94% word accuracy (WER). Expect up to several hours of training for an accuracy above 98% in the case of complex spoken text!

9.4.3.1 Speaker Group Separation

The fourth parameter in the PrepareData() function is called *idtype*. Depending on this parameter the software will automatically identify the speakers and save their ID's into the utt2spk file (used later).

The possible values for idtype are the following:

- *idtype* $= 0 \rightarrow$ directory name is the speaker ID (spaces replaced with underscore),
    NOTE: all wav files in the directory are from a specific speaker
- *idtype* $= 1 \rightarrow$ utt_id derived from the file name is the speaker ID,
    NOTE: this will not identify separate speakers (DEFAULT)
- *idtype* $> 1 \rightarrow$ the first *idtype* number of characters from the file name is the speaker ID, and the wav files can be in the same directory. In order for this to work all ID's must have the same length! Shorter ID's must be padded with, e.g., zeros!

9.4.4   Developing Your Own ASR Software with VoiceBridge

The two previous examples demonstrate how you can develop your own automatic speech recognition (ASR) software based on VoiceBridge. The examples combined the training and the inference in one module. You will need to create a separate module for inference which can be used to recognize speech by using the trained ASR model, since you will not want to train a model every time you wish to recognize speech (see Sect. 9.4.2). Separating the training from inference is a straightforward task, but you will have to be careful to include all of the necessary sub-modules. Inference uses very similar sub-modules to the training because, for example, feature extraction must be part of both. You may of course also modify some of the internal functions in the VoiceBridge DLL. The entire source code is at your disposal!

9.5   VoiceData

VoiceData can be used for generating data for training ASR models (see Fig. 9.15). The generated data includes both the transcription synchronized audio (the input text is read by a machine-trained human-sounding synthesized voice; male or female) and transcription files for training ASR models *in several languages* for any text you desire. There are many languages and voices (female and male) available. VoiceData has a special multiprocessor/core support which makes the toolkit much faster, especially when processing several audio files or images at once.

9.5.1   Text Normalization

Before we can generate audio recordings for training an ASR model we must make sure that the input text is normalized. Normalization means that all non-text elements (e.g. 1/1/2018 or 10.5 kg) are converted to text. This is a complex task because the normalization must use language dependent grammar definitions in order to detect

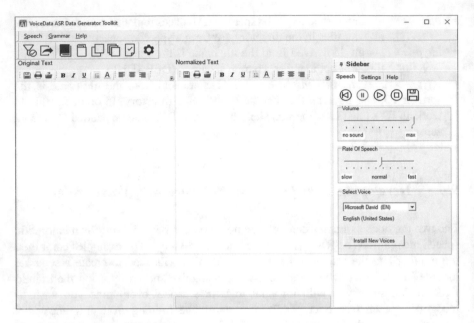

Fig. 9.15 VoiceData user interface

and normalize such elements automatically. You may also decide to provide the normalized text directly. If the text is not normalized well, then the transcriptions will not contain the correct words and may even contain characters which are not allowed while training an ASR model.

VoiceData contains grammar definitions used for text normalization for two languages, English and Dutch. You can develop your own text normalization grammars for any language with the help of the built-in GrammarEditor (Grammar menu). You can even replace the distributed English and Dutch grammars by replacing the language directories with your own files.

You can learn more about the VoiceData text normalization and the grammar definition format from the complementary book *VoiceData Text Normalization* [12] (included in the VoiceData software distribution in PDF format). The book contains extensive explanations about how to develop language dependent grammars and about many other related subjects. It is a complex subject, and therefore it is recommended to read the book before making grammar definitions yourself. Or if you take care of the normalization yourself, then you do not need to learn the built-in normalization module and grammars.

Grammar definitions are stored in the 'normdata' directory under the program folder. Each language has its own subdirectory designated with the ISO two letter language code (e.g., 'en' is English, 'nl' is Dutch). The root directory per language contains the following required configuration files:

- norm_config.tnm—this contains the normalization (classification and verbaliza-tion) settings for the specific language.

- grammar_config.tnm—this contains the list of root grammar definitions to compile. Grammar dependences will be detected automatically.

Please read the complementary book VoiceData Text Normalization [12] for more information!

9.5.1.1   How to Normalize Text

The procedure for normalizing text is as follows:

- Provide the original text by copy-pasting it into the 'Original Text' editor or typing it in.
- Select the desired language in the sidebar's Speech tab. Click 'Install New Voices' if you need another language.
- Make sure that grammar definitions are available for the chosen language in the language subdirectory (two letter ISO language code) of the 'normdata' directory under the program folder. Normalization can only be performed if the grammar definitions are available!
- Use the 'Normalize Original Text' command to normalize the text.
- The normalized text will appear in the normalized text editor where you can still adjust the text.

> NOTE: Check the distributed English grammar definitions for an example about how to develop your own grammars!

9.5.1.2   The 'en_ex' Example Grammar

After installing VoiceData you will notice a subdirectory called "en_ex" under the "normdata" directory, which is normally located in:

C:\Program Files\VOICEDATA\normdata\en_ex

This is an example (incomplete) English grammar containing all grammar source codes (grm files). In order to use this grammar in VoiceData you will have to rename the current "en" folder to, for example, "en_final" and the "en_ex" folder to "en". In this way you will be able to use the example grammar as the English grammar in VoiceData. Make sure that you do not delete the full English grammar but rename it so that you can restore it later! Please note that the full English grammar does not contain the source grm files!

The 'en_ex' grammar contains a simplified grammar which can be used for experimentation. It also contains all of the examples explained in the VoiceData Text Normalization book.

VoiceData text normalization is based on Sparrowhawk, an open source (Apache 2.0 license) implementation of the Google's Kestrel text-to-speech (TTS) text normalization system reported in Ebden and Sproat (2015). More information about Sparrowhawk can be found on the following website: https://github.com/google/sparrowhawk.

9.5.2   Export ASR Data (Audio and Transcriptions)

After you have the normalized text and you have chosen the desired language on the sidebar's Speech tab, use the 'Export Normalized Audio & Transcriptions' command to export the audio and the accompanying text transcriptions. The generated data includes both the transcription synchronized audio (the input text is read by a machine-trained human-sounding synthesized voice; male or female) and transcription files for training ASR models.

IMPORTANT: The normalized text must contain each sentence on a new line. The normalization module will do this automatically but if you enter the text yourself then make sure that each sentence is on a new line! Do not use very long sentences because that may decrease the accuracy of the ASR model later.

9.5.3   AI Text Synthesizer/Speaker

The synthesizer can read the text present in the normalized text editor in several languages. You can select the languages/voices on the sidebar's Speech tab. You can also install other languages/voices. You can operate the synthesizer from the Speech command center on the sidebar's Speech tab.

You can also adjust the volume and the rate of speech; you can start, pause, resume and even save the speech in an audio file, which you can use to listen to the text later on your computer or on another device (smart phone, MP3-player, etc.). You can easily convert the saved WAV file to MP3 with the audio editor (9.2.9) in the AI-TOOLKIT.

For installing new languages and voices follow the instructions which appear when you press the 'Install new Voices' command!

9.6 Document Summary

The DocumentSummary application can be used to create a short summary from any text document. You can summarize text in simple text files, in PDF files, in HTML files, etc. The text documents can be located on your computer or even on the internet (internet connection is needed).

DocumentSummary uses machine learning (ML) powered language models (natural language processing) in the summarization process. You can train your own language model by using a lot of text in any language. You can add specialized text specific to your discipline (law, medicine, chemistry, etc.). The ML model will learn specific language features (synonyms, related words, frequent words, etc.) which will be used to make the summary. You can also choose the number of sentences in the summary. If you are summarizing a longer document then you can increase the number of summary sentences. The summary may contain several subjects and DocumentSummary will choose these subjects from the whole document. You may also decide to subdivide your document (e.g., book) and request a summary from each part.

DocumentSummary contains a module for training a language model (for any language), a module to extract the summary from your documents and also a server which can be used to automate the summary extraction process. There are also several useful built-in utilities for extracting text from PDF documents and from HTML pages, for editing large text files (up to several GB's), for combining several text files together, for annotating PDF files with the summary, etc.

The software is very easy and intuitive to use and there is also a built-in help feature for each module.

9.6.1 How Does It Work?

This section will briefly describe how the summary extraction process works.

First a vocabulary is created from the input text with all of the words in the text and with their occurrence counts. Then excluded words, non-important words and posterior strings are removed from the vocabulary (all of these are defined by the user).

A trained language model is used for automatically removing non-important words from the vocabulary, and for the automatic detection of the synonyms for all of the words in the vocabulary. Synonyms are important because they must be grouped together and not taken as separate subjects while extracting the summary.

Next the sentences are scored according to how many times important words occur in them, and finally the requested number of sentences with the maximum score are extracted.

9.6.2 Creating a Summary

IMPORTANT: The DocumentSummary setup does not contain a trained language model because a well-trained language model may take up several hundreds of megabytes. You must first train a language model with a large text file of your choice! You may use the so called Wikipedia text dumps for this purpose in any available language (which can be freely downloaded from the internet). These are text files with several gigabytes of text. You may first decide to clean up these files in order to remove text with a different language or erroneous symbols.

Start by selecting the language for the document you want to summarize. If you want to summarize a local document then click on the 'Summarize Document' button (move the mouse cursor over a button to see a tooltip), and you will be presented with a file selection dialog. If you want to summarize a web document then enter the web address and click on the 'Summarize Web Document' button.

In the right sidebar you can select the desired number of summary sentences, you can define a PDF password (if it is needed for opening and extracting text from the PDF) and you can change the language directory.

NOTE: Only languages for which there is a trained language model are shown.

9.6.2.1 Frequent Words

You can also set the Frequent Words Limit in the settings, which determines how many of the most frequent words in the main language model vocabulary (lm.voc) will be removed from the summary vocabulary during the summary creation. These are unimportant words identified by the ML model during the training of the language model (e.g., 'the', 'and', etc.). *In order for this list to be accurate, the input text for the training of the language model must be large enough (several gigabytes)!*

IMPORTANT: If you have used a small text file for training the language model then the default value of this parameter may be too high for your application! Always check the 'lm.voc' file to decide the value of the Frequent Words Limit parameter!

9.6.2.2 Fine Tuning Your Language Model

With the configuration file editor you can add words to the excluded word vocabulary, called 'exclude.voc' on the 'Language Model' tab. These words will not be used when deciding the subject(s) of the summary. In English such words are, e.g., *isn't, you, the,* etc. Most of these unimportant words will be automatically removed (as discussed in the previous section) but, depending on the input text used for the training of the language model, some redundant words may not be removed and should be added to the excluded word list. The format of the 'exclude.voc' file is as shown in Table 9.38.

First indicate the word to exclude followed by a space and then a number (the number is not important here, it is just needed because of the vocabulary format).

IMPORTANT: While extracting the summary from a document the software will present you with a short list of the most frequent words in the document. You should make sure that this list does not contain unimportant words. If it contains unimportant words then add these to the 'exclude.voc' list before recreating the summary (*the 'exclude.voc' file is not used during the training of the language model*).

The configuration file 'nonword.cfg' contains a list of symbols which should <u>not</u> be evaluated as actual words. You may further extend this list. *This list is used both in the language model training and in the summary creation.*

The configuration file 'posterior.cfg' contains a list of symbols which should be removed from the end of words. In English this list could, for example, contain the symbol 's (as in "software's"). *This list is used in the language model training and also in the summary creation, and therefore it must be the same with both*!

The configuration file 'norm_config_sp.tnm' contains the 'sentence_boundary_regexp' parameter which is used for defining the sentence boundaries in your document. The default value is "[\cr.:!;\cr?] ".

Table 9.38 'exclude.voc' file format	
	one 1
	two 1
	three 1

IMPORTANT: Please note that the period '.' and the question mark '?' must be entered with '\cr' in front of them because these are special characters used in the software! The whole definition must be constrained by the square brackets [ ].

The configuration file 'norm_config_sp.tnm' also contains the 'sentence_boundary_exceptions_file' parameter which defines the text file containing symbols which should *not* be evaluated as sentence boundaries. The default value is "langdata\cren\crsentence_boundary_exceptions.tnm". Look at the "langdata\cren\crsentence_boundary_exceptions.tnm" file under the DocumentSummary installation directory for an example (default install location: C:\Program Files\DocumentSummary).

IMPORTANT: All parameter values must be placed between quotation marks (i.e., "..."). Folder paths must be relative to the language directory ('langdata' in this case) and note also the '\cr' as a folder separator! The language directory is defined by the international two letter language code! For example, for English, 'en' is used. In some cases some extra characters may also be present after the two letter language code (more information about this can be found in the language model help feature).

9.6.3 Annotating Your PDF

DocumentSummary generates a script file with the extension ".yaml" which can be used for automatically annotating (marking words and sentences) a PDF document. Use the 'Annotate PDF' command and select the yaml file stored in the same directory as the new PDF summary. If you want to clear all annotation from a PDF file then use the 'Clear Annotation PDF' command and select the PDF file in which you want to clear all annotations.

You may need to modify the text in the automatically generated annotation scrip file if the annotation does not appear well in the PDF document (e.g., because of complex symbols or because the text extraction was not 100% correct owing to a complex PDF structure). In this case just remove all annotations from the PDF with the 'Clear Annotation PDF' command, change the summary text in the script file and run the 'Annotate PDF' command again.

9.6.4   Training a Language Model

Let us start with training a language model for English. All necessary files are included in the DocumentSummary distribution except for the trained language model and vocabulary. Select English in the Select Language Model list and click Train Language Model. You will then have to select a text file containing the text you want to use for training the language model. This text file should be several hundreds of MB's in order to provide enough text for the training of the model. You can edit this text file with the built in big text editor (Edit Text File button). Please note that files of this size cannot be opened by standard Windows text editors (e.g. Notepad). You can cancel the training any time with the Cancel button.

TIP: The software first creates a vocabulary from the input text. You may check when this vocabulary is ready in the language directory and then cancel the training process in order to clean the vocabulary first (there can be unwanted symbols and foreign words in it). After cleaning the vocabulary start the training process again and make sure that you answer the question of whether to use the pre-existing vocabulary with YES!

Next you may add a new language model on the Language Model tab. Select the desired language, add a Language Model Extension (e.g., for English, you may add "final" as an extension and have the language signature "en_final". In this way you may train several English language models) and click the 'Add' button. The new language will be added under the selected language directory and all configuration files will be created automatically! You will have to edit these configuration files (see below)! After you have added the desired language model then select it in the 'Select Language Model' list and click 'Train Language Model'. Follow the same procedure as explained above.

You can cancel the training any time with the Cancel button.

Remember also Sect. 9.6.2.2 about how to finc-tune your language models!

IMPORTANT: All input text files must be in UTF-8 without signature format! The built-in text editor will save automatically in this format if the text contains special characters (UTF-8) and you can also select this option in the editor.

NOTE: There are many text files on the internet which can be used for the training of a language model. One of the sources is, e.g., the Wikipedia Text Dumps, which are available in several languages and contain thousands of pages text on many subjects. You may also add your own specialized text to these files, containing specific subjects and words.

9.6.4.1   Settings

In the Settings in the right control panel you can adjust the minimum word occurrence count. Words with fewer occurrences in the input text will not be taken into account while training the language model.

The language directory can also be set to another location with the Language Directory setting.

IMPORTANT: Please note that you will have to add languages to the directory if you change the main language directory location because language directories will not be copied. You may decide to copy the language directories manually, but in that case you will have to modify some configuration files manually in the text editor (as they contain directory information). Folder paths in configuration files must be relative to the language directory ('langdata' as default) and with '\cr' as a folder separator! The language directory is defined by the international two letter language code! Look at the example included in the setup.

9.6.5   Extracting Text from PDF and HTML Files

With the 'Extract Text' button you can extract text from any PDF or HTML files (local or internet). This can be interesting for different purposes; for example, you could assemble a text file for training the language model by using extracted text from different sources.

9.6.6   Combining Text Files

With the 'Join Text Files' button you can combine several text files together, for example, to make a larger text file for training a language model.

9.6.7   Big Text Files Editor

The built-in big text files editor will allow you to edit text files up to several gigabytes in size! Files which are larger than 1GB will be split automatically for editing and then will be re-combined when you click the save button. The built-in AI-TOOLKIT big text files editor can be used for editing text files up to 2GB.

Just open any text file and then double click the file(s) in the presented list to edit them with the AI-TOOLKIT text editor. Click save when you are ready.

Appendix: From Regular Expressions to HMM

This appendix will introduce regular expressions, finite state transducers (FST) and the hidden Markov model (HMM) in simple lay terms. A HMM is a kind of efficient searching mechanism, one of the most important parts of natural language processing (NLP) and automatic speech recognition (ASR), and it is based on regular expressions.

A.1 Introduction

Because there is an immense number of possible word combinations in any regular language we need an efficient way of searching for sentences which have a high probability of matching the input sentence (or word sequence). This efficient searching mechanism is one of the most important parts of natural language processing (NLP) and automatic speech recognition (ASR), and it is based on an extended form of so-called regular expressions the hidden Markov model (HMM).

This appendix will introduce regular expressions, finite state transducers (FST) and the hidden Markov model (HMM) without going into too much detail. Modeling natural language features is a very complex subject. The aim of this appendix is to explain in simple lay terms why we need HMM's, how they work and how they relate to the ASR process. For a more in depth description of natural language processing (NLP) and the related subjects of grammatical modeling (orthography and morphophonology), please refer to the reference section for an example selection of books.

Regular expressions are the backbone of how we model and characterize text sequences, or in other words, regular languages. In fact, if you know regular expressions already, then you know a lot about HMM's.

Based on the knowledge of how regular expressions work we will look at the so-called finite state automaton (or automata) (FSA), which is just one step away from the finite state transducer (FST). An finite state automaton is a kind of

Z. Somogyi, *The Application of Artificial Intelligence*,
https://doi.org/10.1007/978-3-030-60032-7

mathematical device to implement and visualize regular expressions and an important tool in computational linguistics.

An HMM is a special version of the so-called weighted finite state transducer which is an extension of FST. You probably already have a sense of the complexity from just reading all of these names, but don't worry all will be clear after reading this appendix!

A.2   Regular Expressions

A text in any language may contain a sequence of letters, numbers, spaces, punctuation marks, etc. We call one segment of such a text a string. Regular expressions define how we select one or more strings out of a text. We call this selection process a 'search'. In all applications of regular expressions we are searching for strings in a text. In order to distinguish regular expressions from other text we enclose them between slashes '/'. The syntax of regular expressions is a very simple kind of textual programming language.

The simplest regular expression is that which searches for a specific word in a text. For example, the regular expression /fst/ searches for any string containing the substring 'fst'. So if we have a sentence 'fst is very useful' then the regular expression /fst/ will match this sentence (string).

Square braces [ ] are used to specify several characters to match. It is also possible to combine this with the string you are searching for. For example, the regular expression/[fb]ar/will match a sentence with both the words 'far' and 'bar', and thus both of the following sentences will be matched: 'the bar is open' and 'the shop is far'. The expression [fb] thus means that we accept 'f' and also 'b' as the first letter.

In order to match all upper case letters you can use [A-Z], and for all single digits use [0–9], etc. Regular expressions are case sensitive.

The '*' star symbol is used to match *zero* or more occurrences of the character or substring which immediately precedes the symbol, and the '+' plus symbol is used to match *one* or more occurrences in the same way; for example, /ho*/ will match 'h', 'ho', 'hoo', etc. whereas /ho+/ will match 'ho', 'hoo', etc. The '?' question mark symbol means that the preceding character may be missing (zero or single occurrence), e.g., /ho?/ will match 'ho' and 'h'. The '.' period (dot) symbol is used for matching any character, e.g., /f.r/ will match 'far', 'fir', etc.

We can also choose parts from different strings to match with the pipe 'l' symbol. For example /far|bar/ matches both 'far' and 'bar'.

The repetition and wild card symbols together with the pipe symbol are called operators because they make an operation on two or more objects (strings). The star operator has highest precedence followed by plus and the question mark operators, and then the pipe operator. You can also use parentheses ( ) to indicate your choice of precedence.

This short introduction to regular expressions is not an exhaustive list of possibilities (there are many more rules which are not mentioned) but it gives you just enough information to understand how the backbone of natural language processing (string matching) works.

A.3 Finite State Automaton (FSA)

A finite state automaton (FSA) is a kind of mathematical device to implement and visualize regular expressions and an important tool in computational linguistics. The FSA and the strings it matches can be illustrated with a simple diagram.

The best way to understand how an FSA works is by example. We will draw the FSA for the regular expression /hoo+/. Remember that this regular expression matches the strings 'hoo', 'hooo', 'hoooo', etc. The '+' sign means one or more 'o' characters (the character immediately in front of the plus sign).

Each FSA diagram consists of several nodes (one node per state; we will see a bit later what state means in this context) and links (often called arcs) between the nodes ending in arrows to illustrate the direction of the processing.

An FSA recognizes a set of strings (as does the regular expression /hoo+/), and thus each FSA can be described with a regular expression and vice-versa.

The FSA for this example can be seen on Fig. A.1. It has four states denoted with circles, of which one is indicated with a double circle, the final state. This is important because in the final state the FSA decides whether it 'accepts' the input string or not.

The FSA processes the string it receives symbol by symbol, and each of these processing steps ends in a new state. The last or final state is also the accepting state, where the FSA decides if it accepts (matches) the incoming string or not.

Each state may also have a transition arc, which arrives back at itself (self-loop), in order to indicate the acceptance of several symbols (corresponding to the regular expression operator '*' star).

Fig. A.1 Finite state automaton

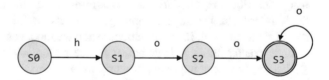

Table A.1 Example 1

Example 1: Let the input string be 'booo'. The following table indicates what is happening in each state and between them.

State	Decision	Description
S0	$h \neq b$	The FSA compares the first symbol of the input 'b' with the first symbol on the arc going out of state S0. Because the symbols do not match, the FSA cannot move to state S1 and it fails to recognize, or in other words, it rejects the input string.

Table A.2 Example 2

Example 2: Let the input string be 'horse'. The following table indicates what is happening in each state and between them.

State	Decision	Description
S0	h = h	The FSA compares the first symbol of the input 'h' with the first symbol on the arc going out of state S0 'h'. Because the symbols do match the FSA can proceed to state S1 and it also moves to the next input symbol.
S1	o = o	The same as in the previous step. The FSA can advance to the next step.
S2	o ≠ r	The third input symbol does not match the letter 'o', which is expected by the FSA, and therefore the machine can not proceed further and rejects the input.

Table A.3 Example 3

Example 3: Let the input string be 'hoooo'. The following table indicates what is happening in each state and between them.

State	Decision	Description
S0	h = h	OK! The FSA proceeds to the next state and reads the next input symbol.
S1	o = o	OK! The FSA proceeds to the next state and reads the next input symbol.
S2	o = o	OK! The FSA proceeds to the next state and reads the next input symbol.
S3	o = o	OK! The FSA proceeds to the next state and reads the next input symbol.
S4	o = o	OK! This is the final input symbol. The FSA accepts the whole input string.

Fig. A.2 Non-deterministic finite state automaton with self-loop

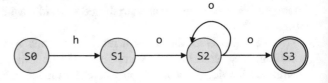

Let us examine several examples in order to understand how our FSA works in Tables A.1, A.2, and A.3.

In the case of this FSA (/hoo+/) example we could say that the different accepted strings describe a regular language containing an infinite number of words such as 'hoo', 'hoooo', etc. This regular language consists of a finite set of symbols which are 'h' and 'o' (only two). In real situations when we model an existing natural language we have exactly the same logic but of course with many more symbols and rules!

There is one more important property of an FSA which must be mentioned—its deterministic nature. We call an FSA deterministic if it does not need to make decisions about which state comes next. Our /hoo+/ example is a deterministic FSA because at each step the next state is known (without doubt).

If we slightly change our example to /ho+o/ (notice that the plus sign has moved one symbol to the left) then it becomes a non-deterministic FSA. You can see this easily if you draw the FSA diagram (see Fig. A.2).

In state S2 the machine will have to make a decision, if it receives the symbol 'o', on whether to stay in state S2 (which produces an 'o') or to move on to state S3 (which also produces an 'o'). The choice is of course obvious (if there is only one

Fig. A.3 Non-deterministic finite state automaton with epsilon arc

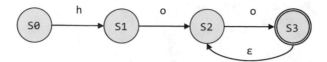

letter 'o' left then it must move to S3 otherwise it must stay in S2) but it has to be made!

The same non-deterministic behavior can be modeled with a so-called epsilon (ε) arc instead of a self-loop. The above diagram should then be modified as shown on Fig. A.3.

The epsilon arc means that S3 may not be evaluated and processing can return to S2. This is exactly the same as the FSA on Fig. A.2.

Non-deterministic FSAs occur often in natural language processing. According to [13] there are three standard decision making strategies in the case of non-deterministic FSAs:

1. **Backup**: "Whenever we come to a choice point, we could put a marker to mark where we were in the input, and what state the automaton was in. Then if it turns out that we took the wrong choice, we could back up and try another path." [13] For example, if we arrive at the wrong string we can go back and try another path through the FSA.
2. **Look-ahead**: "We could look ahead in the input to help us decide which path to take." [13] For example, we decide what to do by checking the next letter.
3. **Parallelism**: "Whenever we come to a choice point, we could look at every alternative path in parallel" [13]

The previous two sections contain the most important part of the information needed to understand how an FSA works. The following sections will extend this information in order to finally arrive at how a hidden Markov model (HMM) works.

A.4   Finite State Transducer (FST)

The word 'Transducer' indicates that an FST converts symbols from one form to another, or in other words, it maps between two sets of symbols. A finite state transducer is a special form of FSA, and therefore they both function very similarly and they can also be visualized similarly.

Mapping between two sets of symbols is indicated by a ':' colon. For example, a: b means the mapping of 'a' to 'b', or cycle:bike maps the word 'cycle' to 'bike'. This is exactly the same notation as we have used previously to map between two characters or words. You could, for example, map one word in a specific language to the same word in another language (automatic machine translation).

The visualization of an FST which transforms any number of 'a's to the same number of 'b's can be seen on Fig. A.4.

Fig. A.4 Simple FST for a:
b mapping

Fig. A.5 Epsilon arc:
modified from [14]

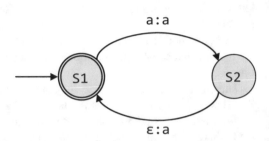

Everything is the same as with an FSA but now instead of one symbol we have two symbols separated with a colon on each arc. The left side of the colon we call the input string and the right side of the colon the output string (the input is mapped to the output).

Blackburn and Striegnitz [14] summarize the possible uses of an FST as follows:

1. **Generation Mode**: The FST in Fig. A.4 writes a set of 'a's as the input string and a set of 'b's as the output string. Both set of symbols have the same length.
2. **Recognition or Acceptance Mode**: The FST accepts the input only if the input string has the same number of 'a's as the output string has 'b's.
3. **Translation Mode (left to right)**: Very intuitively the FST reads 'a's from the input and writes the same number of 'b's to the output string.
4. **Translation Mode (right to left)**: This is the exact inverse of the previous case (and this is also what we call the function which produces the inverse of an FST), and thus it reads 'b's from the output and writes 'a's to the input. We could also call this reverse translation.

These possible uses indicate already several linguistic fields in which FST's are used with great success.

We have seen already that in an FSA an epsilon arc can be used to return processing to a previous node without evaluating the current state. The same functionality is possible in case of FSTs. Both the input and the output may be replaced by an epsilon arc as shown on Fig. A.5.

This example does the following very intuitively:

1. In Generation Mode it first reads an 'a' from the input and writes an 'a' to the output. Then in state S2 it returns to state S1 without doing anything with the input but it writes an 'a' to the output. S1 is the final state indicated by the double circle. This behavior can also be illustrated with two tapes which symbolize the input and output strings as it is shown on Fig. A.6.

Fig. A.6 Generation mode

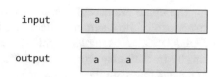

Fig. A.7 FST for
normalizing English
numbers; modified from
[14]

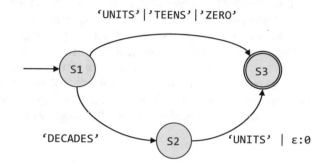

Table A.4 FST lexicon

Symbol	Regular expression representation
UNITS	("1":"one")|("2":"two")|("3":"three")|("4":"four")|("5":"five")|("6":"six")| ("7":"seven")|("8":"eight")|("9":"nine")
TEENS	("10":"ten")|("11":"eleven")|("12":"twelve")|("13":"thirteen")|("14":"fourteen")| ("15":"fifteen")|("16":"sixteen")|("17":"seventeen")|("18":"eighteen")| ("19":"nineteen")
DECADES	("2":"twenty")|("3":"thirty")|("4":"forty")|("5":"fifty")|("6":"sixty")|("7":"seventy")| ("8":"eighty")|("9":"ninety")
ZERO	("0":"zero")

2. In Recognition Mode the FST only accepts the input when the output has twice as many 'a's as the input. Remember that the output is generated by following the FST rules, defined in the FST diagram, from the input state until the output state!

3. In Translation Mode the FST translates (writes) twice as many 'a's to the output than there are 'a's in the input.

4. In Reverse Translation Mode (right to left) the FST reads at least two 'a's from the output and writes half as many to the input.

In a very similar manner FST's can also use string labels containing several symbols.

The following example from [14] illustrates the conversion between English numbers (integers) and their names. And this is exactly how the normalization of numbers works in text normalization systems based on FST's (for example, in VoiceData in the AI-TOOLKIT). Normalization replaces non-text elements with their text representation (Fig. A.7).

The lexicon (collection of words and symbols) for this FST is shown in Table A.4.

The upper arc accepts numbers in ranges of 1–9 and 10–19 along with 0. The lower part of the diagram produces the two digit numbers built from the sub-lexicon DECADES, for example, 20, 30, 40, etc. and also 21, 22 ... 31, ... 42, etc. It does this by transitioning from state S2 with an epsilon arc (when 20, 30, 40...etc.) and concatenating with the UNITS lexicon to obtain 21, 22, 31...etc.

The zero after the epsilon (ε:0) means that at the output a zero will be written. This is necessary because the DECADES lexicon contains the decades as 2, 3, and 4, etc., and thus we form the numbers 20, 30...etc. by adding a '0'.

For example, if we have a sentence "10 meters is too long" then this FST could be used to normalize (transform) this sentence to the following "ten meters is too long". The regular expression ("10":"ten") replaces '10' with 'ten' according to FST rules. Normalization is important in several NLP and speech applications, for example, when we want to synthesize text (e.g., in VoiceData in the AI-TOOLKIT).

A.5   Weighted Finite State Transducer (WFST)

The FSAs and FSTs can be extended by including weights on the arcs between the nodes. Weights can also be seen as 'costs' on traversing the different paths through an FSA or FST. If there are several acceptable paths possible (due to the input) through an FST, then the path which has the lowest total cost (lowest weight) will be chosen.

In Fig. A.8 we show the weights next to the arcs between < > symbols.

The weights can be any number depending on the software implementation. The aim is to order the weights according to their magnitude.

The FST in Fig. A.8 will prefer (accept) the word 'far' instead of the word 'bar', whenever both are possible due to the input, because the weight on 'f' is lower than the weight on 'b'.

There are several operations possible with (between) FSTs such as inversion (switching the input with the output), composition (connecting several FSTs in series, one after the other), etc. In practice several FST's are connected together which form the acoustic model or define a special language dictionary.

Fig. A.8 Weighted finite state transducer (WFST)

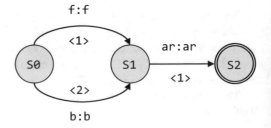

NOTE: An FST diagram is only a graphical device to design and explain the FST model we are using. Each FST and several connected FST's can be modeled with a mathematical algorithm which is used, for example, in speech recognition as part of the acoustic model.

A.6 The Hidden Markov Model (HMM)

"The accuracy of automatic speech recognition remains one of the most important research challenges after years of research and development. There are a number of well-known factors that determine the accuracy of a speech-recognition system. The most noticeable ones are context variations, speaker variations, and environment variations. Acoustic modeling plays a critical role to improve the accuracy. It is not far-fetched to state that it is the central part of any speech-recognition system. Acoustic modeling of speech typically refers to the process of establishing statistical representations for the feature vector sequences computed from the speech wave-form. HMM (Baum 1972; Baker 1975; Jelinek 1976) is one of the most common types of acoustic models" [19].

"Acoustic modeling also encompasses 'pronunciation modeling', which describes how a sequence or multi-sequences of fundamental speech units (such as phones or phonetic feature) are used to represent larger speech units such as words or phrases that are the object of speech recognition" [19].

"In HMM-based speech recognition, it is assumed that the sequence of observed vectors corresponding to each word is generated by a Markov chain" [19]. As shown in Fig. A.9, "an HMM (Hidden Markov Model) is a *finite state machine* that changes

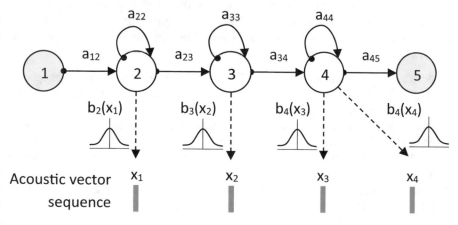

Fig. A.9 An illustration of a five-state left-to-right HMM; modified from [19]

state once every time frame, and at each time frame t when a state j is entered, an observation vector x_t is generated from the emitting probability distribution $b_j(x_t)$. The transition property from state i to state j is specified by the transition probability a_{ij}. Moreover, two special non-emitting states are usually used in an HMM. They include an entry state, which is reached before the speech vector generation process begins, and an exit state, which is reached when the generative process terminates. Both states are reached only once. Since they do not generate any observation, none of them has an emitting probability density... For each emitting state, the HMM is only allowed to remain at the same state or move to the next state." [19].

"Generally speaking, each emitting distribution characterizes a sound event, and the distribution must be specific enough to allow discrimination between different sounds as well as robust enough to account for the variability in natural speech. *Numerous HMM training methods are developed to estimate values of the state transition probabilities and the parameters of the emitting probability densities at each state of the HMM.*" [19].

Notice that the HMM on Fig. A.9 is very similar to the Weighted Finite State Transducer (WFST) discussed in the previous section except that instead of weights there are transmission probabilities on the arcs. If there are several arcs leaving a state then the sum of all transition probabilities leaving a state is 1.

"The basic unit of sound represented by the acoustic model is the *phone*. For example, the word 'bat' is composed of three phones /b/ /ae/ /t/. About 40 such phones are required for English." [19].

"Decomposing each vocabulary word into a sequence of context-independent base phones fails to capture the very large degree of context-dependent variation that exists in real speech. For example, the base form pronunciations for "mood" and "cool" would use the same vowel for "oo," yet in practice the realizations of "oo" in the two contexts are very different due to the influence of the preceding and following consonant. Context independent phone models are referred to as monophones. A simple way to mitigate this problem is to use a unique phone model for every possible pair of left and right neighbors. The resulting models are called tri-phones and if there are N base phones, there are N^3 potential triphones.

The core acoustic models of a modern speech recognizer typically consist of a set of tied three-state HMMs with Gaussian output distributions. This core is commonly built in the following steps:

1. A flat-start monophone set is created in which each base phone is a monophone single-Gaussian HMM with means and covariances equal to the mean and covariance of the training data.
2. The parameters of the Gaussian monophones are re-estimated using 3 or 4 iterations of EM.
3. Each single Gaussian monophone q is cloned once for each distinct triphone $x - q + y$ that appears in the training data.
4. The resulting set of training-data triphones is again re-estimated using EM and the state occupation counts of the last iteration are saved.

5. A decision tree is created for each state in each base phone; the training-data triphones are mapped into a smaller set of tied-state triphones and iteratively re-estimated using EM. The final result is the required tied-state context-dependent acoustic model set." [20]

NOTE: EM is the so-called expectation-maximization algorithm. The variations of the EM algorithm are used most often for training an HMM. The so-called Baum-Welch or forward-backward algorithm is also an extension of EM. Both the transition probability matrix (a_{ij}) and the parameters of the emitting probability densities $(b_j(x_t))$ are trained/learned with this algorithm.

Most state-of-the-art speech recognition systems are triphone based but they first create a monophone model which is then extended/further trained to accommodate a triphone model. VoiceBridge in the AI-TOOLKIT works also like this.

References

1. Batch normalization: accelerating deep network training by reducing internal covariate shift, Christian Szegedy (Google), Sergey Ioffe (Google), arXiv:1502.03167v3 [cs.LG]. 2 Mar 2015
2. Bradley, P.S., Fayyad, U.M.: Refining initial points for k-means clustering. In: ICML'1998: Proceedings of the Fifteenth International Conference on Machine Learning. Morgan Kaufmann, San Francisco, CA
3. Cheng, Y.: Mean shift, mode seeking, and clustering. IEEE Trans. Pattern Anal. Mach. Intell. **17**(8), 790 (1995)
4. Comaniciu, D., Meer, P.: Mean shift: a robust approach toward feature space analysis. IEEE Trans. Pattern Anal. Mach. Intell. **24**(5), 603–619 (2002)
5. The 5 essential components of a data strategy (White Paper), SAS (2018)
6. Albright, S.C., Winston, W.L., Zappe, C.: Data Analysis for Managers, 2nd edn. Brooks/Cole, Belmont, CA (2004). isbn:0-534-38366-1
7. George, M., Rowlands, D.: The Lean Six Sigma Pocket Toolbook. McGraw-Hill, New York (2004)
8. Albright, S.C., Winston, W.L., Zappe, C.: Data Analysis for Managers, 2nd edn. Brooks/Cole, Belmont, CA (2004). isbn:0-534-38366-1
9. Curtin, R.R., Edel, M., Lozhnikov, M., Mentekidis, Y., Ghaisas, S., Zhang, S.: mlpack 3: a fast, flexible machine learning library. J. Open Source Softw. (2018). https://doi.org/10.21105/joss.00726
10. NIST: NIST/SEMATECH e-Handbook of Statistical Methods. U.S. Department of Commerce, Washington, DC (2012)
11. Jolliffe, I.T.: Principal Component Analysis, 2nd edn. Springer, New York (2002)
12. Somogyi, Z.: VoiceData Text Normalization, Automatic Speech Recognition (ASR) Data Generator Toolkit, AI-TOOLKIT (free PDF book, included in the VoiceData software distribution) (2018)
13. Jurafsky, D., Martin, J.H.: Speech and Language Processing, An Introduction to Natural Language Processing, Computational Linguistics, and Speech Recognition. Prentice-Hall, Upper Saddle River, NJ (2008)
14. Blackburn, P., Striegnitz, K.: Natural Language Processing Techniques in Prolog. Union College Computer Science Department. http://cs.union.edu/striegnk/courses/nlp-with-prolog/html/node13.html.
15. Lei Ba, J., Kiros, J. R., Hinton, G. E.: Layer Normalization, arXiv: 1607.06450v1 [stat.ML] 21 Jul 2016
16. Goodfellow, I., Bengio, Y.: Deep Learning. MIT Press, Aaron Courville (2016)

© Springer Nature Switzerland AG 2021
Z. Somogyi, *The Application of Artificial Intelligence*,
https://doi.org/10.1007/978-3-030-60032-7

17. Russell, S.J., Norvig, P.: Artificial Intelligence – A Modern Approach, 3rd edn. Prentice Hall, Hoboken, NJ (2010)
18. Bonaccorso, G.: Machine Learning Algorithms. Packt Publishing, Birmingham (2017)
19. Indurkhya, N., Damerau, F.J.: Handbook of Natural Language Processing, 2nd edn. Chapman & Hall/CRC, New York (2010)
20. Gales, M., Young, S.: The application of hidden Markov models in speech recognition. Found Trends Signal. Process. **1**(3), 195–304 (2008)
21. Everest, F.A.: The Master Handbook of Acoustics, 4th edn. McGraw-Hill, New York (2001)
22. Chazan, D., Hoory, R., Cohen, G., Zibulski, M.: Speech Reconstruction from Mel Frequency Cepstral Coefficients and Pitch Frequency. IBM Research Laboratory, Haifa
23. Gelfand, S.A.: Hearing: An Introduction to Psychological and Physiological Acoustics, 5th edn. Taylor and Francis, London (2010)
24. Palmer, H.E., Martin, J.V., Blandford, F.G.: A dictionary of English pronunciation with American variants (in phonetic transcription). W. Heffer & Sons, Cambridge, MA (1926)
25. Davis, S., Mermelstein, P.: Comparison of parametric representations for monosyllabic word recognition in continuously spoken sentences. IEEE Trans. Acoust. Speech Signal Process. **28** (4), 357–366 (1980)
26. Haeb-Umbach, R., Ney, H.: Linear Discriminant Analysis for Improved Large Vocabulary Continuous Speech Recognition, Proceedings ICASSP, pp. 13–16. IEEE, San Francisco, CA (1992)
27. Omar, M.K., Hasegawa-Johnson, M.: Model enforcement: a unified feature transformation framework for classification and recognition. IEEE Signal Process. **52**, 2701–2710 (2004)
28. Saon, G., Padmanabhan, M., Gopinath, R., Chen, S.: Maximum likelihood discriminant feature spaces. IEEE International Conference on Acoustics Speech and Signal Processing. **2**(2000), II-1129–II-1132 (2000)
29. Povey, D., Kuo, H-K. J., Soltau, H.: Fast Speaker Adaptive Training for Speech Recognition (2008)
30. Rendle, S., Freudenthaler, C., Gantner, Z. and Schmidt-Thieme, L. BPR: Bayesian Personalized Ranking from Implicit Feedback, UAI (2009)
31. Michael P. Holmes, Alexander G. Gray and Charles Lee Isbell, Jr.: QUIC-SVD: fast SVD using cosine trees. (2008)
32. Musco, C., Musco, C.: Randomized block Krylov methods for stronger and faster approximate singular value decomposition. (2015)
33. Halko, N., Martinsson, P.G., Tropp, J.A.: Finding structure with randomness: probabilistic algorithms for constructing approximate matrix decompositions. Survey Rev. Sect. **53**(2), 217–288 (2011)
34. Stolfo, S. J., Fan, W., Lee, W., Prodromidis, A., Chan, P. K. Cost-based modeling and evaluation for data mining with application to fraud and intrusion detection: results from the JAM project. (2000)
35. Ivan, T.: An experiment with the edited nearest-neighbor rule. IEEE Trans. Syst. Man Cybern. (6), 448–452 (1976)
36. Chawla, N.V., Bowyer, K.W., Hall, L.O., Kegelmeyer, W.P.: SMOTE: synthetic minority over-sampling technique. J. Artif. Intell. Res. **16**, 321–357 (2002)
37. Han, H., Wang, W.-Y., Mao, B.-H.: Borderline-smote: a new over-sampling method in im-balanced datasets learning. In: Huang, D.S., Zhang, X.P., Huang, G.B. (eds.) International Conference on Intelligent Computing, pp. 878–887. Springer, Berlin (2005)
38. Kingma, D. P.: Adam: A Method For Stochastic Optimization, Jimmy Lei Ba, ICLR (2015)
39. Matthew D. Zeiler, Adadelta: An Adaptive Learning Rate Method, CoRR (2012)
40. Duchi, J., Hazan, E., Singer, Y.: Adaptive subgradient methods for online learning and stochastic optimization. J. Mach. Learn. Res. **12**, 2121–2159 (2011)
41. Nguyen, L. M., Liu, J., Scheinberg, K., Tak, M.: SARAH: a novel method for machine learning problems using stochastic recursive gradient. ArXiv e-prints, (2017)

42. Kolmogorov, A.N.: On the representation of continuous functions of many variables by superpositions of continuous functions of one variable and addition. Doklay Akademii Nauk USSR. **14**(5), 953–956 (1957) Translated in: Amer. Math Soc. Transl. 28, 55–59 (1963)

43. Hecht-Nielsen, R.: Kolmogorov's mapping neural network existence theorem. In: Proceedings IEEE International Conference on Neural Networks, vol. II, pp. 11–13, New York, IEEE Press (1987)

44. Castillo, P.A., Arenas, M.G., Castillo-Valdivieso, J.J., Merelo, J.J., Prieto, A., Romero, G.: Artificial neural networks design using evolutionary algorithms. In: Benítez, J.M., Cordón, O., Hoffmann, F., Roy, R. (eds.) Advances in Soft Computing. Springer, London

45. Goldberg, D.E.: Genetic Algorithms in Search, Optimization, and Machine Learning. Addison-Wesley, Boston, MA (1989)

46. Kaufman, L., Rousseeuw, P.J.: Finding Groups in Data: an Introduction to Cluster Analysis Wiley Series in Probability and Statistics. Wiley, New York (1990)

47. Caliński, T., Harabasz, J.: A dendrite method for cluster analysis. Academy of Agriculture, Poznań, Poland. Published online 27 June 1974

48. Rousseeuw, P.J.: Silhouettes: a graphical aid to the interpretation and validation of cluster analysis. J. Comput. Appl. Math. **20**, 53–65 (1987)

49. Xu, L.: Bayesian ying–yang machine, clustering and number of clusters. Pattern Recogn. Lett. **18**(11), 1167–1178 (1997)

50. Manning, C.D., Raghavan, P., Schütze, H.: An Introduction to Information Retrieval. Cambridge University Press, London (2009)

51. Ultsch, A.: Clustering with SOM: U*C, In: Proc. Workshop on Self-Organizing Maps, Paris, France, pp. 75–82 (2005)

52. Kirchgässner, W., Wallscheid, O., Böcker, J.: Deep Residual Convolutional and Recurrent Neural Networks for Temperature Estimation in Permanent Magnet Synchronous Motors. Empirical Evaluation of Exponentially Weighted Moving Averages for Simple Linear Thermal Modeling of Permanent Magnet Synchronous Machines. (2019)

53. Cardiovascular diseases (CVDs), World Health Organisation Report. (2017)

54. AML workshop proceedings and data, Microsoft Corporation (MIT license). (2017)

55. Postoperative dataset. Sharon Summers, School of Nursing, University of Kansas Medical Center, Kansas City, KS 66160 Linda Woolery, School of Nursing, University of Missouri, Columbia

56. Olah, C.: Understanding LSTM Networks, Github Blog

57. Sutton, R.S., Barto, A.G.: Reinforcement Learning: An Introduction. The MIT Press, London

58. Szepesvari, C.: Algorithms for Reinforcement Learning. Morgan & Claypool, Sand Rafael, CA (2009)

59. Florian, R. V.: Correct equations for the dynamics of the cart-pole system. (2007)

60. Alaa Tharwat Classification assessment methods. Applied Computing and Informatics. (2018)

61. Sparrowhawk/Kestrel, the Google-internal TTS text normalization system reported in Ebden and Sproat. Copyright 2015 and onwards Google, Inc., Apache 2.0 license, Website: https://github.com/google/sparrowhawk (2014).

62. The Simd Library. http://ermig1979.github.io/Simd, Ihar Yermalayeu

63. The Kaldi project. http://kaldi-asr.org

64. Aggarwal, C.C.: Recommender Systems. Springer, Heidelberg (2016)

65. Johnson, R.W.: Body Fat Dataset. Department of Mathematics & Computer Science, South Dakota School of Mines & Technology, Rapid

66. Penrose, K.W., Nelson, A.G., Fisher, A.G.: Generalized body composition prediction equation for men using simple measurement techniques. Med. Sci. Sports Exerc. **17**(2), 189 (1985)

67. Harper, R., Tee, P.: The Application of Neural Networks to Predicting the Root Cause of Service Failures, FIP/IEEE IM 2017 Workshop.

68. Fisher, R.A.: The Use of Multiple Measurements in Taxonomic Problems, Annual Eugenics, 7, Part II, 179–188 (1936); also in "Contributions to Mathematical Statistics". Wiley, New York (1950)

69. Somogyi, Z.: The AI-TOOLKIT. https://ai-toolkit.blogspot.be, https://ai-toolkit.github.io
70. Somogyi, Z.: VoiceBridge. https://github.com/AI-TOOLKIT/VoiceBridge
71. Somogyi, Z.: VoiceData Text Normalization, Automatic Speech Recognition (ASR) Data Generator Toolkit. (2018)
72. Somogyi, Z.: Business Process Improvement Handbook for Office & Services (for Experts & Managers, an Advanced Guide), ISBN 9789090296203. (2016)
73. Kaiming He, Xiangyu Zhang, Shaoqing Ren, Jian Sun: Deep Residual Learning for Image Recognition, Microsoft Research. (2015)
74. Schroff, F., Kalenichenko, D., Philbin, J.: FaceNet: A Unified Embedding for Face Recognition and Clustering. Google Inc., (2015)
75. Kazemi, V., Sullivan, J.: One Millisecond Face Alignment with an Ensemble of Regression Trees. (2014)
76. Vahid Kazemi and Josephine Sullivan: Face Alignment with Part-Based Modeling (2011)
77. Li, S.Z., Jain, A.K.: Handbook of Face Recognition. Springer, London (2005)
78. Facial Recognition Technology: Fundamental Rights Considerations in the Context of Law Enforcement, FRA – European Union Agency for Fundamental Rights. (2019)
79. Dalal, N., Triggs, B.: Histograms of Oriented Gradients for Human Detection. (2005)
80. Beigi, H.: Fundamentals of Speaker Recognition. Springer, Boston, MA (2011)
81. Reynolds, D.A.: Automatic speaker recognition using Gaussian mixture speaker models. Lincoln Laboratory J. **8**(2), 173 (1995)
82. Reynolds, D. A., Quatieri, T. F., Dunn, R. B.: M.I.T. Lincoln Laboratory. (2000)
83. Makhoul, J.: Linear Prediction: A Tutorial Review. Proc. IEEE. **63**(4) (1975)
84. Amrutha, R., Lalitha, K., Shivakumar, M., Michahial, S.: Feature extraction of speech signal using LPC. Int. J. Adv. Res. Comp. Commun. Eng. **5**(12) (2016)
85. Dehak, N., Dehak, R., Kenny, P., Brummer, N., Ouellet, P., Dumouchel, P.: Support Vector Machines versus Fast Scoring in the Low-Dimensional Total Variability Space for Speaker Verification. Interspeech (ISCA). (2009)
86. Senoussaoui, M., Kenny, P., Dehak, N., Dumouchel, P.: An i-vector Extractor Suitable for Speaker Recognition with both Microphone and Telephone Speech. (2010)
87. Kenny, P.: Joint Factor Analysis of Speaker and Session Variability: Theory and Algorithms. (2006)
88. Dehak, N., Kenny, P., Dehak, R.'e., Dumouchel, P., Ouellet, P.: Front-end factor analysis for speaker verification. IEEE Trans. Audio Speech Lang. Process. **2010**, 1–11 (2010)
89. Garcia-Romero, D., Zhou, X., Espy-Wilson, C. Y.: Multicondition training of gaussian plda models in i-vector space for noise and reverberation robust speaker recognition. ICASSP. (2012)
90. Khosravani, A., Glackin, C., Dugan, N., Chollet, G., Cannings, N.: The intelligent voice 2016 speaker recognition system. (2016)
91. Snyder, D., Garcia-Romero, D., Povey, D., Khudanpur, S.: Deep Neural Network Embeddings for Text-Independent Speaker Verification. (2017)
92. C++ Mathematical Expression Toolkit Library, Arash Partow (1999–2019)
93. The Boost C++ project, https://www.boost.org
94. Kingma, D.P. (Google), Welling, M. (Universiteit van Amsterdam, Qualcomm).: An Introduction to Variational Autoencoders. (2019)
95. Creswellx, A., White, T., et al.: Generative Adversarial Networks: An Overview. IEEE-SPM (2017)
96. Alqahtani, H., Kavakli-Thorne, M., Kumar, G.: Applications of Generative Adversarial Networks (GANs): An Updated Review. Archives of Computational Methods in Engineering
97. Baker, J.: Stochastic modeling for automatic speech recognition. In: Reddy, D.R. (ed.) Speech Recognition. Academic, New York (1975)
98. Baum, L.: An inequality and associated maximization technique occurring in statistical estimation for probabilistic functions of a Markov process. Inequalities. **III**, 1–8 (1972)
99. Jelinek, F.: Continuous speech recognition by statistical methods. Proc. IEEE. **64**(4), 532–556

Index

A

Accuracy, 6–8, 90, 92, 169, 170, 224–227
Acoustic modeling, 158
Acrobot, 329, 334
Action reward, 71
Action selection policy, 64, 70
Activation function, 22, 33, 46, 139
Activation layer, 20
Adaptive learning rate optimizer: Adam, Sarah
    Plus, RMS Prop, Adagrad, 31
Autoencoders, 82
AI-TOOLKIT, 311
Alternating least squares (ALS), 220
Anomaly detection, 241, 247–265
Area Under The ROC Curve (AUC), 97
ARPAbet, 166
Audio editor, 372
Autoencoder, 82
Automatic speech recognition (ASR), 145
Average pooling, 44

B

Backpropagation, 29
Batch normalization, 26, 27, 319
Bayesian analysis, 220
Bayesian personalized ranking (BRP), 220
Bias, 5, 21
Binary encoding, 128
Binomial distribution, 132
Borderline synthetic minority oversampling,
    136
Business process automation, 68–75

C

Calinski-Harabasz Index (CHI), 101
Cart-pole, 77
Class imbalance, 134–137, 259, 356
Classification, 8, 88, 90
Clustering, 10, 52, 99
Cohen's kappa, 92
Collaborative filtering, 216, 345, 348, 351
Confusion matrix, 89, 90, 97
Content-based recommendations, 230
Context-sensitive recommendations, 228
Contingency table, 108
Control chart, 120
Convolution layer, 37
Convolutional filter, 39
Cross entropy loss, 29

D

Data cleaning, 118–123
Data collection, 140
Data transformation, 123–129
DBScan clustering, 58, 59, 61
Decoding, 147
DeepAI educational, 374
Deep Q-learning, 66
Dimensionality reduction,
    157, 210–215, 344
Discounted cumulative future
    reward, 64
Document summary, 405
Drop connected layer, 27
Dropout layer, 26

Printed in the United States
by Baker & Taylor Publisher Services